FUNDAMENTALS OF ROTARY DRILLING

FUNDAMENTALS OF ROTARY DRILLING

EDITED BY
W. W. MOORE

THE ROTARY DRILLING SYSTEM: A PROFESSIONAL AND PRACTICAL TRAINING GUIDE TO ITS EQUIPMENT, PROCEDURES AND TECHNOLOGY.

ENERGY PUBLICATIONS
A DIVISION OF HARCOURT BRACE JOVANOVICH
800 Davis Building, Box 1589, Dallas Texas 75221

Library of Congress Catalog Card Number 81-66583

Table of Contents

Table of Contents

DRILLING PROCEDURES AND TECHNOLOGY

Foreword

Man's custom of digging holes into the earth is believed to be as old as civilization itself. One reasonably can suppose that his need for drinking water would have caused prehistoric man to scoop out depressions in stream beds in search of water during a dry spell. With the development of better hand tools, such depressions or holes would be dug deeper and deeper.

Manually dug deep holes (or wells) had to be big enough to permit a man to work. Wells dug with hand tools eventually were deep enough to encounter subterranean water flows, even though the well site was far removed from a stream bed. Dependable fresh water supply from subterranean sources did much to expand the frontiers of civilization.

For centuries, manually dug wells provided much of the potable water used by mankind. Well digging became an art, however, when means were developed to perform all the work from the surface, making it unnecessary for man to work downhole. The hole or well could be much smaller and less actual work need be performed. The percussion system of well drilling slowly evolved. The basic elements of excavation remained the same: break up earthen materials and remove debris from the hole.

The earliest recorded history of a percussion drilling system comes from China where it was used first for digging brine wells. The system developed by ancient Chinese employed the same principles later used by cable tool rig for drilling the Drake well in 1859. Materials used and power sources were different but a string of tools, on a cable, was reciprocated in a manner which pulverized an interval of earth downhole and mixed it with water, making a sludge which was bailed from the hole.

The Drake well in Pennsylvania was drilled expressly in search of oil. From that time on through the rest of the 19th century, great progress was made in cable tool drilling methods and equipment. Very little advancement, if any, has been made since about 1930. Cable tools were used exclusively in oilwell drilling in this country prior to 1920, except in California, Louisiana and Texas where the rotary drilling system was undergoing rapid development.

1. Crown
2. Mast
3. Catline Boom
4. Racking Platform
5. Drill Line
6. Traveling Block
7. Hook
8. Swivel
9. Rotary Hose
10. Standpipe
11. Drawworks
12. Drillers Console
13. Pipe Setback
14. Drill Floor
15. Rotary Table
16. Substructure
17. Blow Out Preventer Stack
18. Dog House
19. Choke Manifold
20. Gas Flare
21. Mud Gas Separator
22. Shale Shaker
23. Degasser
24. Desander
25. Mud Cleaner
26. Mud Guns (Bottom Type)
27. Mud Agitators
28. Mud Tanks (3)
29. Mud Sack Storage
30. Mud Mixing Hopper
31. Mud Mixing Pumps
32. Mud Pumps (2)
33. Pulsation Dampeners
34. Shock Hoses
35. Mud Discharge Lines
36. Brake Water Tank
37. Mud Lab
38. Trip Tank
39. Mud Return Line
40. Drilling Water Tank
41. S.C.R. House
42. Cable Tray
43. Cable Elevator
44. Engines & Generators
45. Engines & Air Compressors
46. Parts Storage
47. B.O.P. Closing Unit
48. Work Shop
49. Pump Parts Storage
50. Fuel Tank
51. Junk Bin
52. Personnel Elevator
53. Wire Line Stand
54. Stairway w/Pipe Ramp
55. Catwalks
56. Drill Pipe
57. Pipe Rack
58. Auxiliary Brake

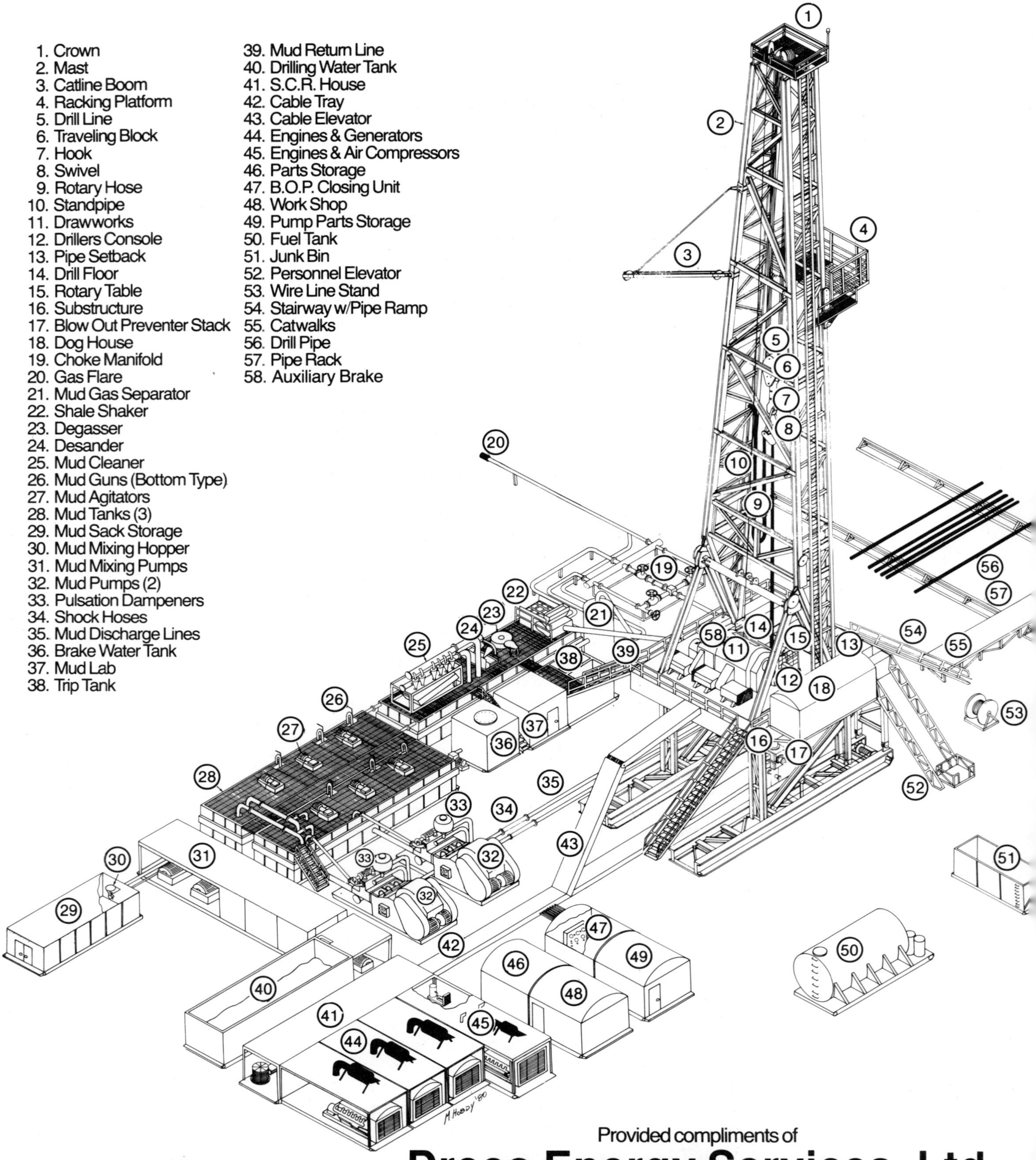

Provided compliments of

Dreco Energy Services Ltd.

Manufacturers of full service equipment for the drilling and production industry.

Chapter 1
Evolution Of The Hydraulic Rotary Drilling System

The hydraulic rotary drilling system was an outgrowth of core drilling practices. Early use of rotary systems in drilling for oil was developed mainly at Corsicana, Texas in the late 1800s but gained its greatest impetus after being used to drill the Lucas well at Spindletop in 1901.

This famous gusher marks the beginning of the petroleum industry as we have come to know it. Also it was a broad step in development of the rotary drilling system which is undergoing continuous improvement in both equipment and techniques.

Basic principles used in this sytem are much the same as in other methods of well digging: break up earth particles and remove from hole. Methods or mechanics employed, however, are altogether different.

The bit, with a downward thrust, digs by rotation rather than by reciprocation. Some bit types also produce percussive forces which are helpful.

Resulting debris is washed from under the bit and circulated out of the hole in a continuous stream of drilling fluid, rather than being removed intermittently, by bailing, as in the cable tool system.

The rotary rig used to drill the Lucas well in 1901 consisted of a small locomotive-type oilfield boiler and a single cylinder steam engine with chain drive to the countershaft of a 2-shaft, single speed, single brake drawworks. The hoist drum with jaw-clutched sprocket was driven by chain from the countershaft.

The rotary table was chain driven from countershaft to clutched sprocket on the pinion shaft of the rotary. The fluid circulating system used a conventional duplex steam pump designed for general industrial use, capable of pumping 150 gpm at 175-200 psi with steam pressure of 75-100 psi. The swivel had an exposed ball-type thrust bearing to support the drill string weight. Below that was the packed-off rotating fluid connection.

The Lucas well drill string consisted of ordinary lapweld pipe and couplings with one long length of "extra strong" or "double extra strong" pipe to serve as griefstem at the top of the string, which could withstand grooving produced by grip ring rollers. Fishtail rotary bits were used exclusively in drilling gumbo and soft sand formations encountered.

The rotary table had open bevel gear drive and a grip ring rotating the round griefstem as it moved downward through the table while drilling ahead. Drilling fluid was natural mud derived from the clay and gumbo formations being drilled.

The men involved in this operation deserve special credit for drilling to slightly below 1000 ft with this primitive equipment. The Lucas well launched not only the oil industry but an effective drilling system that pushed petroleum into first place among all energy sources.

Notable improvements in rotary drilling rigs and practices so far in this century have been continuous and substantial. Let's look at some highlights.

The rotary table with grip rings to drive the round griefstem, as used on the Lucas well, had gained preference over another type using posts and clamps to turn the drill string. Both drive systems, however, were developed simultaneously in the late 1880s.

Next development was a fluted kelly used with grip rings (1914) but the grip rings were soon replaced by kelly drive bushings. The use of a square kelly with matching drive bushings appears to have been developed in California about 1915 and gained wide acceptance during the early 1920s. Other than the development of roller-type drive bushings and hexagon kellys, no major change has been made in these two items.

The rotary machine or "table" also was improved continuously. The rotary clutch was moved from pinion shaft to drawworks countershaft in the early 1920s and soon thereafter came the roller bearing pinion shaft with drive sprocket mounted outboard. The fully enclosed oil bath rotaries came into general use in the 1930s.

Until the late 1960s the largest rotary machine in general use weighed about 10,000 lb and had a round opening of 27½ in., whereas the rotary used to drill the Lucas well weighed about 2000 lb and had a round opening of perhaps 14 in. In the 1970s, to meet the needs of offshore operations, 37½ in. rotaries weighing approximately 13,000 lb were developed to pass 36-in. conductor pipe. More recently larger machines with 49½-in. round openings and weighing approximately 25,000 lb have been placed on the market and are proving popular.

Drill pipe tool joints replaced ordinary pipe couplings about 1910 or 1912, and seamless drill pipe replaced the lapweld or "merchant's pipe" in the early 1920s. The first seamless drill pipe had only a very light internal upset but a few years later the internal upset was increased sufficiently to satisfy requirements for the next 10 to 15 years.

Improvements in metallurgy, however, were required as drilling depths increased. There is a continuing battle going on between greater well depth and the development of drill strings strong and durable. Alloy steels now are used extensively in drill string components.

During the steam rig era the philosophy seemed to be "the bigger the better". The rotary rig used on the Lucas well (1901) had only one boiler, such as those used on cable tool rigs, with a normal rating of perhaps 35 or 40 boiler horsepower and a working pressure of 75 psi. The single cylinder steam engine is reported to have been 9-in. bore by 12-in. stroke. The hoisting equipment and rotary were likewise designed for light duty.

About 1915-16, heavy duty combination steam rigs were developed for California operations. Such rigs used two 40-hp boilers, one 14-in. x 14-in. single cylinder slide valve steam engine, a double brake 2-speed drawworks and a 12-in. x 6¾-in. x 14-in. duplex slush pump. The 23-in. rotary table was chain drive with jaw-clutch on the pinion shaft and open bevel gears. Square kellys with matching drive bushings were preferred in most instances, but grip ring drive was also available.

The roller cutter rock bit was invented by Mr. Hughes in 1908 but did not come into general use until 1915 and then only to a limited extent in hard rock areas. With continued improvement in tooth and bearing designs the roller cutter rock bit soon dominated bit business throughout the entire world.

Steel derrick usage increased during the early 1920's in areas where good rig timber was not available. Wooden derricks were used in California perhaps later than elsewhere in the U.S. The first cantilever mast, 87 ft high, was fabricated in 1938 at Tulsa, Okla. These masts were enlarged as time passed until they took over nearly all the onshore market by mid-1950.

Steam-powered rigs were getting better through the 1920s and 1930s. Twin cylinder steam engines came along about 1918, requiring at least two boilers. Twin cylinder engines thereafter were upgraded through use of piston valves and roller bearings.

Size was increased to 14-in. x 14-in. and designed for steam pressures of 350 psi plus. Usually four boilers of 125 hp each were used with these large steam rigs. Steam slush pumps likewise were being enlarged. Notably the size 14-in. x 7¼-in. x 18-in., good for about 600 gpm, came into general use in the late 1920s. Even larger pumps, 15-in. x 7¾-in. x 20-in., came along in the early 1930s.

Steam-powered drawwork size also increased during this period. Three-speed, 3 post, 2-shaft drawworks were popular in the 1920s. The drum shaft diameter in these new machines was 6 or 7 in., compared to 5 in. on 2-speed hoists and to 3½ in. for earlier machines.

Subsequently, steam-driven drawworks with 9-in. and 10½-in. drum shafts were skid-mounted and, in some instances, fully enclosed oil bath units. The first unitized, skid-mounted drawworks (1932) were field shop conversions of old jackpost-type rigs. The steam era, so far as new developments were concerned, ended in the 1930s, although steam rigs continued to dominate deep drilling in the Gulf Coast area through the 1940s. In this area good boiler water was plentiful and fuel was relatively cheap.

The 1930s marked the turn away from steam powered rigs. West Texas and other arid regions provided very limited quantities of good boiler water and sometimes no gas for fuel. Oil had to be hauled in for boiler fuel for widely scattered operations.

Cable tool operations in these arid regions had been successfully changed over to internal combustion power. To accomplish this conversion the steam engine was replaced by a gasoline (or oil) engine and a reversing clutch. It was quite natural, therefore, that the first powered rotary rigs were also similarly converted. Power driven slush pumps, in small sizes, had already been developed for electric drive. As power for the drawworks, two or more engines, compounded, driving through a heavy duty reverse clutch, provided a reasonable substitute for the steam engine.

In the early 1930s newly designed small powered rigs were used for drilling shallow wells. Some of these grew from well servicing units. The old steam rig conversion was most common for moderate depth wells. After the mid-1930s several independents, including manufacturers of well servicing equipment, pioneered the development of consolidated power rigs. During the next several years rapid progress was made, both in consolidated rigs and in engines to power them. Power driven mud pumps were upgraded and so were other components such as masts, blocks, swivels, hooks, etc. There has been an ongoing progress in power rig development since the 1930s and around 1950 capacity equalled conventional steam rigs.

Auxiliary equipment evolution was also making important contributions to the art. The roller-type rock bit was continuously improved and no doubt ranks most important in development of the rotary drilling system.

Rig instrumentation, starting with weight indicators in the 1920s, took a lot of guesswork out of rig running. Dependable mud pump pressure gauges came along about the same time. The Wigle spring hook, Wilson tongs, center-latch elevators and drill pipe slips, introduced in the 1920s, did much to lighten the toil and provide crew safety.

The hydromatic brake, introduced in the early 1930s, has helped solve the brake wear problem while tripping in at great depths. Ram-type blowout preventers were first used in the 1920s and have become almost indispensable.

Chapter 2
Downhole Tools

Bits

The bit at the bottom of the drill string is the indispensable tool. Without it all other items of the conventional rotary drilling rig are useless. Other rig items merely support bit functions. Bits of many designs have been used in rotary drilling but all require downward thrust plus rotary motion in order to make hole. The rotary rig is similar to an out-size drill press.

The forerunner of all rotary drilling bits is the fishtail, so called because the cutting blade resembles the tail of a fish. These bits plow through the earth and are not suitable for drilling hard formations. Fig. 2-1 shows typical early-day bits. The outlet for drilling mud is near the bit shank, with considerable length of blade below.

After a period of use in drilling, the blade became dull and had to be dressed or sharpened. This usually was done at the drill site, using forge, anvil and manpower at hand. The long blade of a new bit provided enough steel for many sharpening jobs. Later the mud outlet was moved down on the blade, as shown in Fig. 2-2, and improved results were noted. Sharpening these bits was done by building up and hardfacing cutting edges.

Fishtail bits were first referred to as drag bits. Included were diamond point bits used for sidetracking junk and chisel bottom bits used for spudding through thin hard shells. Four-bladed drag bits (Fig. 2-3) remained popular for many years especially after

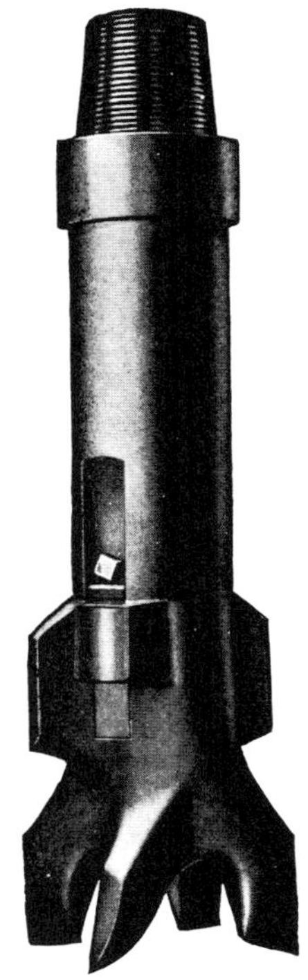

Fig. 2-3. Four-way Drag Bit.

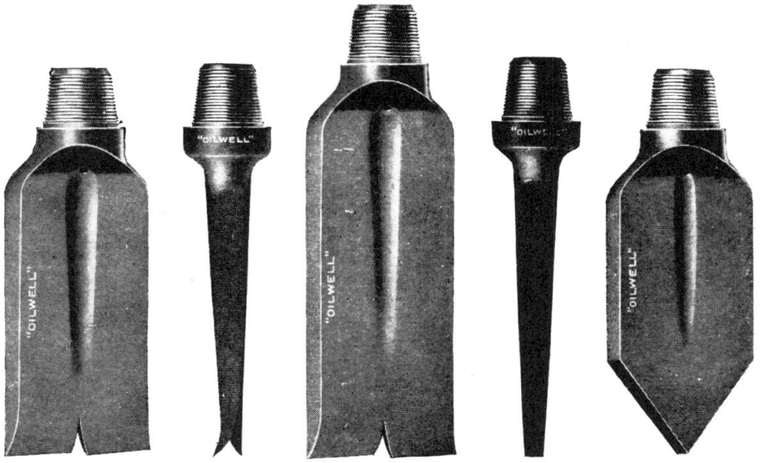

Standard Length Fish Tail | **Extra Long or Reaming Fish Tail** | **Standard Length Diamond Point**

Fig. 2-1. Fishtail bits. Long Blade and Diamond Point.

Fig. 2-2. Fishtail bits. Short blade.

Fig. 2-4a. Diamond bit.

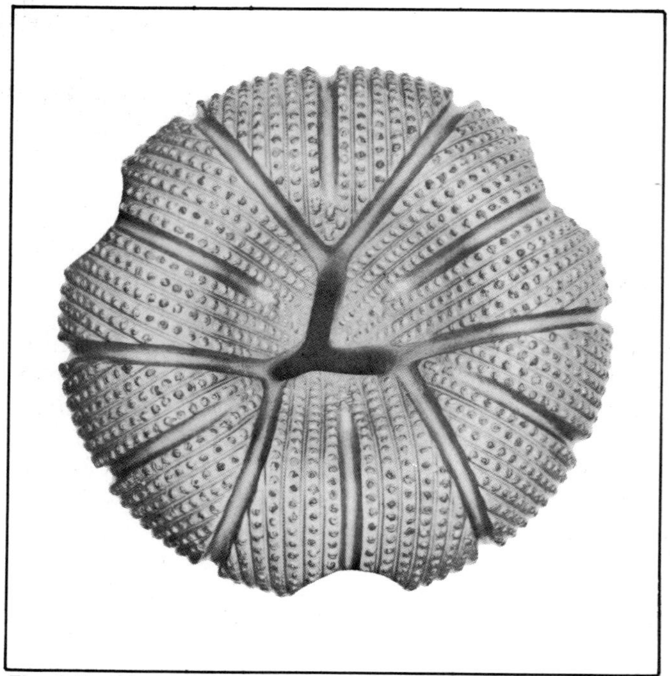

Fig. 2-4b. Diamond bit.

jet nozzles were used. Four-way drag bits are dressed by applying hardfacing metal to the cutting structure.

Diamond bits (Fig. 2-4) currently in use are in fact drag bits, but are designed expressly for the hardest formations likely to be encountered. A modified diamond bit combining the basic design of a 4-way drag bit with diamond studded wear surfaces has proved useful in drilling medium-to-soft formations. Because fewer carats of diamonds are used, it is much cheaper than the conventional diamond bit.

In 1909 Hughes brought out a 2-cone rolling cutter bit (Fig. 2-5) that proved effective in hard rock formations, hence the term rock bit was coined. This early model did not have intermeshing (self-cleaning) teeth so it was poor in gumbo or shale which caused the cones to ball up or lock. Cross rolling cutter-type bits came out later but rock bits were not a commercial success for several years, and seemed to have limited application.

In the late 'teens Hughes brought out the famous simplex two-cone bit (Fig. 2-6) with long split body having side reamers and lubricated journal-type bearings. These bits could be dressed on the job by replacing cones and reamers when needed. In the early 1920s the bit was improved by offering optional acme self-cleaning cones with long intermeshing teeth. This design proved effective in drilling shales and soft formations. Reed's cross rolling cutter bit was improved about this same time and drag bits had real competition for the first time ever.

In the early 1930s both Reed and Hughes went back to short sport model bits which were unitized with much smaller bodies and were leased as a unit. The Hughes bit was a tricone roller bearing whereas the Reed bit retained the cross roller arrangement.

With the expiration of some Hughes basic patents, all rolling cutter rock bits became similar. They were tricone roller bearing assemblies with a great variety of tooth structures, including both milled and tungsten carbide insert types. Some six U.S. manufacturers make them.

The jet bit principle was first applied to drag bits and proved effective. In the late 1950s jet nozzles were applied to rock bits, even with tungsten carbide in-

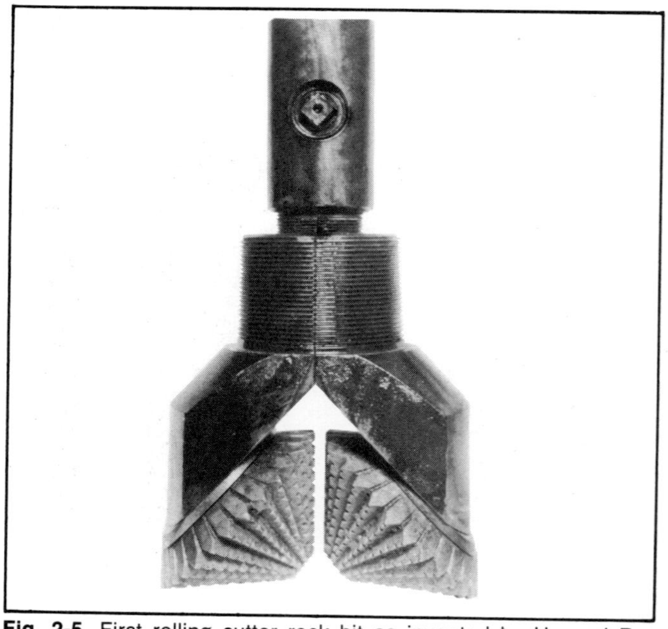

Fig. 2-5. First rolling cutter rock bit as invented by Howard R. Hughes, Sr. and run at Goose Creek, Texas in 1909.

Fig. 2-6. Early model Hughes Simplex rock bit with lubricated journal-type rolling cones. Subsequent models with acme self-cleaning cones — intermeshing teeth — were effective in drilling shale intervals.

serts. Higher cost of jet bits was justified in deep drilling because bottomhole scavenging could be done with less drilling fluid being circulated and less friction loss in drill string.

Substituting fluid velocity for fluid mass was effective in cleaning debris from under the bit. At great depth, if small drill pipe is required, it is illogical to attempt to circulate large fluid volumes. The surface

Fig. 2-7. Fishtail bit with removable blade.

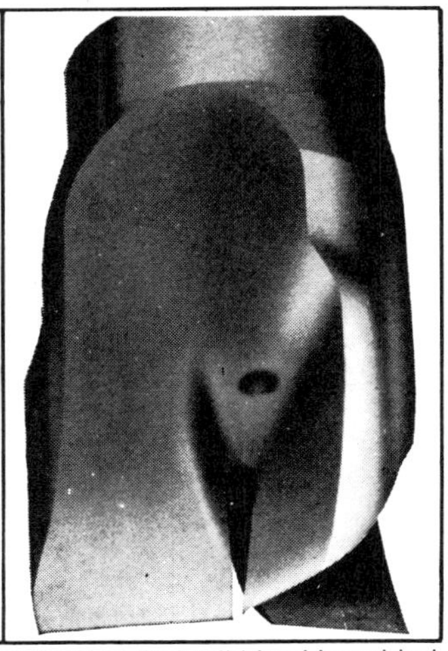

Fig. 2-8. Fishtail bit with welded-in blade.

Fig. 2-9. Guiberson disc bit.

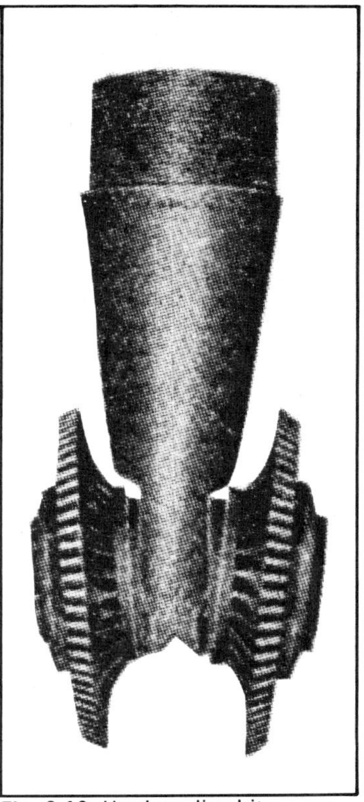

Fig. 2-10. Hughes disc bit.

equipment (pump and prime mover) would soon be considerably overloaded.

The most recent improvement in rock bits is lubricated bearings, particularly lubricated journal-type bearings. Remember the Hughes simplex bit of the 1920s? It had lubricated journal-type bearings.

The concluding portion of this section will be confined exclusively to tricone rock bits since they account for a large portion of bit usage today. But before doing so, something must be said about specialty bits that have made significant contributions to the art.

In the drag bit category, fishtail bits with inexpensive replaceable alloy steel blades (Fig. 2-7) made it possible to use hardfacing material more effectively and expedited bit dressing at the rig. Bit expense was lowered since only the blade was expendable. One other approach to the problem was replaceable, welded-in alloy steel blades used with a cast steel body (Fig. 2-8). (Available in 2-blade, 3-blade and 4-blade models.)

The Guiberson disc bit (Fig. 2-9) was used extensively in Texas during the early 1930s. Hughes offered a bit of similar construction (Fig. 2-10) except, instead of using four plain discs, two discs were used, each with teeth around the outer circumference. The Grant differential bit (Fig. 2-11) employs a single cutter wheel assembly mounted 15° from vertical. The small cutter discs on the wheel rim are free to turn as the whole assembly rotates. It is used primarily as a hole opener.

The Zublin bit, (Fig.2-12) sometimes referred to as

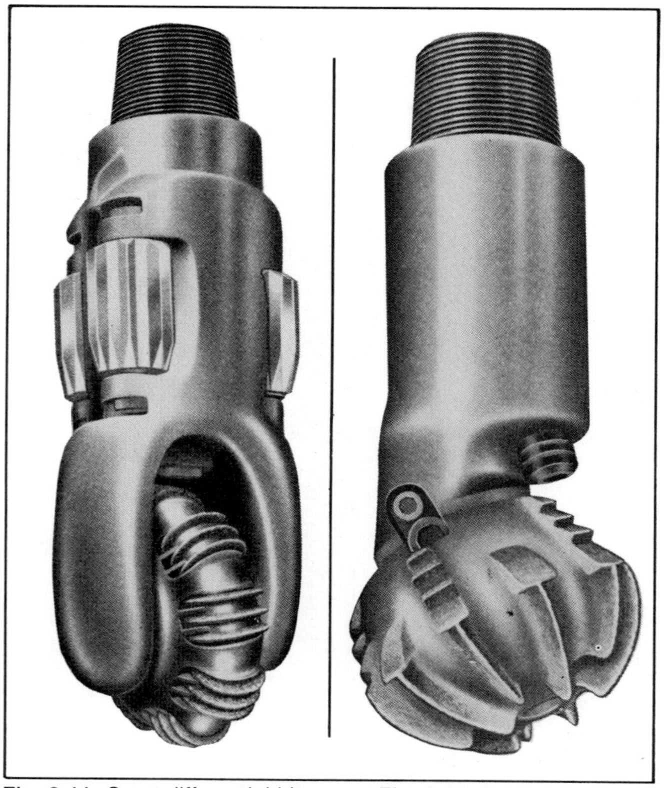

Fig. 2-11. Grant differential bit. Fig. 2-12. Zublin bit.

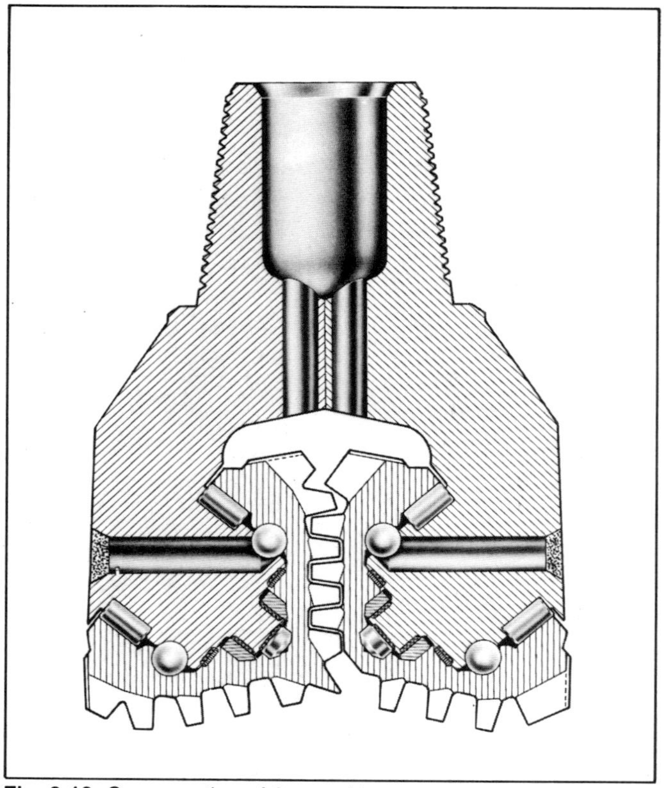

Fig. 2-13. Cross section of 2-cone bit.

the Simplex bit, and now marketed by Grant Oil Tools, deserves special mention. The design is unique in that the single cutter head is angle-mounted and is retained on its spindle with balls in a bearing race, very much like cone-type bits. It might be referred to as a 1-cone bit.

Four different types of cutter heads are available for specialized use such as rolling out collapsed casing, drilling off whipstocks, drilling hard sand, drilling soft formations and reaming out ratholes. Zublin bits were first used extensively in California and foreign fields. They were used to some extent in East Texas in the early 1930's and have been improved continuously since.

The rolling cone rock bit has become the most commonly used type, world-wide. In many respects

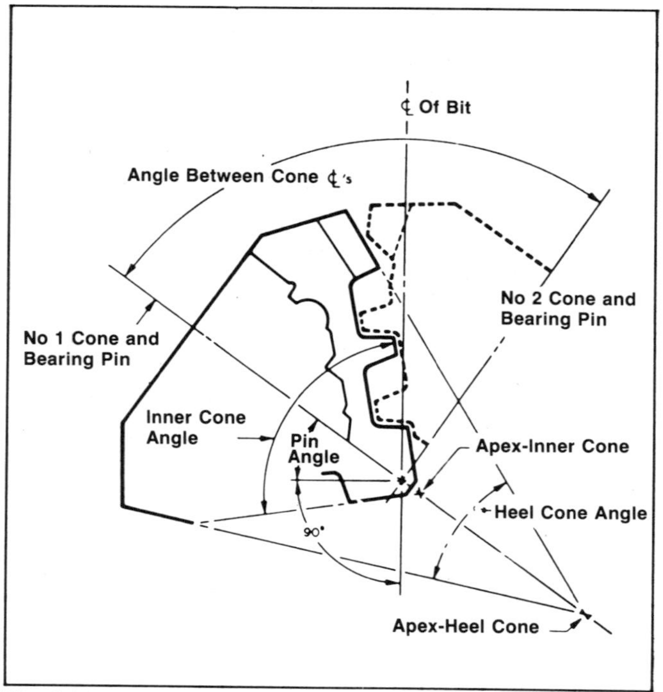

Fig. 2-14a. Three cone design.

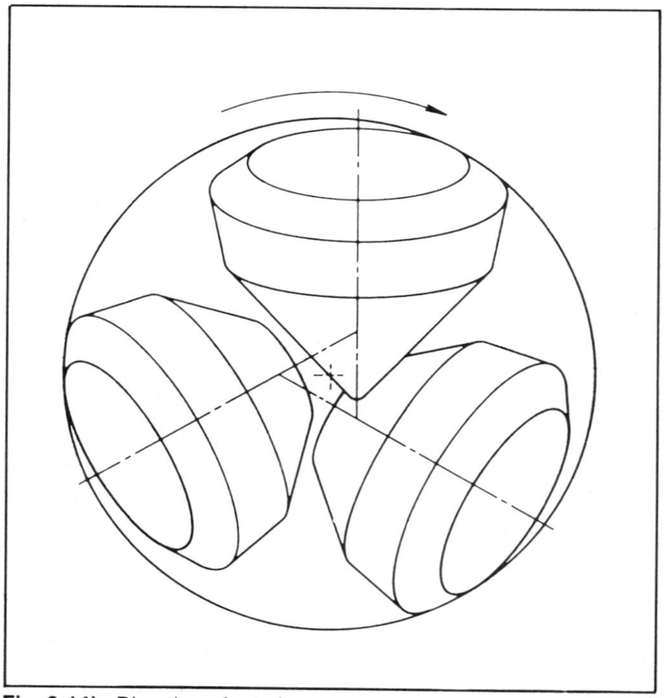

Fig. 2-14b. Direction of rotation

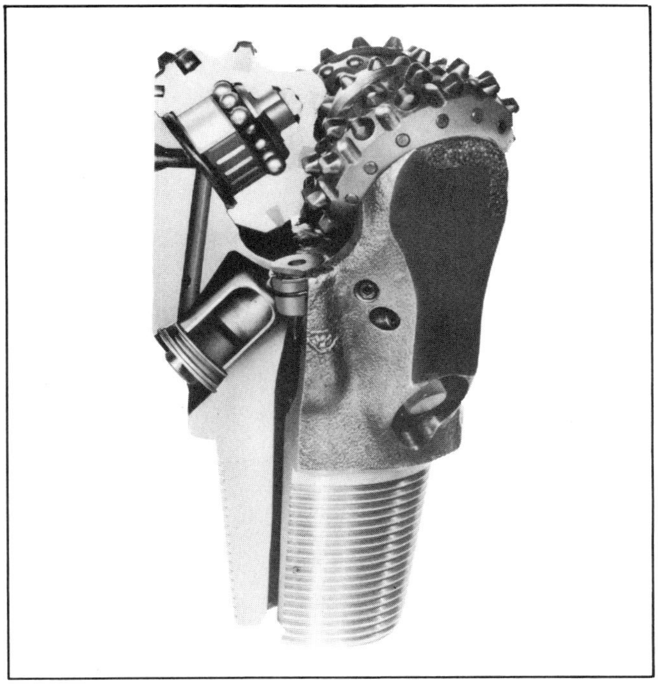

Fig. 2-16. Cutaway of Hughes Tool Co. journal bearing bit. Hughes journal bearings match the life of carbide cutting structures, have no inserted parts that can collapse.

The surfaces in the pressure area — where journal bearing pin meets cone interior — are carburized mated surfaces.

They are extremely hard and won't shatter under heavy impact. Thick noncarburized steel core underneath supports carburized case and acts as a shock absorber to soak up heavy drilling shocks.

Mated bearing surfaces in Hugheset bits are inlaid with special metals that add extra wear resistance and protection against galling.

This marriage of different metals, plus structural clearances and Hughes' sealed lubrication system, minimizes friction, heat and turning torque.

Hughes' patented O-ring seals prevent drilling fluid from entering the journal bearing and retain special grease in the bearing.

Patented pressure relief valves automatically release excessive pressure build-up inside the bearings, a feature that protects the O-rings and the compensator system from possible damage by unbalanced pressure.

Fig. 2-17. Hughes J-22 Bit for soft formations.
FORMATIONS: Soft with low compressive strength (shales, clays, red beds, salt, soft limestone, sands).
CUTTING STRUCTURE: Large diameter Hugheset tungsten carbide compacts with relatively sharp, chisel-shaped crests have maximum projection and intermesh.
GAUGE PROTECTION: Maximum number of flat-ended tungsten carbide compacts installed in cone gauge surfaces to retard gauge wear. Hughes-X hardfacing on shirttail area of heads.
DRILLING ACTION: Fast penetration achieved by gouging and scraping action due to large cone offset and wide spacing of Hugheset chisel-shaped compacts.
NORMAL DRILLING WEIGHT: 3000 to 4000 lb/in. of bit diameter.
NORMAL ROTARY SPEED (rpm): 100 to 35 (lower rpm with higher weight).
BEARINGS: Hughes journal bearings with silver inlays in cone and hard alloy inlays on bearing pin. Special lubricants in patented, O-ring-sealed pressure compensating system.

the basic construction employed by several manufacturers is similar. Fig. 2-13 shows a cross-section of a bit with roller bearing cones. Figs. 2-14a and 2-14b show 3-cone bit geometry. Fig. 2-15 shows terminology used in reference to tricone rock bit elements. Figs. 2-16 through 2-50 shows views of many types of roller cone bits.

Widely spaced long teeth are most effective in soft formations while closely spaced short teeth are more effective in hard formations. The chert bit, with tungsten carbide inserts, is the ultimate in this respect, but the higher cost cannot always be justified.

Fig. 2-13, which is a cross-section of a conventional 2-cone bit, illustrates the relationship of the various elements in cone-type bit design. The 2-cone type is used to show the relationship of all parts in a single plane. What will be said about various parts in this design and their relationships also applies to the 3-cone design.

Note in Fig. 2-13 that cutters, or cones, are mounted on bearing pins which are an integral part of the bit body. This provides necessary strength to withstand loads imposed on bits in present-day drilling. Radial loads are taken by the roller bearing near the cutter base and by plain bearing near the nose.

The ball bearing retains the cutter on bearing pin and absorbs all inward thrust loads. The primary capacity for outward thrust is obtained by providing a plain thrust bearing at the ball race inner flange face and at the inner end of the bearing pin.

This 5-element bearing design is typical for most types of 3-cone bits in sizes from 4¾ in. to 12¼ in. A recent trend in bearing design is to replace roller bearings with a plain or journal bearing as primary radial load element. In larger bits additional space available permits use of two or more roller bearings for increased capacity. Because of space limitations in most types of bits smaller than 4¾ in., a ball bearing in combination with an outer and inner plain bearing is used.

Steel Tooth Bits

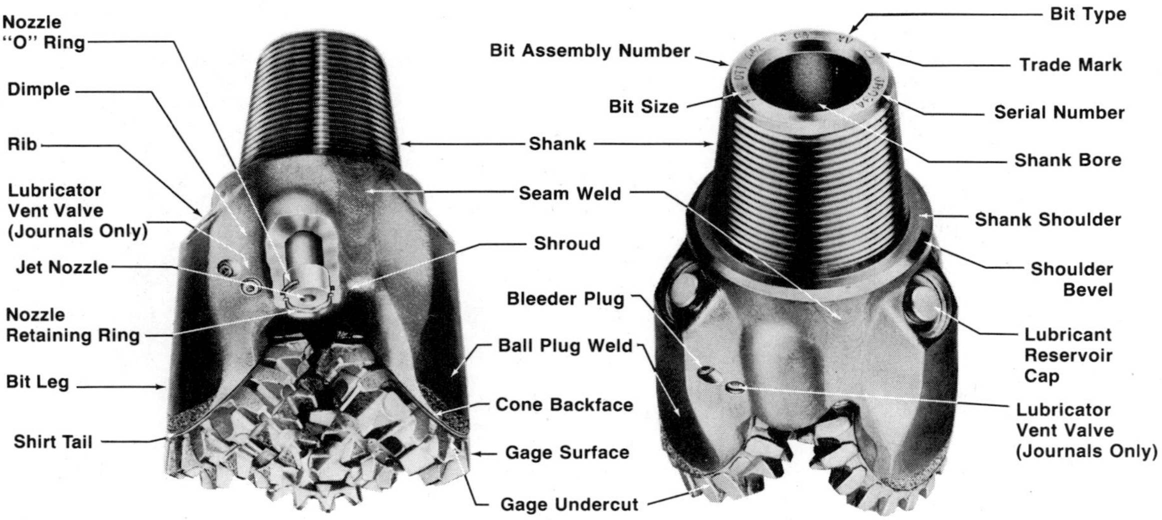

Basic outside geometry of all steel tooth bits is the same. Differences in tooth arrangement, tooth length and angle of teeth in each bit type provide the most efficient action on bottom for the particular formation to be drilled.

For softer formations, chisel-shaped teeth are deeply cut and provide a twisting, tearing, gouging action for rapid penetration. Teeth for harder formations are shorter and stronger to carry the higher loads necessary for efficient penetration.

Cone action in relation to bit rotation is changed from scraping and gouging more to true rolling motion for more chipping and crushing in bits designed for the harder formations.

Scientifically directed jet streams combined with correct nozzle sizes clean the cones of the bit and provide the most efficient removal of cuttings from bottom.

Rows of teeth on the inner portion of each cone interfit with the grooves on adjacent cones. This provides positive self-cleaning action and permits the best use of available space for both teeth and bearings.

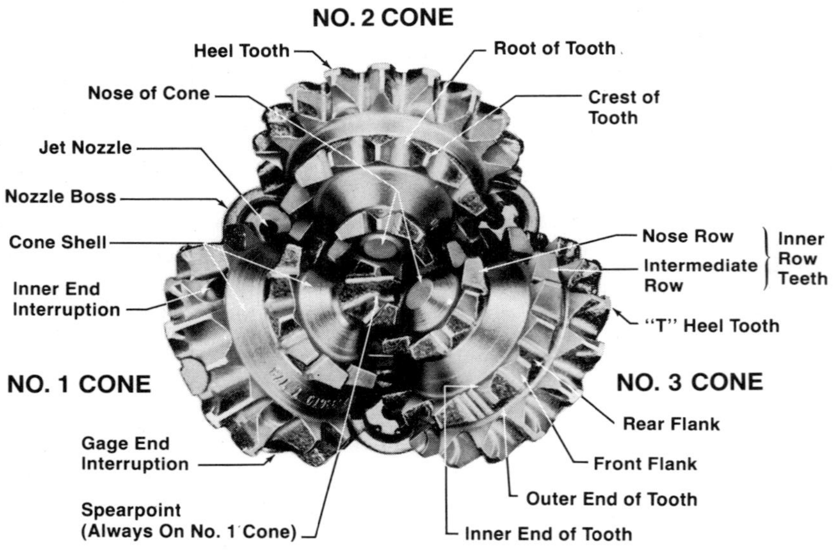

Fig. 2-15. Three cone bit terminology. *—Courtesy of Hughes Tool Co.*

Typical Steel Tooth Cutting Structures

For Soft Formations

For Medium Formations

For Hard Formations

Fig. 2-18. Hughes J-33 bit for medium soft formations.
FORMATIONS: Medium-soft, low strength, with harder, more abrasive streaks.
CUTTING STRUCTURE: Large diameter Hugheset tungsten carbide compacts with chisel-shaped crests. Less projection and intermesh of compacts than on J22. Cutting action is similar to Hughes X1G steel tooth bit.
GAUGE PROTECTION: Maximum number of flat-ended tungsten carbide compacts installed in cone gauge surfaces to retard gauge wear. Hughes-X hardfacing on shirttail area of heads.
DRILLING ACTION: Primarily gouging and scraping with moderate crushing and chipping. Slightly shorter compacts than J22 provide increased durability in slightly harder formations.
NORMAL DRILLING WEIGHT: 3500 to 5000 lb/in. of bit diameter.
NORMAL ROTARY SPEED (rpm): 80 to 35 (lower rpm with higher weight).
BEARINGS: Hughes journal bearings with silver inlays in cone and hard alloy inlays on bearing pin. Special lubricants in patented, O-ring-sealed pressure compensating system.

Fig. 2-19. Hughes J-44 bit for medium hard formations.
FORMATIONS: Medium hardness with high compressive strength, especially those with many thick, hard streaks (limestones, sands, dolomites, hard shales).
CUTTING STRUCTURE: Hugheset tungsten carbide compacts with blunt, chisel-shaped crests. Less projection and intermesh than J33. Cutting action similar to Hughes XV steel tooth bit.
GAUGE PROTECTION: Maximum number of flat-ended tungsten carbide compacts installed in cone gauge surfaces to retard gauge wear. Hughes-X hardfacing on shirttail area of heads.
DRILLING ACTION: Balance of gouging-scraping and crushing-chipping action achieved by small degree of cone offset plus increased bearing pin angle.
NORMAL DRILLING WEIGHT: 4000 to 5000 lb/in. of bit diameter.
NORMAL ROTARY SPEED (rpm): 60 to 35 (lower rpm with higher weight).
BEARINGS: J44 has Hughes journal bearings with silver inlays in cone and hard alloy inlays on bearing pin. Sealed lubrication system is provided.

In order to determine the cone contour, a decision must be made on the proper pin angle. This is the angle between the centerline of the bearing pin and cone and the horizontal plane (Fig. 2-14a). This pin angle determines the outside contour of all three cones because each cone is designed to fill com-

Fig. 2-20. Hughes J-55 bit for medium hard formations.
FORMATIONS: The hard side of medium, generally abrasive and homogeneous (limestone, dolomite, chert, sandstone).
CUTTING STRUCTURE: Hugheset tungsten carbide compacts of 60° conical design, fairly blunt ends, moderate projection and intermesh. Cutting action is similar to Hughes J7 steel tooth bit.
GAUGE PROTECTION: Maximum number of flat-ended tungsten carbide compacts installed in cone gauge surfaces to retard gauge wear. Hughes-X hardfacing on shirttail area of heads.
DRILLING ACTION: Almost entirely crushing and chipping. Minimum scraping. Cones are non-offset to minimize compact motion relative to formation.
NORMAL DRILLING WEIGHT: 4000 to 60000 lb/in. of bit diameter.
NORMAL ROTARY SPEED (rpm): 50 to 35 (lower rpm with higher weight).
BEARINGS: J55 has Hughes journal bearings with silver inlays in cone and hard alloy inlays on bearing pin. Sealed lubrication system is provided.

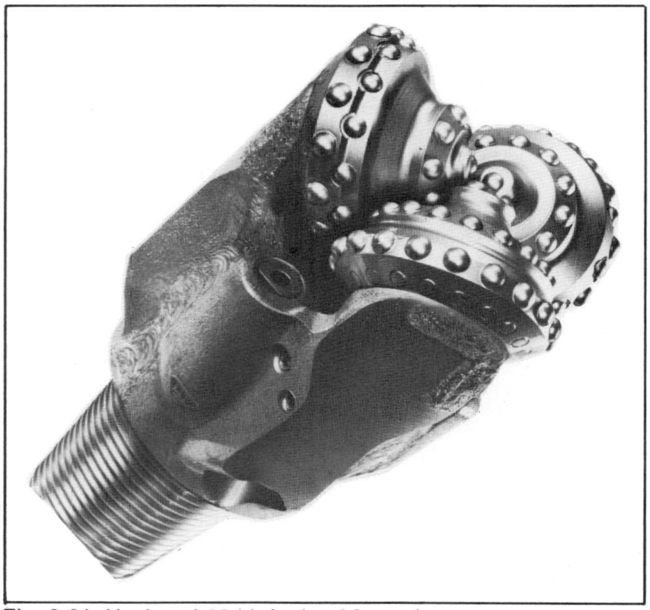

Fig. 2-21. Hughes J-88 bit for hard formations.
FORMATIONS: Hard semi-abrasive (limestone, dolomite and hard, sandy shales).
CUTTING STRUCTURE: Hugheset tungsten carbide compacts. Hemispherical ends, small projection and slight intermesh provide maximum resistance to breakage.
GAUGE PROTECTION: Maximum number of flat-ended tungsten carbide compacts installed in cone gauge surfaces to retard gauge wear. Hughes-X hardfacing on shirttail area of heads.
DRILLING ACTION: Primarily crushing and chipping. Reaming action from gauge compacts help maintain full hole line.
NORMAL DRILLING WEIGHT: 4000 to 6000 lb/in. of bit diameter.
NORMAL ROTARY SPEED (rpm): 50 to 30 (lower rpm with higher weight).
BEARINGS: J88 has Hughes journal bearings with silver inlays in cone and hard alloy inlays on bearing pin. Sealed lubrication system is provided.

Fig. 2-22. Reed FP 63 Bit (Cutaway view). The "FP" series is the top of the Reed rock bit line. It was designed for use in situations that demand long hours of cutting under heavy weights and at fast penetration rates. The lubricated journal-type bearing employs a floating beryllium copper bushing with radial seal recessed in and rotating with the cone. Bits of the "FP" series are available in insert type only.

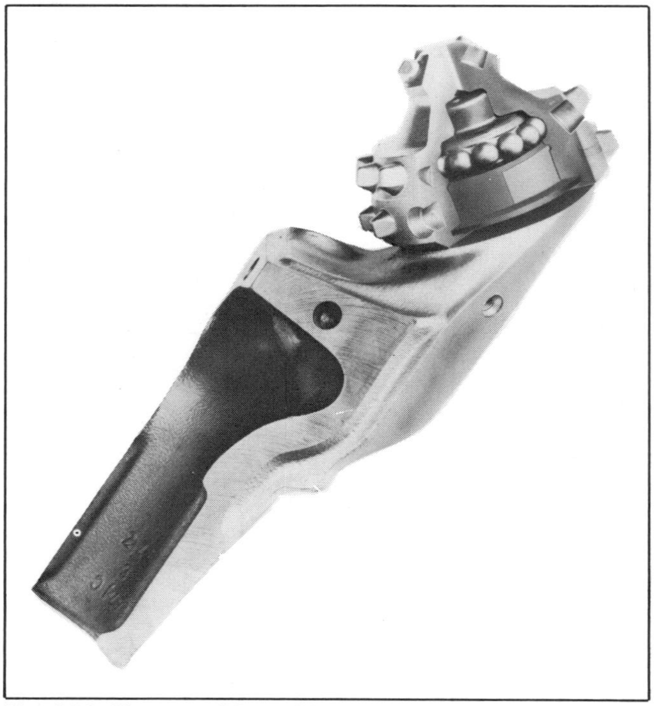

Fig. 2-23. Close-up of Reed FP63 Lug. This cutaway section of cone reveals the floating bushing, the radial seal and the retaining balls.

Fig. 2-24. Reed FP51 Insert Bit. Designed for drilling soft top hole formation of low compressive strength and high drillability. The skew angle of the cutters and positioning of the long conical inserts provide gouging, tearing and scraping action.

pletely its allotted space. Selection of the proper pin angle is determined after careful consideration of the desired depth of interfit, tooth depth, heel surface length and angle, gauge contact and length, cone diameter, cone shell thickness, available space for bearings and bit leg strength.

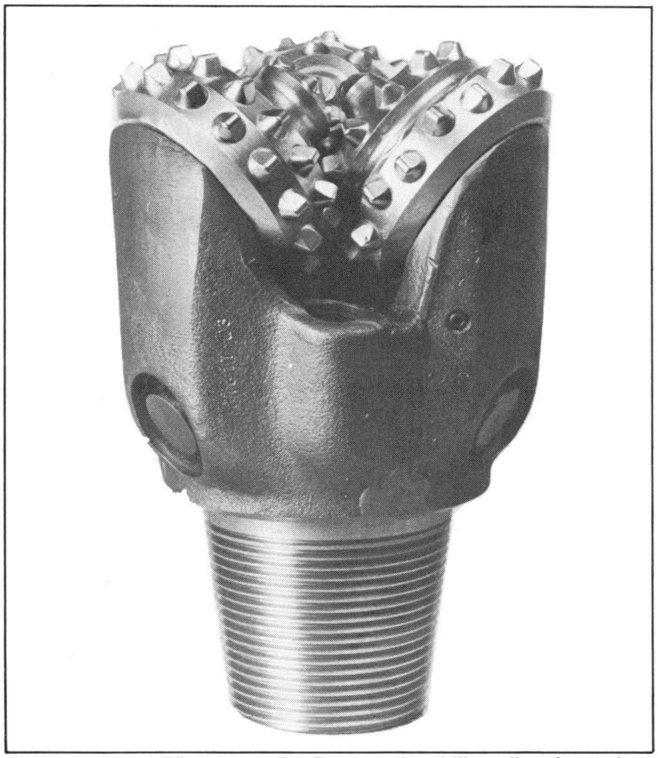

Fig. 2-25. Reed FP53 Insert Bit. Designed to drill medium formation having both hard and soft streaks. Recommended for weights of 4000 to 5000 lb/in. of bit diameter at rotary speeds of approximately 50 rpm.

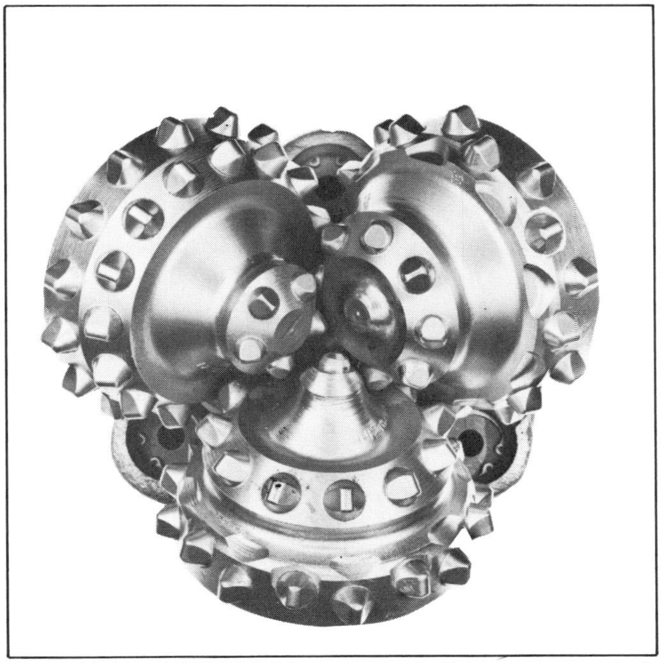

Fig. 2-26. Reed FP53 bit (close-up view). Note arrangement of chisel-shaped inserts.

All of these elements must be balanced in order to achieve the most desirable compromise for drilling.

One of these elements is the depth of interfit, which is the distance one cone's tooth crests extend into the grooves of the adjacent cone. Four advan-

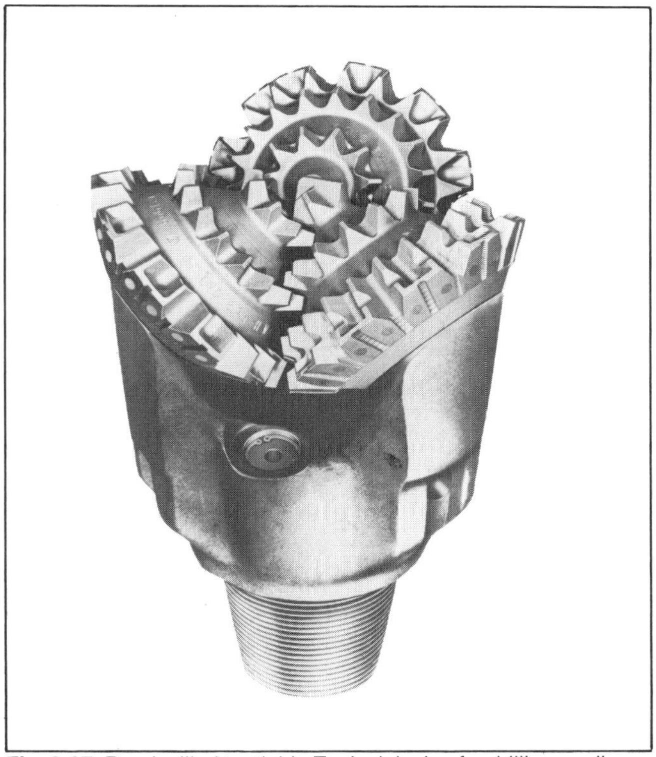

Fig. 2-27. Reed milled tooth bit. Typical design for drilling medium-hard formations. Note webbing and gauge protection. These bits are often used in mining operations.

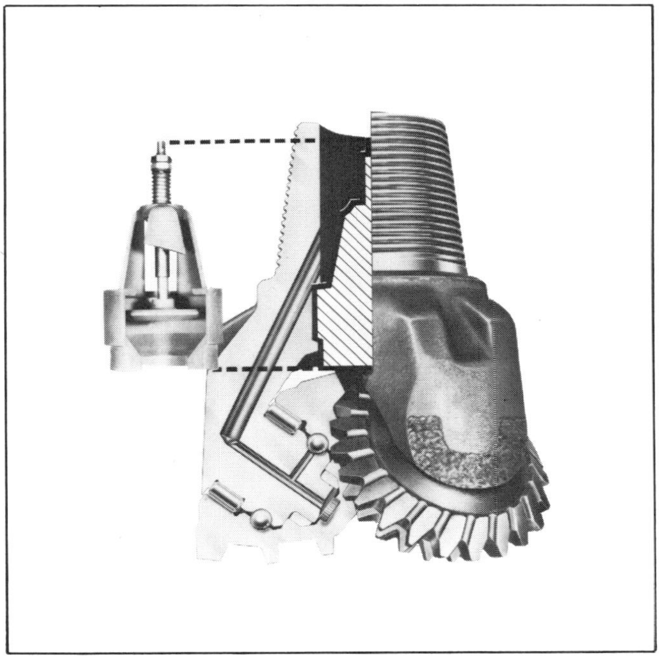

Fig. 2-28. Reed milled tooth bit for ore mining usually employs air circulation. This bit is provided with an air handling valve that controls air diverted to bearing surfaces for cleaning and cooling.

tages are gained from interfit:

1. It provides room for larger cone contour, allowing increased space to allot to tooth depth, cone shell thickness, bearing size.
2. It provides mechanical groove cleaning as cutters rotate, which assists drilling fluid in

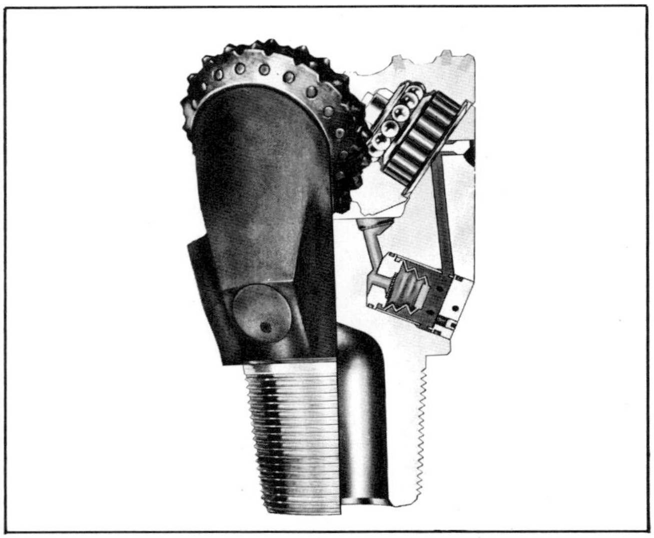

Fig. 2-29. Smith Roller-Ball-Friction Bearing System. Rock-bit bearing must withstand high unit impact loading.

Conventional rolling cutter rock bits have two different bearing structure arrangements. The first, a roller-ball-friction (RBF) bearing, is usually used on bit sizes up through 12¼-in. diam.

On bits larger than 12¼-in. the bearing arrangement is roller-ball-roller (RBR) type. On RBF-type bearings, radial loads on the bits are supported by the outer and inner bearings. The outer bearing is the roller bearing and the inner bearing is a friction-type bearing.

The friction portion of this bearing consists of a special case-hardened bushing pressed into the inner end of the cone and a male piece or pilot pin on the leg, hardfaced with material which, in combination with the bushing, displays a low coefficient of friction to resist seizure and wear.

The middle bearing on both bearing types is composed of a row of ball bearings that resist some thrust or longitudinal loading.

The majority of the thrust loading on a bit is resisted by a tool steel thrust button pressed in the cone that rides or rotates against the end of hardfaced pilot pin.

On RBR bearing design the friction bushing is replaced by roller bearings in the nose section.

keeping the bit clean.

3. It controls part of the action on bottom because cones cannot roll about the bottom of the hole without slippage. If a cutter has a single conical surface with its apex at the bit's center of rotation, it will not interfit with the adjacent cone and will roll around bottom in a true rolling motion without slippage or scraping action. The greater the interfit, the greater is the amount of slippage, thus meaning a greater amount of action on bottom.
4. It provides space for extending one cone across the center of the bit to resist coring action.

The length and the conical angle of the heel surface are also important factors. Change in conical angle at the heel teeth brings this row out of interfit and permits a wider selection of heel teeth that work near the wall of the hole. The difference between the conical angle of the heel surface and inner row surface is another factor which controls the slippage, or action, on bottom.

The conical heel surface would like to roll about a center at its theoretical apex, whereas the inner row surface would also like to roll about a center at

Fig. 2-30. Smith 3-cone bit cutaway lubricated journal bearing. Purpose of bearing arrangement is to increase capacity for the bearing structure to withstand unit and impact loading and to dissipate heat rapidly.

In solid journal bearing design, roller bearings are eliminated. In this area of the bearing structure a proprietary hard metal is inlaid in the leg journal and a high heat conductive, bearing material is inlaid in the cone.

The bearing material inlaid in the cone aids in rapid dissipation and transfer of heat from the sliding surfaces of the cone and journal to the outer cone.

The solid journal bearing design retains the same type of ball bearing structure for the middle bearing and a similar type of friction bearing structure for the inner bearing.

Other design improvements in the Smith Tool "F" Series solid journal type bearing are as follows:

1. An improved bearing sealing arrangement compared to conventional roller-ball-friction bearing design is used.
2. An improved lubrication system and grease reservoir has been employed. Bearing lubrication in the field or on the drilling rig is made easier.
3. A pressure compensation system has been incorporated in the solid journal bearing design.

Fig. 2-31. Smith type DS milled tooth bit for soft formations drills effectively in red bed, salt, softer shales and clays. Deep, widely spaced tooth design allows maximum penetration and cleaning. Interrupted tooth spacing prevents bottomhole tracking and increases drilling efficiency while partial tooth deletions help provide additional cleaning in critical tooth rows.

To provide adequate gauge protection and faster penetration rates, DS bit has flare crest gauge teeth. Drilling rates are also increased by large amount of cone offset, which produces optimum gouging-scraping action.

Normal operating weights range between 2000 and 6000 lb/in. of bit diameter, with corresponding rotary speeds of 75 to 250 rpm. Speeds should be decreased as weight is increased.

Fig. 2-32. Smith DT bit. Developed for soft, sticky shales, soft shales, clays, red beds, salts and soft limestone. Harder formations are often drilled at an effective rate when crooked hole problems dictate use of extremely light weights.

Rapid penetration is provided by deeply cut teeth and cone offset for maximum gouging-scraping action. Flare crest gauge teeth give adequate-gauge protection and assure a full gauge hole.

As with type DS bit, maximum resistance to formation packing is achieved through widely spaced teeth and ample tooth deletions. Interrupted tooth spacing prevents bottomhole tracking.

When using the DT bit, operate with weights between 2000 and 6000 lb/in. bit diameter, with corresponding rotary speeds of 75 to 250 rpm. Speed should be decreased as weight increases.

Fig. 2-33. Smith DG bit operates best in medium soft formations or soft formations interspersed with harder streaks. The geometric design, incorporating maximum gouging-scraping action and deeply cut teeth, makes the Type DG applicable for a wide variety of formations, including unconsolidated shales, medium soft shales, red bed, slate, anhydrite, gypsum, softer limes and tough, waxy shales. Tungsten carbide gauge hardfacing, applied to the cutter teeth, resists abrasion and increases bit lift. Added protection is provided by flare crest teeth with a "T" configuration on No. 2 cutter gauge teeth.

When using the DG bit, operate with weights between 3000 and 7000 lb/in. of bit diameter with corresponding rotary speeds of 60 to 175 rpm. Rotation speed should decrease as weight increases.

Fig. 2-34. Smith Type V milled tooth bit was designed to replace the softer formation bits in strata where shales alternate with streaks of limestone and sand, or where gypsum tends to ball up between teeth.

A combination drilling action of chipping/rushing and some gouging/scraping is provided because of reduced cutter offset. Teeth are shorter and more closely spaced than in soft and medium soft formation bits.

Gauge row teeth are deleted and interrupted. This ventilation of gauge row teeth contributes to faster penetration rates.

The most effective weight range varies from 3000 to 8000 lb/in. of diameter, while rotary speeds of 50 to 120 rpm give best performance. Speed should be decreased as weight increases.

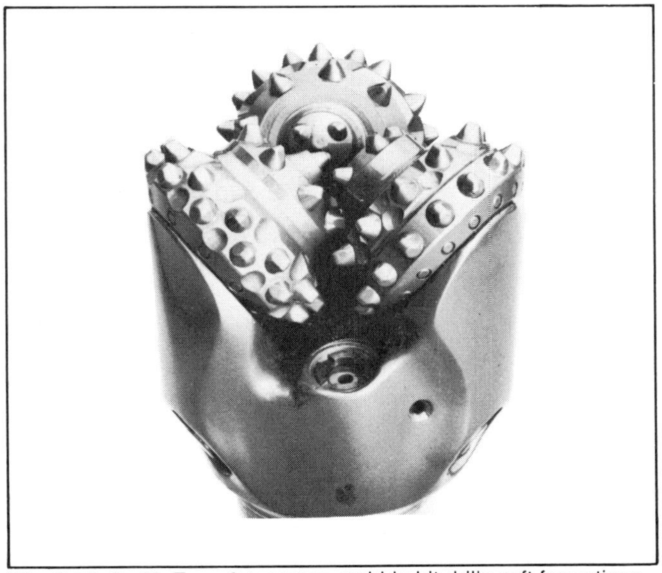

Fig. 2-35. Smith Type 2 tungsten carbide bit drills soft formations, such as unconsolidated sticky shales, soft shales, clays, red beds, salts, soft limestones, anhydrites, gypsum and tough, waxy shales. It is especially effective where heavier weight drilling muds are used.

Large diameter tungsten carbide inserts offer maximum strength and resistance to wear. The chisel-crest configuration of the inserts, plus maximum cutter offset, provides efficient gouging-scraping action and fastest penetration in soft formations.

Precise insert spacing insures smooth rotation and prevents tracking. Bottom hole cleaning is aided by deeper relief grooves on three cutters.

Recommended weight ranges between 2000 and 5000 lb/in. of bit diameter with corresponding rotary speeds of 40 to 70 rpm.

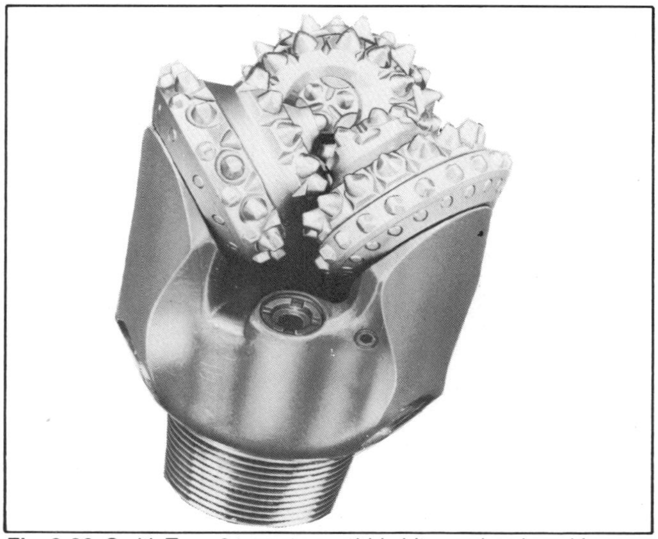

Fig. 2-36. Smith Type 3 tungsten carbide bit was developed for use in medium hard formations. These include sandy shales and medium hard shales with some softer streaks. Penetration rates are as high as with medium soft formation milled tooth bits because of the chisel-crest inserts with maximum extension from cone body.

Improved bottomhole cleaning and flushing is made possible by use of long inserts and deep relief grooves on the cones. Inserts are made of selected grades of tungsten carbide to provide balanced wear resistance and strength. Cutters are machined from special quality steel for superior strength and insert retention.

Recommended drilling weights are 2000 and 5000 lb/in. of bit diameter with rotary speeds from 40 to 70 rpm.

Fig. 2-39. Smith Type 5 tungsten carbide bit was developed to drill hard formations such as unconsolidated sandy shales, sandy limestone, dolomite and hard brittle shales.

Two grades of chisel-crest shaped tungsten carbide inserts are used on the cutters—one for inner rows and a more wear resistant grade for gauge rows. Extra inserts are placed on the inner end of all three cutters to give added coring protection.

The recommended drilling weights range from 2000 to 6000 lb/in. of bit diameter, with corresponding rotary speeds of 40 to 65 rpm.

Drill collar stabilizers are recommended for maximum bit life. Low density drilling fluids help in obtaining best penetration rates.

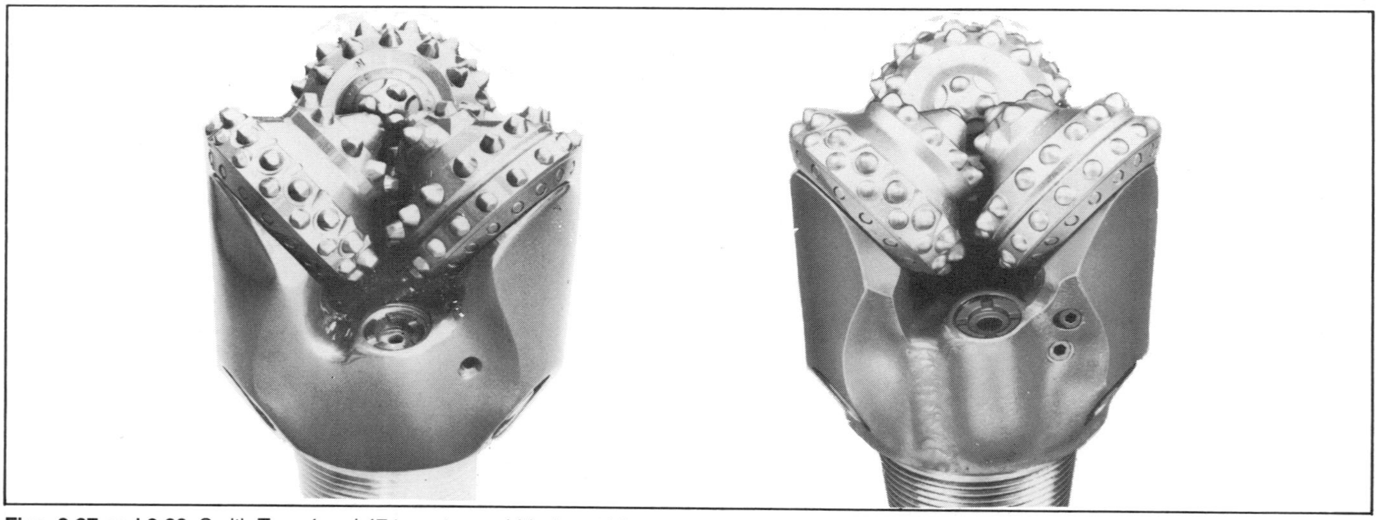

Figs. 2-37 and 2-38. Smith Type 4 and 47 tungsten carbide insert bits were developed to drill in medium hard formations. These include shales and gypsum through harder shales and harder limestone, dolomite and medium hard sand. Chisel-crest shaped inserts on the Type 4 offer deep penetration and fast drilling rates.

Where broken formations are encountered and greater toughness is required, Smith Type 47 is suggested. Its projectile-shaped inserts provide the strength necessary to drill through formations.

Medium gouging-scraping action is provided by medium cone offset. Varied insert pitch in all drive rows reduces tracking while deep relief milling speeds up chip removal.

Both the Type 4 and 47 bits have built-in ruggedness based upon a larger journal diameter and a special quality cutter steel for superior strength and insert retention.

Operating weights suggested are 2000 to 5000 lb/in. of bit diameter with rotary speeds of 35 to 65 rpm.

its apex. Obviously, the cone cannot do both, so there must be some slippage (Fig. 2-14a).

The length of the gauge surface must be designed to provide enough gauge area to prevent cones from wearing undersize. The length of this gauge surface is restricted not only by cone diameter but also by bearing diameter in the cone.

When additional action is needed on bottom, one method is to offset cones by twisting their centerlines to avoid intersection at a common point on the bit axis (Fig. 2-14b). Cone offset increases drilling speed in most formations due to additional scraping action it gives cone teeth. The amount of desired offset, however, depends on type of formation

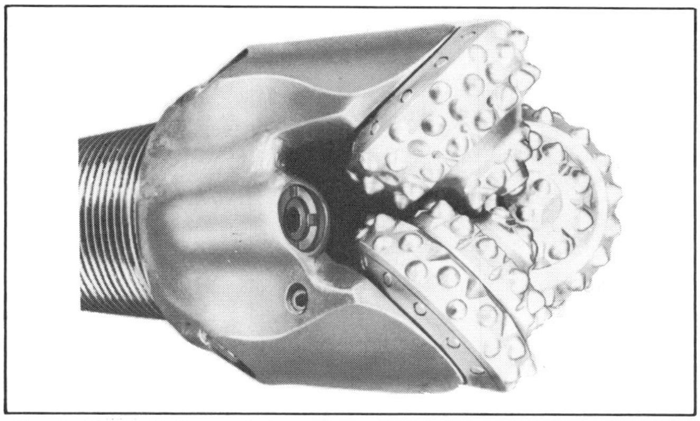

Fig. 2-40. Smith Type 57 tungsten carbide bit was designed to drill hard semi-abrasive formations. These include hard sandy limestone, shale, limestone and hard dolomite with cherty streaks.

Two grades of tough, projectile-shaped or conical-shaped tungsten carbide inserts are used on the cutters.

Inserts with extra high abrasion resistance are used on the cutter gauge rows to provide added gauge protection and insure full gauge hole.

Extra inserts of a different grade are added to inner end of cutting structure to prevent coring.

Bit is designed to have true cone roll to provide maximum chipping-crushing action.

Recommended weights and speeds range from 3000 to 8000 lb/in. of bit diameter and from 35 to 60 rpm.

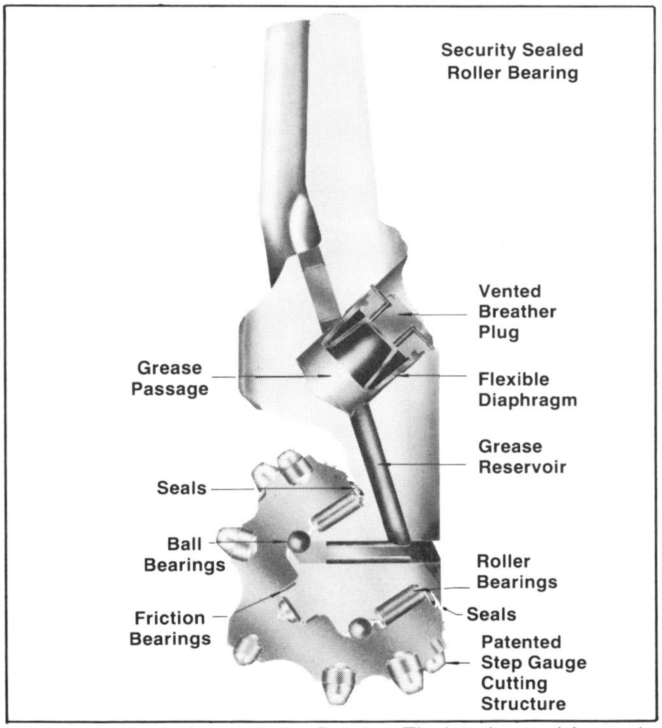

Fig. 2-42. Security Sealed Roller Bearing. The bearing seal, located between the back face of the cone and the arm journal, retains the lubricant and prevents drilling fluid and abrasive cuttings from entering the bearing cavities.

Bonded to the arm, the seal is permanent. Cutting cannot damage rubber by getting trapped between the seal and the shirttail. Contaminating well fluids and abrasive cuttings cannot get past the seal into the bearing cavities. At the other end of the system is a vented breather plug which allows fluid pressures downhole to push against the lubricant-filled flexible diaphragm, so that pressures surrounding the bearing seals are equalized.

The diaphragm-controlled reservoir, which supplies lubricant to the bearing cavity, acts as a surge chamber to allow for volumetric changes within the bearing cavity during operations.

Fig. 2-41. Smith Type & tungsten carbide bit is a very hard formation bit developed to provide maximum penetration and bit life through harder or higher compressive strength formations. These include granite, quartzite, chert, flint, novaculite, taconite, basalt, quartzitic sand and pyrite.

Conical shaped inserts withstand high loads and provide good penetration rates. All three cones have multiple inserts toward center of the hole. The true roll feature of Type 9 gives maximum chipping-crushing action.

On the gauge surfaces, closely spaced inserts of sintered tungsten carbide give added gauge protection and insure a full gauge hole.

Normal operating weight ranges between 3000 and 8000 lb/in. of bit diameter with rotary speeds of 30 to 55 rpm.

Drill collar stabilizers are recommended for maximum bit life.

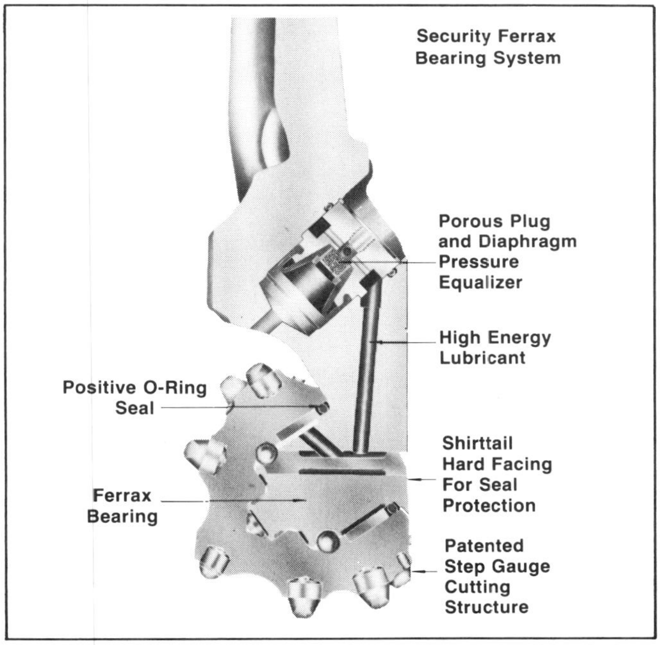

Fig. 2-43. The FerrAx Bearing System is Security's journal-type bearing combined with a special lubrication system. These features are well illustrated in this cutaway view.

being drilled, because this scraping action is more conducive to bit wear in abrasive formations.

Grooves in cones form inner and outer sides of cone teeth and determine tooth crest length. Lands between the grooves are then notched to form circumferential rows of chisel-shaped teeth. It is desirable to make these crest lengths short enough to assure easy penetration of the formation, but long

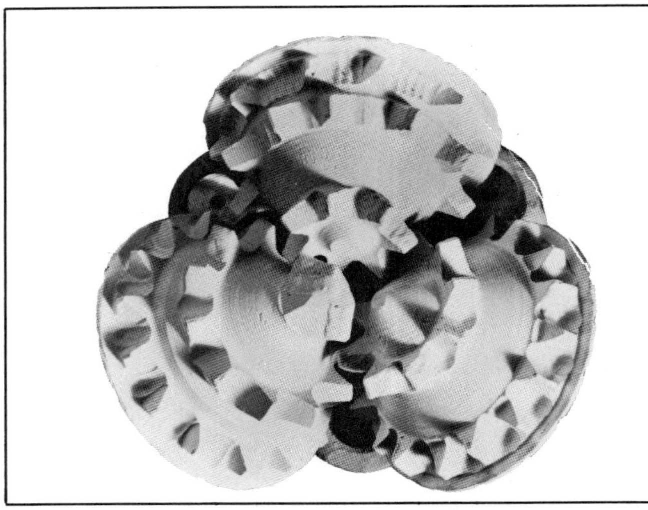

Figs. 2-44 and 2-45. DS and DM Deviation Control Bits. For soft and medium formations, these bits are specially designed to stay on course in directional drilling and to hold or correct angle in crooked holes. Special bit geometry plus unique, patented "Guide Track Groove" eliminates tendency to drill into the side wall.

Available, on order, in most popular sizes. DS design is also available incorporating sealed bearing, the DSS.

Fig. 2-46. Security Type S84 Sealed Bearing Insert Bit is designed for use in sandy shales, medium hard shales and unconsolidated shales and sands with some streaks for softer formations. S84 combines lower journal angle with largest cone offset for maximum gouging, scraping action; larger diameter tooth-shaped inserts with longer extensions are widely spaced to provide maximum penetration.

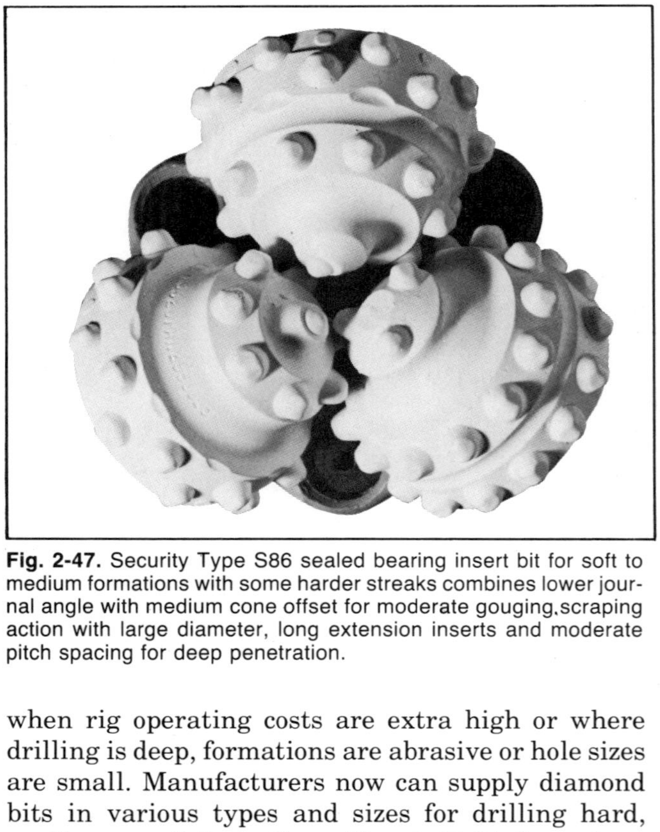

Fig. 2-47. Security Type S86 sealed bearing insert bit for soft to medium formations with some harder streaks combines lower journal angle with medium cone offset for moderate gouging,scraping action with large diameter, long extension inserts and moderate pitch spacing for deep penetration.

enough to provide strength against breakage.

Cone teeth must have sufficient included angle to provide adequate strength to resist breakage but still not be thicker than necessary. If they are made excessively thick, they will be too blunt in a half-dulled condition to penetrate the formation satisfactorily.

Earlier chert bits would not drill soft formations at an acceptable rate but more recent development of long tooth-like inserts has alleviated this shortcoming. Practically all types and sizes of rolling cone bits are now available with sealed lubrication roller bearings and some are available with sealed lubrication journal-type bearings.

The use of diamond bits for drilling oil wells has become an accepted practice. Despite higher initial cost, they become competitive with roller cone bits when rig operating costs are extra high or where drilling is deep, formations are abrasive or hole sizes are small. Manufacturers now can supply diamond bits in various types and sizes for drilling hard, medium or soft formations. Figs. 2-51-54 show views of diamond bits now in use.

String Reamers

In earlier years of rotary drilling practice, considerable difficulty was experienced in keeping the hole round and out to gauge. Extra long fishtail bits were sometimes used to solve this problem. All earlier roller cutter rock bits included two or more side roller reamers on body of bit, a foot or more above the bottom. With the advent of short bodied roller cutter

Fig. 2-48. Security Type S88 sealed bearing insert bit for medium to medium hard formations combines medium journal angle and medium cone offset with closely spaced long extension tooth-shaped inserts for drilling in sandy shales, dolomites, hard shales and soft limestone with streaks of softer or harder formations.

Fig. 2-49. Security Type M88 sealed bearing insert bit. Mounted in near true-rolling cones, the medium length wedge shaped tungsten carbide inserts are widely pitched and extended from raised lands for deeper bites and maximum fluid cleaning action. Moderate weight and rotary speeds let the M88 penetrate fast in medium hard to hard and abrasive formations. Maximum gauge life is provided for the hard abrasive formations by use of higher strength conical shaped gauge trimmers.

Fig. 2-50. Security Type H88 sealed bearing insert bit. Non-offset, true-rolling cones are studded with closely spaced rows of 90° conical shaped inserts and 120° step-gauge wall trimmers. Together they create a tough, balanced cutting structure that chips out large cuttings and allows them fast exit; provides an even cut on bottom, a full-size hole all the way down; combines moderate rotary speeds with moderately high weight on the bit.

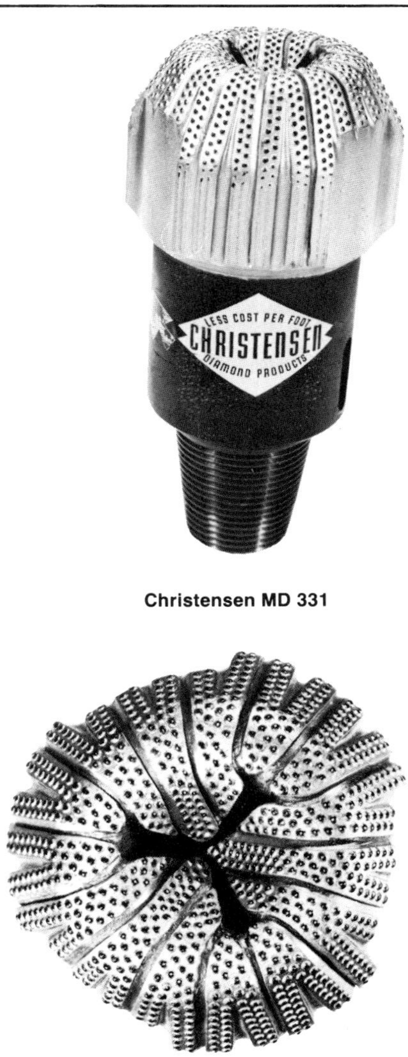

Fig. 2-51. Christensen MD331 diamond bit. Two views.

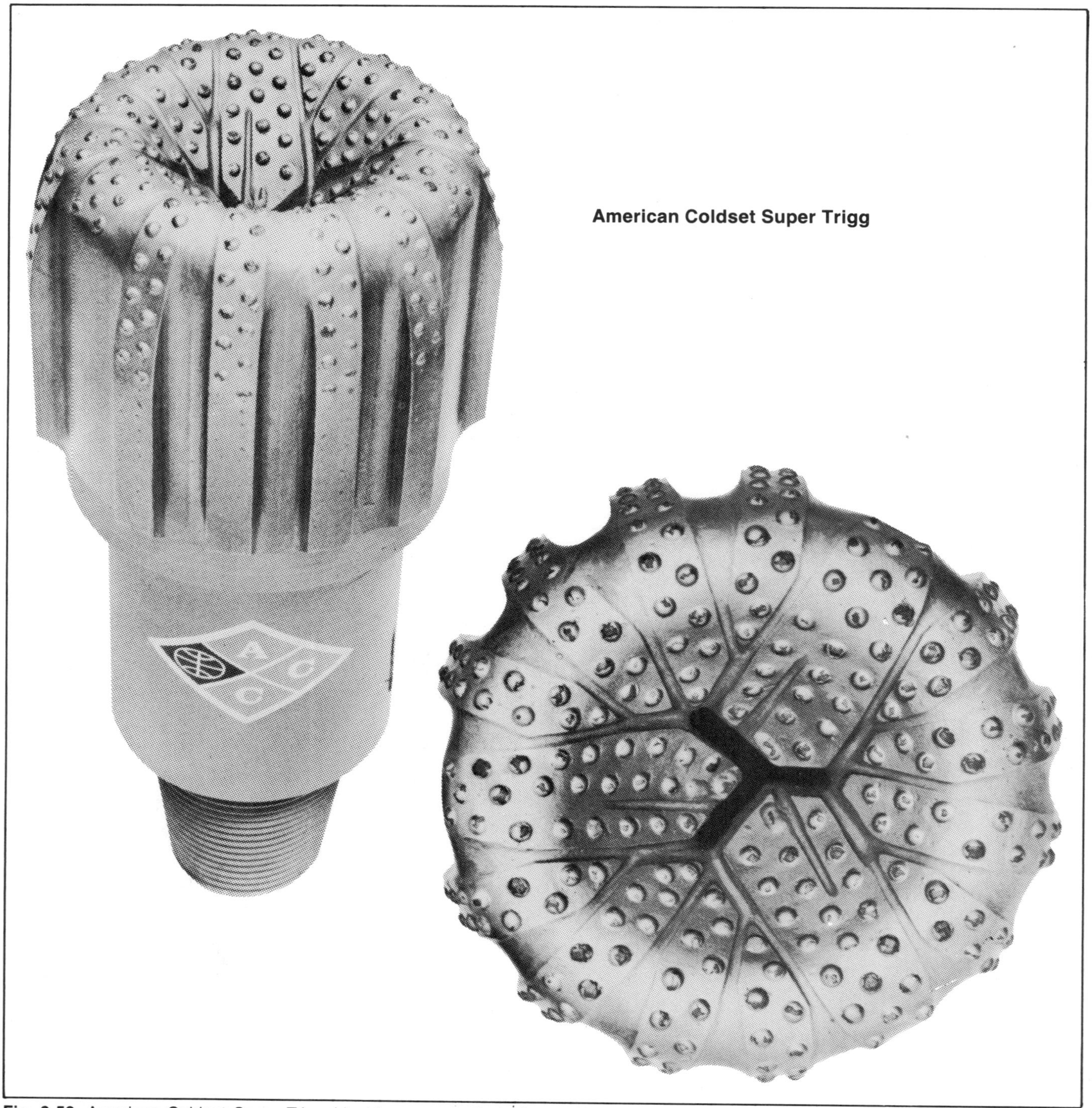

Fig. 2-52. American Coldset Super Trigg (double concentration) diamond bits. Two views.

bits, string reamers came into use, some with three roller-reamers spaced 120° apart around the body while others used two sets of rollers.

String reamers when run just above the bit serve a dual function: first, to assure a round full gauge hole and second, to stabilize the bit and improve effectiveness. One or more roller-type reamers used in the drill collar string, 30, 60, or 90 ft above the bit, tend to limit drill string buckling and prevent doglegs.

Fig. 2-55 illustrates common types of roller-reamers. Figs. 2-56 and 2-57 illustrate some types of hardfaced reamers or stabilizers which are commonly used to prevent buckling of the drill collar string, a technique that has been developed to a high degree of proficiency.

Rotary Percussion Devices

Percussive blows are most effective in rock destruction. The common air driven rock drill has met and survived all competition, for at least a century, in drilling relatively shallow, small diameter holes in

Fig. 2-53. NL Hycalog type MS diamond bit (two views) recommended for drilling medium soft strata such as medium bonded, interbedded sand with shale, granular limestones and most unaltered precipitates. Medium large diamonds with good exposure are used. Three sided configuration minimizes swabbing effect.

Fig. 2-54. Two views of Christensen's experimental bit using bonded poly-crystalline diamond cutting inserts developed by General Electric.

Fig. 2-55a. Three point bit reamer. This type reamer is run between the drill collars and the bit. The distance from bit to reamer is held to a minimum to insure less reaming back to bottom with a new bit.

Fig. 2-55b. Three point string reamer. Run in the drill collar or pipe string this reamer provides centering of drill collars to drill straighter holes in crooked hole areas. When run in the drill pipe, reamer is effective in reaming out doglegs, key seats and ledges in the hole.

Fig. 2-55c. Six point bit reamer. This type reamer is run between drill collars and the bit, as is the three point design, but is used where more stabilization or greater reaming capacity is required. When drilling in crooked hole areas, packing the hole with a six point reamer has been found very satisfactory in preventing sharp changes in hole angle.

rock. It is limited in depth application primarily by its lack of ability to remove the debris.

Solid-rod percussion tools, referred to as Canadian and Galician systems, were used rather extensively for drilling oil wells in Russia during the 19th and early part of the 20th century. Rods permit high frequency downhole percussive blows.

The American adaptation and/or modification of the rod drilling system provides for hollow rods and a circulating system. The rods are reciprocated at high frequency by surface machinery and a continuous circulation of fluid removes the cuttings. The bits employed are not unlike large star bits as used in rock

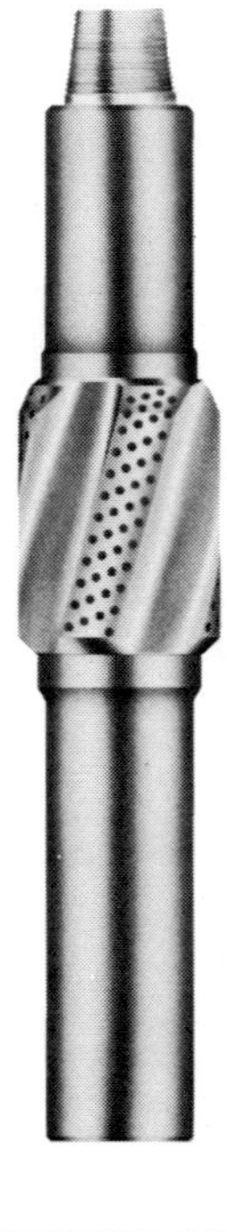

Fig. 2-56. Drilco Integral Stabilizer is a rugged rotating stabilizer generally used in making up close fitting bottomhole assemblies designed to prevent abrupt hole angle changes. One piece construction is used for hole sizes 6⅛ in. thru 7⅞ in. and is furnished for either bottomhole or string application. Replaceable shrunk or sleeve types are available for hole sizes 7⅝ in. through 17½ in. Pressed in large diameter carbide compacts prolong wear life of blades.

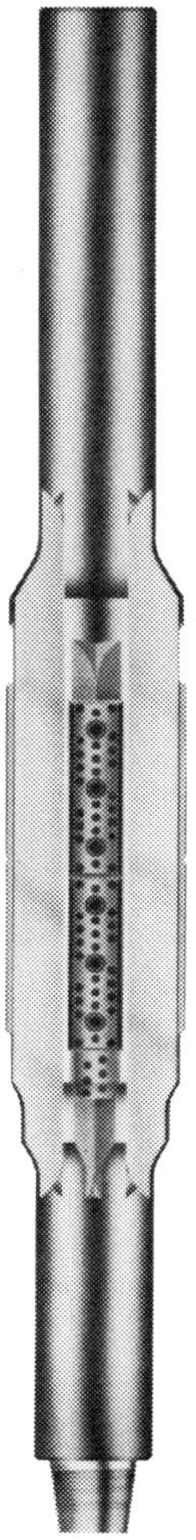

Fig. 2-57. Drilco "Full-Flo" replaceable wear pad stabilizer is designed for use low in the drill collar string or on bottom next to the bit. Long wear pads make it useful for packed hole stabilization and the fluted body provides adequate circulation passage.

drill practice. The hollow rod circulating system has been improved through the years and is still in use in some areas of this country for drilling water wells.

The success of high-frequency percussion drilling methods in shallow wells has led to research and experimentation with percussion devices that can be used with the rotary system to augment drilling action of the rotating bit. The devices which have thus far shown promise employ the old rock drill principle of the reciprocating hammer striking an anvil to which the rotary bit is attached. Mud-powered hammer tools, when perfected, will offer widest possible application. The problems, however, seem to be numerous.

In air-drilling practice the problems are somewhat simpler and as a result the air-hammer percussion tool has shown rather phenomenal increase in penetration rates. Wherever air drilling is used in preference to mud drilling, the hammerdrill (Fig. 2-58) will likely increase both the penetration rate and bit foot-

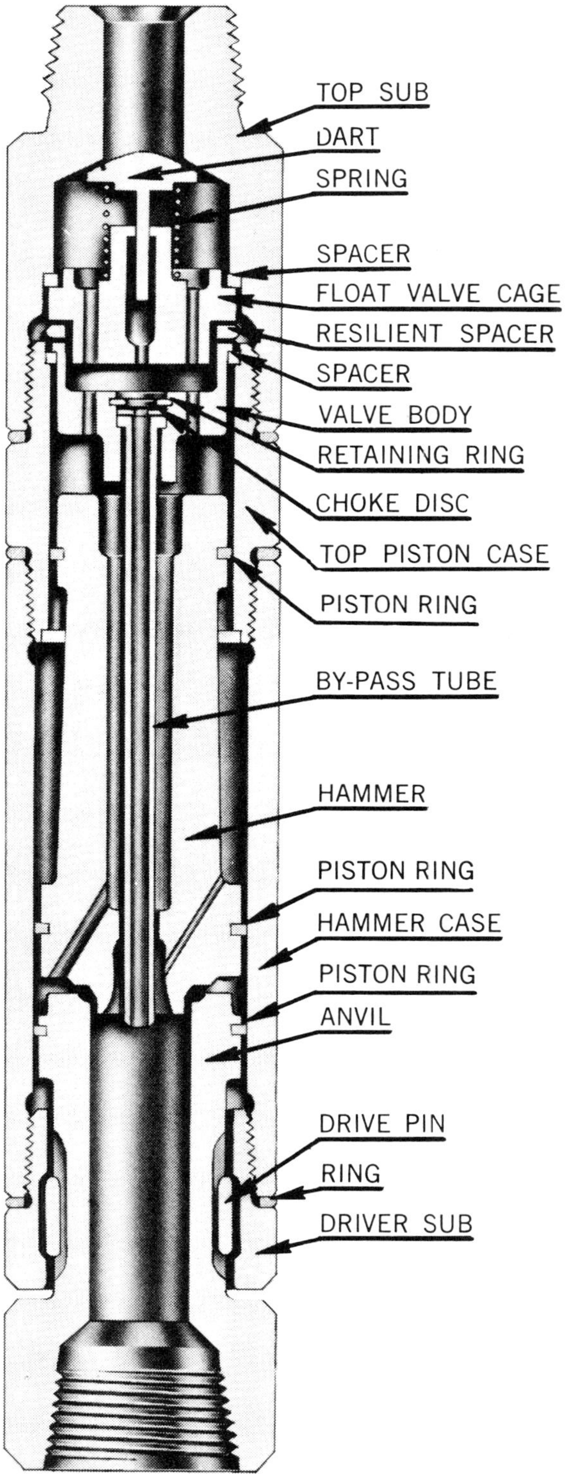

Fig. 2-58. Mission's "Hammerdril" is a pneumatically operated bottomhole device that combines the percussive action of cable tool drilling with the rolling cone action of rotary drilling. It can be used on any rotary rig using an adequate volume of air or gas as the circulating fluid.

The hammer blows are of high frequency and are transmitted directly to the bit, just below the tool. It has proved effective in straight hole drilling of steep or dipping formations.

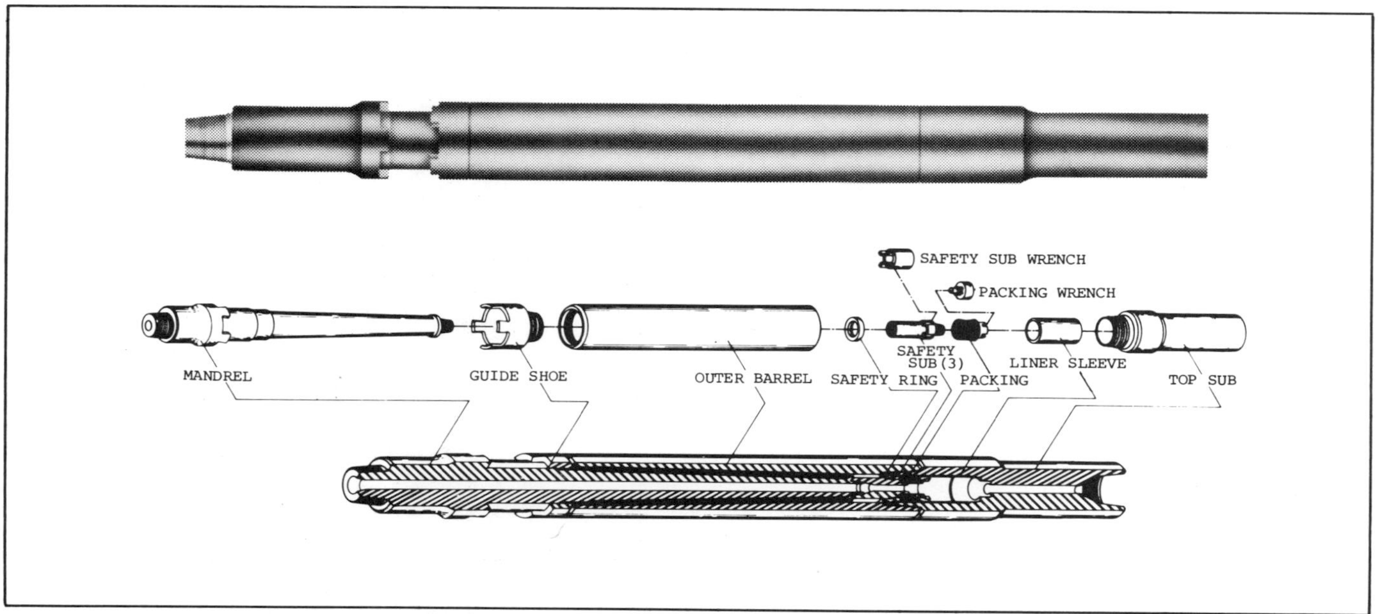

Fig. 2-59 a and b. The Drilco shocksub usually is run between the bit and drill collars to absorb shock loads produced by the bit while drilling.

The device contains a long rubber element which, being bonded (vulcanized) to the ID of the barrel and the OD of the mandrel, transmits torque and weight to the bit. The rubber element is limited to operation in holes not over 200 deg F. For hotter holes the tool with a steel spring should be used (2-59b). In packed hole assemblies, using a square drill collar, the shock-sub usually is run between square collar and the regular collars.

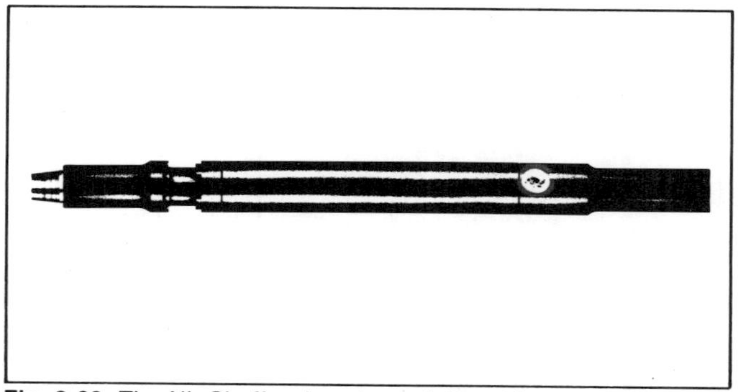

Fig. 2-60. The NL Shaffer damping sub is made up on the string immediately above the bit. Drilling weight is applied to the bit through the shock absorbing element in the tool. Thus bit induced vibrations and impact loads are isolated from the drill collar string and dampened by load deflection characteristics of the shock absorbing element.

Also shown is an enlarged view of the shock absorbing element which is made from stainless steel wire woven into a cloth-like mesh and compressed to the required shape in a forming die. In operation, as the load is applied, the length is compressed and radial thickness is increased.

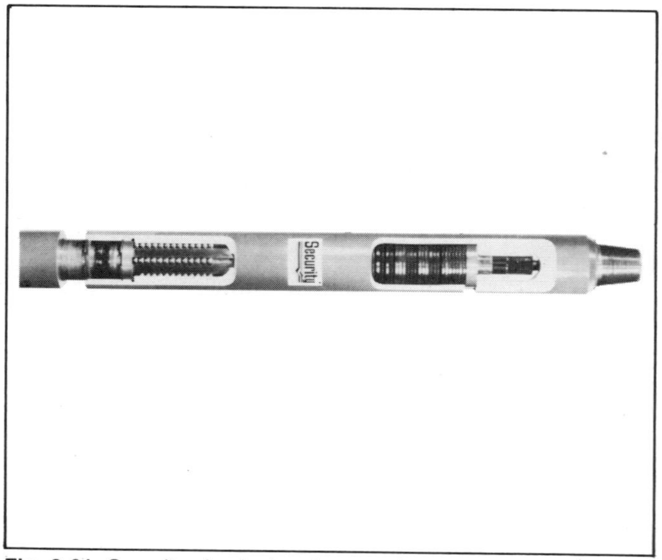

Fig. 2-61. Security shock absorber employs a stack of Belleville-type steel washers as the shock absorbing element.

age. Efforts are continuing to adapt this and similar tools to mud-powered operation.

Downhole Shock Absorbers

High frequency vibrations imparted to the bit by a hammer are highly desirable while natural vibrations developed by the bit can be destructive to drill collar connections. This is particularly true in drilling hard rock formations where the rolling bit action actually produces some percussive blows that help improve the drilling rate.

Obviously a free acting shock absorber between the bit and drill collar string might serve the dual purpose of letting the bit vibrate naturally while at the same time protecting the drill collar string from vibration-induced damage. Fig. 2-59 illustrates one such device that has proved effective in many areas. Fig. 2-59 also shows a damping sub which employs a stainless steel spring absorbing element, suitable for high temperatures. Fig. 2-60 illustrates a damping sub utilizing a stainless steel mesh type shock absorbing element. Fig. 2-61 illustrates a shock absorber which uses stacks of Belleville-type washers as the shock absorbing medium.

Chapter 3
Drill String Components

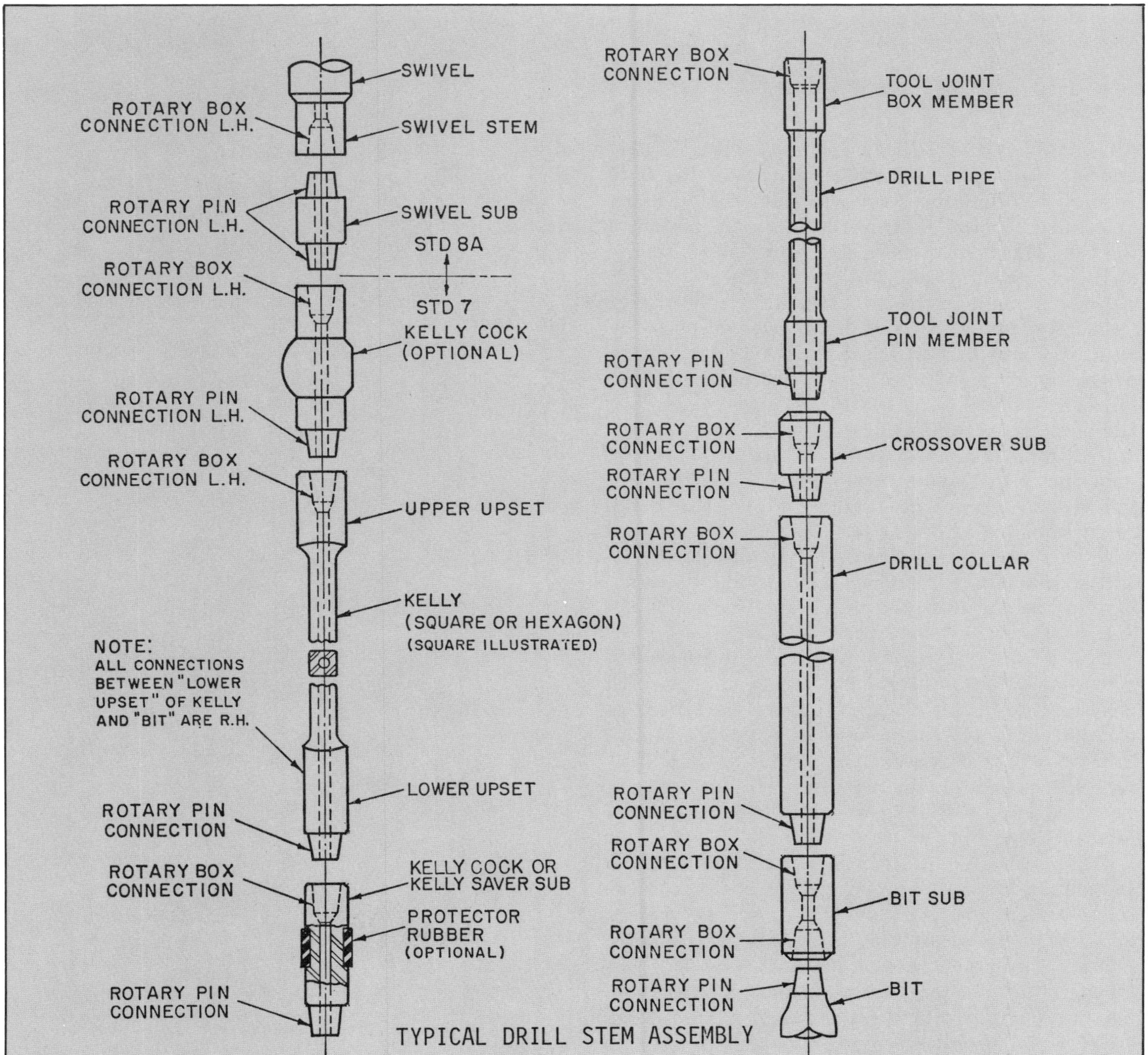

TYPICAL DRILL STEM ASSEMBLY

Drill Collars

The early-day drill collar was a heavy-wall, double-box sub used to connect the bit to the drill pipe. When Hughes simplex bit with a lubricating system arrived it was necessary to employ an over-shot drill collar at least 17 ft long to provide space within its inner bore to accommodate a reservoir for the lubricant. Soon afterwards it was common practice to include at least one additional drill collar which, with one joint of drill pipe, made up a full stand to set back.

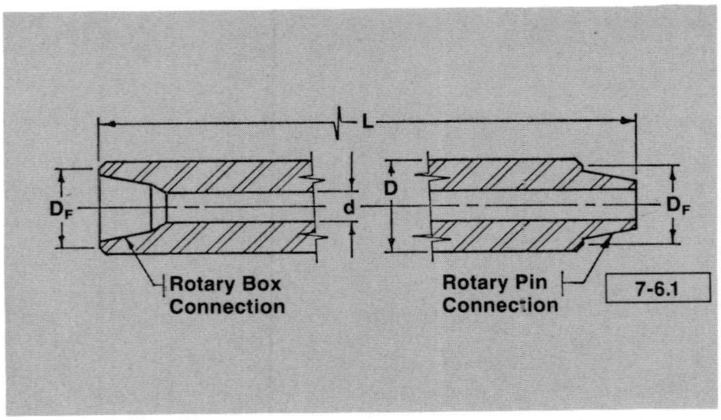

Fig. 3-1. Drill collars.

Use of long drill collar assemblies began in 1934 and has been expanded, in most hard rock areas, to 20 or more 30-ft lengths. Rotary bits, in order to make hole, must be forced downward against the bottom of the hole. This thrust is provided by that portion of the drill string weight not supported by the rig.

At some point in the drill string there is neither tension nor compression. This is referred to as the neutral point which will move up or down the string depending on amount of weight applied to the bit. Below the neutral point the drill string is in compression and subject to buckling forces.

Drill pipe, if rotated while buckled, will not last long due to the severe transverse stresses that are induced. Heavy-wall drill collars, on the other hand, resist buckling much better. It is considered good practice therefore to have more than enough weight in the drill collars to provide the desired bit weight. As long as the neutral point is in the drill collar section all drill pipe will be in tension.

Drill collars are made of alloy steel, heat-treated after machining, and are quite rugged (Fig. 3-1). Even so, they buckle enough to induce heavy stresses in connections and cause wear on the outer surfaces. The drill collar connection, being the weakest part, has been subject of much research and development in recent years with the result that failures now are less frequent.

Further progress surely will be made as time passes. Notable improvements have been in thread form and treatment (Figs. 3-2 and 3-3), stress-relief grooves both on pin and in box (Figs. 3-4 and 3-5) and balance in design bending strengths of pin and box. The bore through the pin is important to joint strength so step bore or reduced bore for a foot or more of the pin-end of a collar will permit increased pin strength without significantly increasing hydraulic pressure losses (Fig. 3-6).

Pin and box threads serve the primary functions of pulling and holding together their respective shoulders. The shoulders must be of sufficient width, strength and finish to resist deformation when subjected to bending forces and also provide an effective fluid seal.

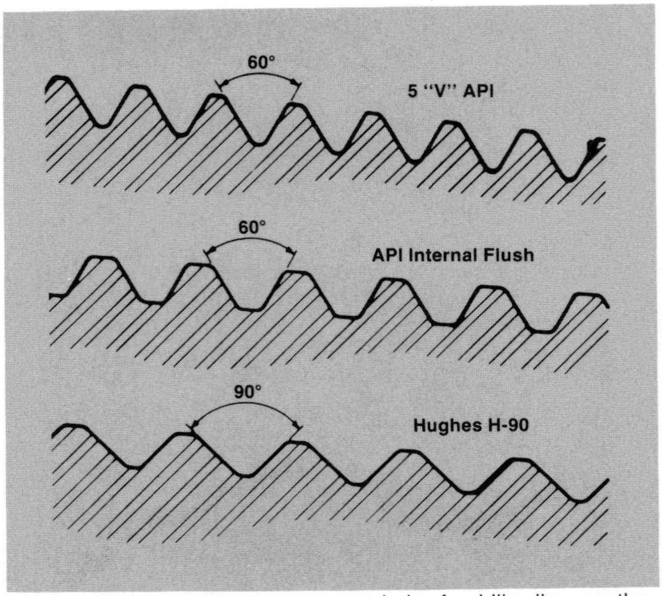

Fig. 3-2. Sharp thread roots are poor choice for drill collar use; the 4½-in. full hole and 4½-in. regular threads have such sharp threads. The internal-flush and H-90 threads are better since they are flatter and have large radii at corners.

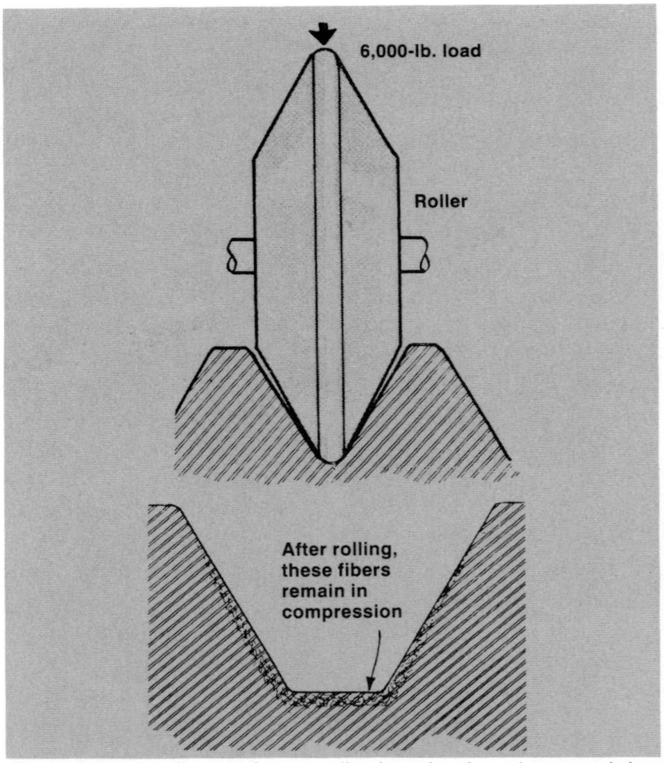

Fig. 3-3. Hydraulic ram forces roller into the thread root and the roller is then moved down the thread spiral. Roller is of the same contour as the thread root and pressure between the two causes a cold working of the metal in the thread root.

Resistance to fatigue failure is property of cold-worked metal fibers. Even after thread rolling is completed, these fibers remain in compression; as a result they can withstand tensile stresses longer without fatiguing.

The conventional drill collar has a cylindrical outer surface. That is, they are round and uniform OD from shoulder to shoulder. There are many specialized de-

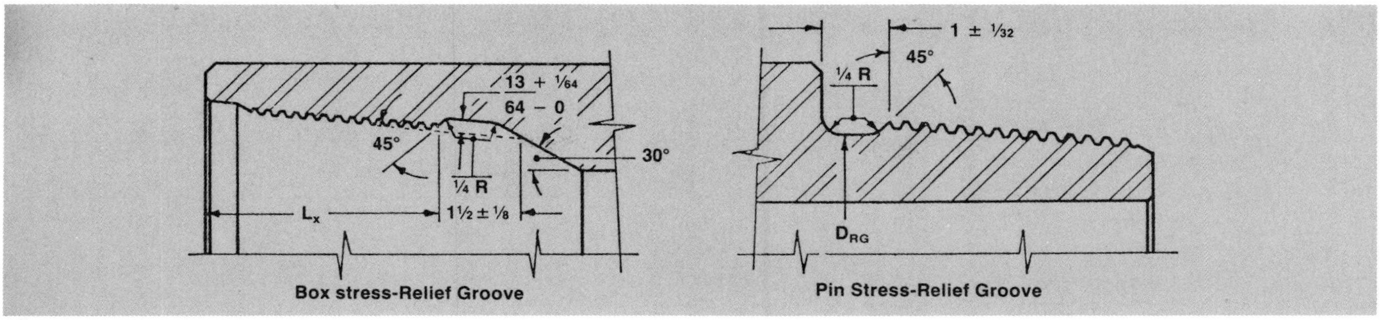

Fig. 3-4.

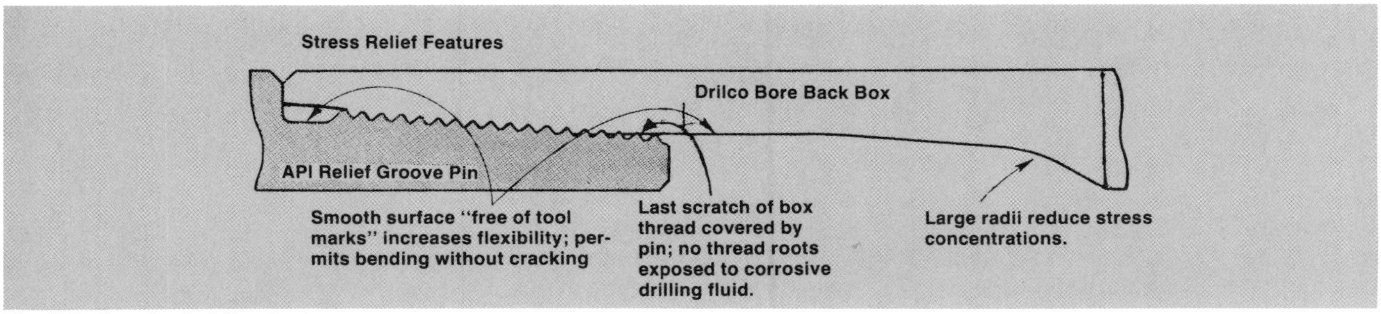

Fig. 3-5. Bore back drill collar box.

signs. One variation is inclusion, below the box, of a slight recess on the OD to provide a shoulder for elevators, thus making lifting subs unnecessary. A second recess further down can be provided for safe use of slips, thus making use of a safety clamp unnecessary (Fig. 3-7).

A second variation is spiral grooving of outer surface, except near the connections (Fig. 3-8). These spiral grooves, usually three in number spaced 120° apart, prove effective in prevention of wall sticking caused by fluid pressure differential when collars are opposite permeable zones.

A third variation is the square drill collar (Fig. 3-9). Normally only one square drill collar is used just above the bit when drilling in crooked hole formations. Its purpose is to reduce rapid change in hole angle. The square drill collar becomes a long stabilizer fitting the hole closely yet allowing free circulation between the flats of the square and the round walls of the hole. The square collar assures a smooth hole free of doglegs. Usually the distance across corners of the square is 1/16-in. less than bit size. Frequently a reamer below and a stabilizer above the square collar are used. For effective use the formation must be firm enough to provide support for the assembly.

Drill Pipe and Tool Joints

An entire chapter could be devoted to this subject. Drill pipe, fitted with tool joints,is, in a sense, an expendable item and as such contributes substantially to the cost of drilling a well. Both drill pipe and

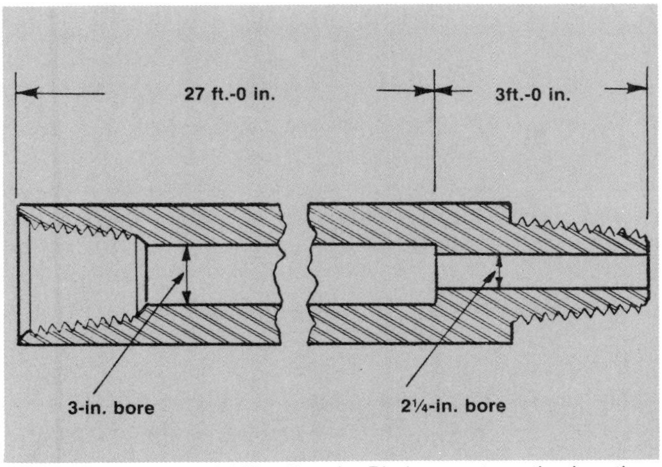

Fig. 3-6. Step bore of drill collar pin. Pin keeps strength when the collar is step-bored as in this drawing. Bore for some 3 ft on pin end is smaller so that there is more metal in joint pin. In remainder of joint, bore is larger to prevent excess pressure drop in mud stream.

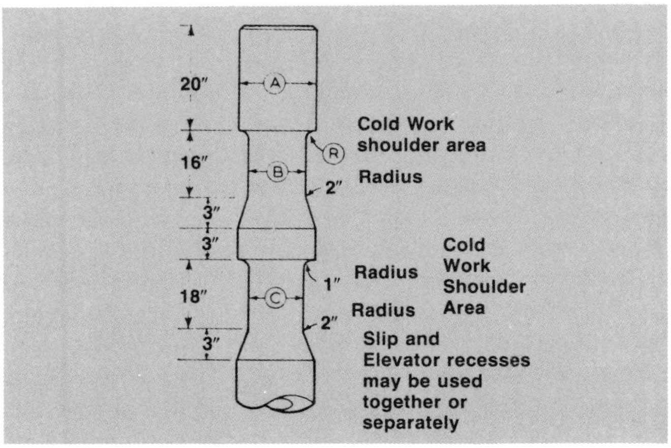

Fig. 3-7. Elevator/slip recess.

TABLE 3-1

Dimensions and Properties of Pipe Body (All Drill Pipe Types)

Drill Pipe				Collapse Resistance*(PSI)				
Size O.D.	Nominal Weight Per Ft.	Wall Thickness	Inside Diameter	Grade D	Grade E	Grade X	Grade G	Grade S
In.	Lb.	In.	In.					
2-7/8	10.40	0.362	2.151	12,110	16,510	20,910	23,110	29,720
3-1/2	9.50	0.254	2.992	. . .	10,040	12,060	13,050	15,720
3-1/2	13.30	0.368	2.764	10,350	14,110	17,880	19,760	25,400
3-1/2	15.50	0.449	2.602	12,300	16,770	21,250	23,480	30,190
4	11.85	0.262	3.476	. . .	8,410	9,960	10,700	12,590
4	14.00	0.330	3.340	8,330	11,350	14,380	15,900	20,170
4	15.70	0.380	3.240	9,460	12,900	16,340	18,050	23,210
4-1/2	13.75	0.271	3.958	. . .	7,200	8,400	8,950	11,070
4-1/2	16.60	0.337	3.826	7,620	10,390	12,750	13,820	16,800
4-1/2	20.00	0.430	3.640	9,510	12,960	16,420	18,150	23,330
4-1/2	22.82	0.500	3.500	10,860	14,810	18,770	20,740	26,670
5	16.25	0.296	4.408	. . .	6,970	8,090	8,610	9,860
5	19.50	0.362	4.276	7,390	10,000	12,010	12,990	15,700
5	25.60	0.500	4.000	9,900	13,500	17,100	18,900	24,300
5-1/2	21.90	0.361	4.778	6,610	8,440	10,000	10,740	12,710
5-1/2	24.70	0.415	4.670	7,670	10,460	12,920	14,000	17,050

Drill Pipe		Internal Yield Pressure*(PSI)					Pipe Body Yield Strength*(Lb)				
Size O.D.	Nominal Weight Per Ft.	Grade D	Grade E	Grade X	Grade G	Grade S	Grade D	Grade E	Grade X	Grade G	Grade S
In.	Lb.										
2-7/8	10.40	12,120	16,530	20,930	23,140	29,750	157,180	214,340	271,500	300,080	385,820
3-1/2	9.50	. . .	9,520	12,060	13,340	17,140	. . .	194,260	. . .	272,000	350,000
3-1/2	13.30	10,120	13,800	17,480	19,320	24,840	199,150	271,570	343,990	380,190	488,820
3-1/2	15.50	12,350	16,840	21,330	23,570	30,310	236,700	322,780	408,850	451,890	581,000
4	11.85	. . .	8,600	10,890	12,040	15,470	. . .	230,750	292,290	323,050	415,350
4	14.00	7,940	10,830	13,720	15,160	19,490	209,260	285,360	361,460	399,500	313,650
4	15.70	9,140	12,470	15,790	17,460	22,440	237,690	324,120	410,550	453,770	583,420
4-1/2	13.75	. . .	7,900	10,010	11,070	14,230	. . .	270,040	342,050	378,050	486,070
4-1/2	16.60	7,210	9,830	12,450	13,760	17,690	242,410	330,560	418,700	462,780	595,000
4-1/2	20.00	9,200	12,540	15,890	17,560	22,580	302,400	412,360	522,320	577,300	742,240
4-1/2	22.82	10,690	14,580	18,470	20,420	26,250	345,580	471,240	596,900	659,740	848,230
5	16.25	. . .	7,770	9,840	10,880	13,990	. . .	328,070	415,560	459,300	590,530
5	19.50	6,970	9,500	12,040	13,300	17,100	290,100	395,600	501,090	553,830	712,070
5	25.60	9,620	13,120	16,620	18,380	23,630	388,770	530,140	671,520	742,200	954,260
5-1/2	21.90	6,320	8,610	10,910	12,060	15,510	320,550	437,120	553,680	611,960	786,810
5-1/2	24.70	7,260	9,900	12,540	13,860	17,830	364,630	497,220	629,810	696,110	895,000

*All values are minimum for the pipe body based on API formulas tool joints are covered in detail by API specifications, which are followed closely around the world.

API specifications cover two grades of drill pipe, based on the physical properties of each grade. They are: Grade D, minimum yield strength 55,000 psi; Grade E, minimum yield strength 75,000 psi. Also available are non-API Grade G, minimum yield strength 105,000 psi; and Grade S, minimum yield strength 135,000 psi.

Drill pipe is furnished internal-upset, external-upset, internal-external-upset and in random lengths as follows: Range 1: 18-ft to 22-ft inclusive; Range 2: 27-ft to 30-ft inclusive; Range 3: 38-ft to 45-ft inclusive.

Note: Internal-external-upset is for weld-on tool joints only. Drill pipe sizes and weights extend from 2⅜-in. OD 4.85 lb/ft to 6⅝-in. OD 25.20 lb/ft. API Spec. 5A covers in detail the various properties of all tubular goods, including drill pipe. (Refer to the latest revised issue for current standards.)

Table 3-1 is an excerpt from Armco's Engineering Data Handbook, 1966 ed. It covers basic specifications of both API and non-API grades of drill pipe.

Tool joints, as used on drill pipe, have much in common with connections on drill collars, or perhaps the other way around. Tool joints usually are not an integral part of drill pipe as are drill collar connections but are semi-permanent attachments to each length of pipe. Much progress has been made in tool joint design both as to relative dimensions and as to metallurgy. Tool joints in new condition are no longer weaker than the body of the pipe but may become so with excessive wear and abuse.

Two methods of attaching tool joints to drill pipe are still in use: weld-on and threaded connection. Weld-on types are rapidly gaining preference in most areas; however, the threaded types have the advantage of being field replaceable. The integral type joint also would seem to offer some advantage but like the weld-on type it too must be produced in a plant or shop.

In addition to these basic types, many variations of each are available in the several drill pipe sizes and styles. Original API tool joint, subsequently referred to as Reg. API, provided for the bore of the pin to be much smaller than the ID of the internal upset of the pipe which, in smaller sizes meant significant restriction to fluid flow, but provided a strong pin.

API FH (full hole) was developed to alleviate this condition somewhat. It was determined that by adding perhaps ¼ in. to OD of pin, the bore could be increased as much as ¾ in. without impairing the strength of the connection. Proprietary designs (non-API) were next developed, having a pin bore equal to the ID of the internal upset portion of the drill pipe. Also available for use on externally upset drill pipe is the IF (internal flush) design which has a

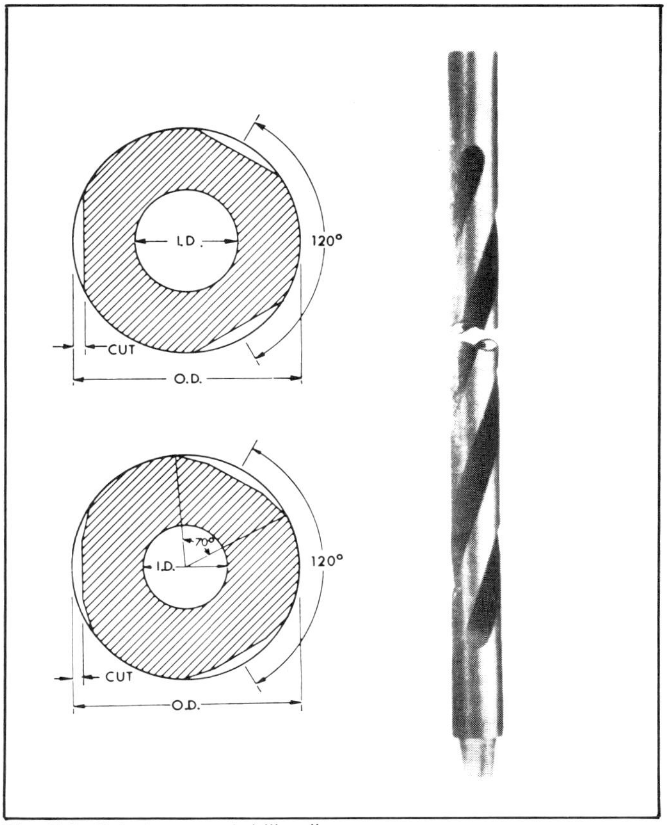

Fig. 3-8. Spiral grooved drill collars.

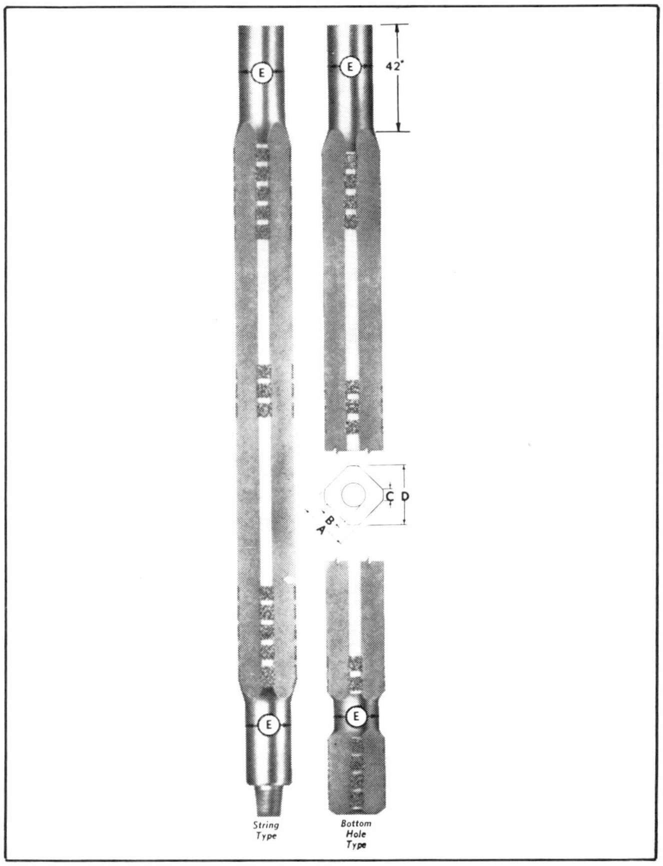

Fig. 3-9. Square Drill Collar

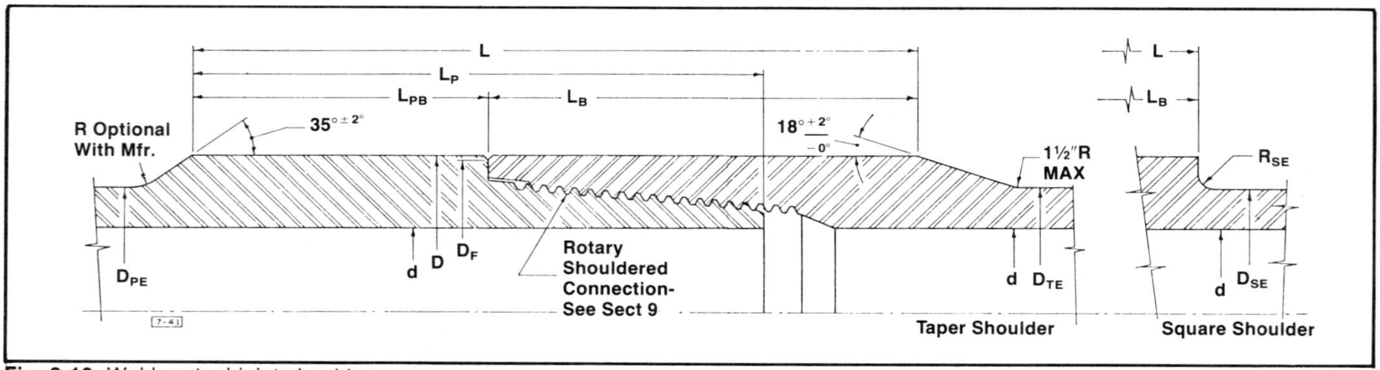

Fig. 3-10. Weld on tool joint shoulders.

pin bore approximately equal to the non-upset pipe section.

All weld-on tool joints are available with either a taper shoulder or a square shoulder for drill pipe elevators (Fig. 3-10). API FH joints can be welded onto internal upset pipe and IF joints can be welded onto external upset pipe. Also available and currently the most popular is an intermediate type joint welded onto IEU drill pipe having a slight external upset plus the internal upset.

API Spec. 7, 30th Edition, April 1975, specifies that "Tool joints shall be of the weld-on type and shall be furnished in the sizes and styles shown in API table 4.2 as specified on purchase order." (See Table 3-2).

Kelly and Drive Bushings

Of all the peculiar terminology found in the oilfield perhaps none finds its origin as remote from this sphere of activity as does the kelly joint, first known as the griefstem.

The latter name was given to the joint because of the trouble drillers had in keeping their grip rings adjusted as it slid through the rotary table.

According to the story told by oil writer and author Jerry Robertson, there was a baseball player named Michael J. (King) Kelly with the Cincinnati team in 1878. He reached his pinnacle of fame with Chicago from 1880 to 1887 as a great base runner who made long slides. King Kelly was encouraged in his trips around the bases by his fans yelling "Slide, Kelly slide." Although he never saw a rotary rig, his name has been given to this square forged joint of pipe, developed by an engineer who doubtless could recall the great King Kelly (Fig. 3-11).

This joint has made great strides through the years in keeping with the advancements made in accurate machining, precision forging and hardening of fine steel. The kelly has become straighter, tougher and made with closer tolerances—truly a precision instrument.

Kelly substitutes, or short, extra hard joints, above and below the kelly have been added to prolong its life

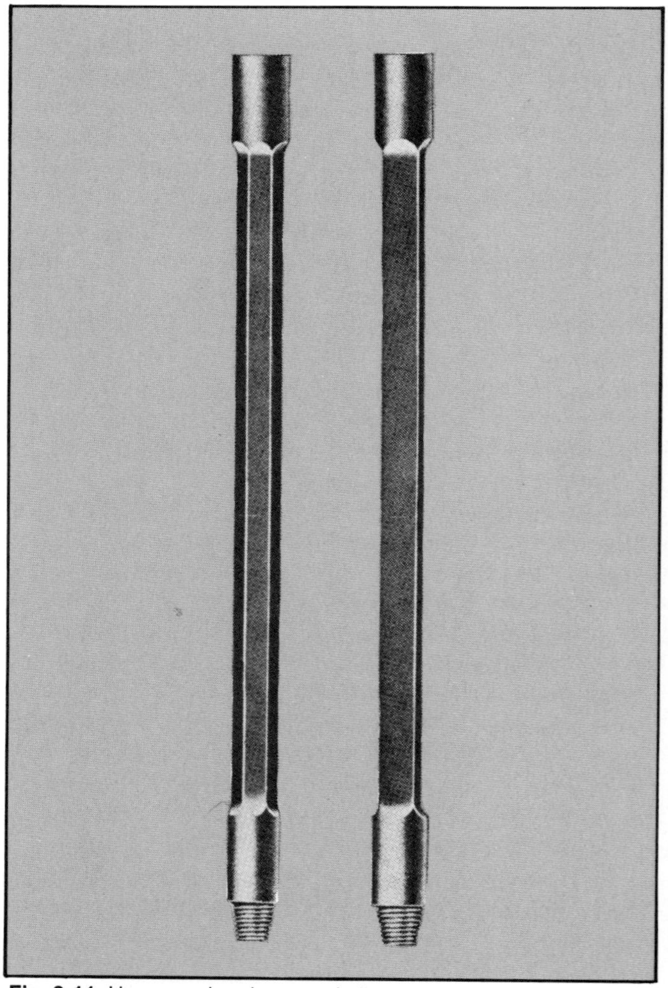

Fig. 3-11. Hexagonal and square kellys.

many times. Kelly drive bushings (Fig. 3-12) have been, and are being, improved. This bushing, as it revolves the kelly joint in the rotary table, holds the key to its success and the safety of the entire drill string. A sudden drop caused by hanging up will be sufficient to ruin a core bit or drill, or cause damage to the wireline. Drill pipe subjected to shock has a short life. If the driller does not control the drill stem at all times an easy fishing job becomes extremely difficult, and a difficult one becomes almost impossible. Because of this, several methods have been tried in an

TABLE 3-2

Tool Joint Dimensions For Grade E, X, G, and S Drill Pipe
All dimensions in inches. See Fig. 4.2

1	2	3	4	5	6	7	8	9	10	11	12	13	14
	Drill Pipe			Tool Joint									
Tool Joint Designation[1]	Size and Style	Nom. Wt.[2] lb per	Grade	Outside Dia of Pin and Box ± 1/32	Inside Dia of Pin[3] + 1/64 − 1/32	Bevel Dia of Pin and Box Shoulder ± 1/64	Total Length Tool Joint Pin + 1/4 − 3/8	Pin Tong Space ± 1/4	Box Tong Space ± 1/4	Com-bined Length of Pin and Box ± 1/2	Dia of Pin at Ele-vator Upset, Max	Dia of Box at Ele-vator Upset, Max	Tor-sional Ratio, Pin to Drill Pipe
				D	d	D_F	L_P	L_{PB}	L_B	L	D_{PE}	D_{TE}	
NC26(2 3/8 IF)	2 3/8 EU	6.65	E75	3 3/8*	1 3/4*	3 17/64	9	6	7	13	2 9/16	2 9/16	1.10
			X95	3 3/8*	1 3/4*	3 17/64	9	6	7	13	2 9/16	2 9/16	.87
			G105	3 3/8*	1 3/4*	3 17/64	9	6	7	13	2/16	2 9/16	.79
NC31(2 7/8 IF)	2 7/8 EU	10.40	E75	4 1/8*	2 1/8*	3 61/64	9 1/2	6	8	14	3 3/16	3 3/16	1.03
			X95	4 1/8*	2	3 61/64	9 1/2	6	8	14	3 3/16	3 3/16	.90
			G105	4 1/8*	2	3 61/64	9 1/2	6	8	14	3 3/16	3 3/16	.82
			S135	4 3/8	1 5/8	3 61/64	9 1/2	6	8	14	3 3/16	3 3/16	.82
NC38[4]	3 1/2 EU	9.50	E75	4 3/4*	3	4 37/64	10 1/2[4]	7	9 1/2	16 1/2	3 7/8	3 7/8	.91
NC38(3 1/2 IF)	3 1/2 EU	13.30	E75	4 3/4*	2 11/16*	4 37/64	11	7	9 1/2	16 1/2	3 7/8	3 7/8	.98
			X95	5	2 9/16	4 37/64	11	7	9 1/2	16 1/2	3 7/8	3 7/8	.87
			G105	5	2 7/16	4 37/64	11	7	9 1/2	16 1/2	3 7/8	3 7/8	.86
			S135	5	2 1/8	4 37/64	11	7	9 1/2	16 1/2	3 7/8	3 7/8	.80
		15.50	E75	5	2 9/16	4 37/64	11	7	9 1/2	16 1/2	3 7/8	3 7/8	.97
			X95	5	2 7/16	4 37/64	11	7	9 1/2	16 1/2	3 7/8	3 7/8	.83
			G105	5	2 1/8	4 37/64	11	7	9 1/2	16 1/2	3 7/8	3 7/8	.90
NC40(4FH)	3 1/2 EU	15.50	S135	5 1/2	2 1/4	5 1/64	11 1/2	7	10	17	3 7/8	3 7/8	.87
	4 IU	14.00	E75	5 1/4*	2 13/16*	5 1/64	11 1/2	7	10	17	4 3/16	4 3/16	1.01
			X95	5 1/4*	2 11/16	5 1/64	11 1/2	7	10	17	4 3/16	4 3/16	.86
			G105	5 1/2	2 7/16	5 1/64	11 1/2	7	10	17	4 3/16	4 3/16	.93
			S135	5 1/2	2	5 1/64	11 1/2	7	10	17	4 3/16	4 3/16	.87
NC46(4IF)	4 EU	14.00	E75	6*	3 1/4*	5 23/32	11 1/2	7	10	17	4 1/2	4 1/2	1.43
			X95	6*	3 1/4*	5 23/32	11 1/2	7	10	17	4 1/2	4 1/2	1.13
			G105	6*	3 1/4*	5 23/32	11 1/2	7	10	17	4 1/2	4 1/2	1.02
			S135	6*	3	5 23/32	11 1/2	7	10	17	4 1/2	4 1/2	.94
	4 1/2 IU	16.60	E75	6 1/4*	3 1/4*	5 23/32	11 1/2	7	10	17	4 11/16	4 11/16	1.09
			X95	6 1/4*	3	5 23/32	11 1/2	7	10	17	4 11/16	4 11/16	1.01
			G105	6 1/4*	3	5 23/32	11 1/2	7	10	17	4 11/16	4 11/16	.91
			S135	6 1/4*	2 3/4	5 23/32	11 1/2	7	10	17	4 11/16	4 11/16	.81
	4 1/2 IEU	20.00	E75	6 1/4*	3	5 23/32	11 1/2	7	10	17	4 11/16	4 11/16	1.07
			X95	6 1/4*	2 3/4	5 23/32	11 1/2	7	10	17	4 11/16	4 11/16	.96
			G105	6 1/4*	2 1/2	5 23/32	11 1/2	7	10	17	4 11/16	4 11/16	.96
			S135	6 1/4*	2 1/4	5 23/32	11 1/2	7	10	17	4 11/16	4 11/16	.81
4 1/2 FH**	4 1/2 IU	16.60	E75	6*	3*	5 23/32	11	7	10	17	4 11/16	4 11/16	1.12
			X95	6*	3*	5 23/32	11	7	10	17	4 11/16	4 11/16	.89
			G105	6*	3*	5 23/32	11	7	10	17	4 11/16	4 11/16	.81
			S135	6 1/4	2 1/2	5 23/32	11	7	10	17	4 11/16	4 11/16	.81
	4 1/2 IEU	20.00	E75	6*	3*	5 23/32	11	7	10	17	4 11/16	4 11/16	.95
			X95	6*	2 1/2	5 23/32	11	7	10	17	4 11/16	4 11/16	.95
			G105	6*	2 1/2	5 23/32	11	7	10	17	4 11/16	4 11/16	.86

*Denotes standard OD or standard ID.
**Obsolescent connection. See Appendix H, I, and J for other dimensional data.
[1]The tool joint designation (Col. 1) indicates the size and style of the applicable connection. See special note, page 3.
[2]Nominal weights, threads and couplings, (Col. 3) are shown for the purpose of identification in ordering.
[3]The inside diameter (Col. 6) does not apply to box members, which are optional with the manufacturer.
[4]Length of pin thread reduced to 3 1/2 inches (1/2 inch short) to accommodate 3 inch ID.

NOTE 1: Neck diameters (D_{PE} & D_{TE}) and inside diameters (d) of tool joints prior to welding are at manufacturer's option. The above table specifies finished dimensions after final machining of the assembly.

NOTE 2: Appendix H contains more dimensions of obsolescent connections and for square elevator shoulders.

NOTE 3: No torsional ratio (tool joint pin to drill pipe) less than 0.80 is shown. In particular areas, tool joints having much smaller torsional yields may prove to be adequate.

—Table courtesy of American Petroleum Institute

TABLE 3-2 (continued)
Tool Joint Dimensions For Grade E, X, G, and S Drill Pipe
All dimensions in inches. See Fig. 4.2

1	2	3	4	5	6	7	8	9	10	11	12	13	14
		Drill Pipe						Tool Joint					
Tool Joint Designation[1]	Size and Style	Nom. Wt.[2] lb per ft	Grade	Outside Dia of Pin and Box ± 1/32	Inside Dia of Pin[3] + 1/64 − 1/32	Level Dia of Pin and Box Shoulder ± 1/64	Total Length Tool Joint Pin + 1/4 − 3/8	Pin Tong Space ± 1/4	Box Space ± 1/4	Com-bined Length of Pin and Box ± 1/2	Dia of Pin at Ele-vator Upset, Max	Dia of Box at Ele-vator Upset, Max	Tor-sional Ratio, Pin to Drill Pipe
				D	d	D_F	L_P	L_{PB}	L_B	L	D_{PE}	D_{TE}	
NC50(4 1/2 IF)	4 1/2 EU	16.60	E75	6 3/8*	3 3/4*	5 59/64	11 1/2	7	10	17	5	5	1.23
			X95	6 3/8*	3 3/4*	5 59/64	11 1/2	7	10	17	5	5	.97
			G105	6 3/8*	3 3/4*	5 59/64	11 1/2	7	10	17	5	5	.88
			S135	6 3/8*	3 1/2	5 59/64	11 1/2	7	10	17	5	5	.81
	4 1/2 EU	20.00	E75	6 3/8*	3 5/8*	5 59/64	11 1/2	7	10	17	5	5	1.02
			X95	6 3/8*	3 1/2	5 59/64	11 1/2	7	10	17	5	5	.96
			G105	6 3/8*	3 1/2	5 59/64	11 1/2	7	10	17	5	5	.86
			S135	6 5/8*	3	5 59/64	11 1/2	7	10	17	5	5	.87
	5 IEU	19.50	E75	6 3/8*	3 3/4*	5 59/64	11 1/2	7	10	17	5 1/8	5 1/8	.92
			X95	6 3/8*	3 1/2	5 59/64	11 1/2	7	10	17	5 1/8	5 1/8	.86
			G105	6 1/2	3 1/4	5 59/64	11 1/2	7	10	17	5 1/8	5 1/8	.89
			S135	6 5/8	2 3/4	5 59/64	11 1/2	7	10	17	5 1/8	5 1/8	.86
	5 IEU	25.60	E75	6 3/8*	3 1/2	5 59/64	11 1/2	7	10	17	5 1/8	5 1/8	.86
			X95	6 1/2	3	5 59/64	11 1/2	7	10	17	5 1/8	5 1/8	.86
			G105	6 5/8	2 3/4	5 59/64	11 1/2	7	10	17	5 1/8	5 1/8	.87
5 1/2 FH**	5 IEU	19.50	E75	7*	3 3/4	6 23/32	13	8	10	18	5 1/8	5 1/8	1.53
			X95	7*	3 3/4	6 23/32	13	8	10	18	5 1/8	5 1/8	1.21
			G105	7*	3 3/4	6 23/32	13	8	10	18	5 1/8	5 1/8	1.09
			S135	7 1/4	3 1/2	6 23/32	13	8	10	18	5 1/8	5 1/8	.98
	5 IEU	25.60	E75	7*	3 1/2	6 23/32	13	8	10	18	5 1/8	5 1/8	1.21
			X95	7*	3 1/2	6 23/32	13	8	10	18	5 1/8	5 1/8	.95
			G105	7 1/4	3 1/2	6 23/32	13	8	10	18	5 1/8	5 1/8	.99
			S135	7 1/4	3 1/4	6 23/32	13	8	10	18	5 1/8	5 1/8	.83
	5 1/2 IEU	21.90	E75	7*	4	6 23/32	13	8	10	18	5 11/16	5 11/16	1.11
			X95	7*	3 3/4	6 23/32	13	8	10	18	5 11/16	5 11/16	.98
			G105	7 1/4	3 1/2	6 23/32	13	8	10	18	5 11/16	5 11/16	1.02
			S135	7 1/2	3	7 3/32	13	8	10	18	5 11/16	5 11/16	.96
	5 1/2 IEU	24.70	E75	7*	4*	6 23/32	13	8	10	18	5 11/16	5 11/16	.99
			X95	7 1/4	3 1/2	6 23/32	13	8	10	18	5 11/16	5 11/16	1.01
			G105	7 1/4	3 1/2	6 23/32	13	8	10	18	5 11/16	5 11/16	.92
			S135	7 1/2	3	7 3/32	13	8	10	18	5 11/16	5 11/16	.86

*Denotes standard OD or standard ID.

**Obsolescent connection. See Appendix H, I, and J for other dimensional data.

[1]The tool joint designation (Col. 1) indicates the size and style of the applicable connection. See special note, page 3.

[2]Nominal weights, threads and couplings, (Col. 3) are shown for the purpose of identification in ordering.

[3]The inside diameter (Col. 6) does not apply to box members, which are optional with the manufacturer.

[4]Length of pin thread reduced to 3 1/2 inches (1/2 inch short) to accommodate 3 inch ID.

NOTE 1: Neck diameters (D_{PE} & D_{TE}) and inside diameters (d) of tool joints *prior* to welding are at manufacturer's option. The above table specifies finished dimensions after final machining of the assembly.

NOTE 2: Appendix H contains more dimensions of obsolescent connections and for square elevator shoulders.

NOTE 3: No torsional ratio (tool joint pin to drill pipe) less than 0.80 is shown. In particular areas, tool joints having much smaller torsional yields may prove to be adequate.

—Table courtesy of American Petroleum Institute

attempt to ease the kelly through the drive bushing. Hard metal liners have been used—set flush against the kelly—in the bushing. Rollers, numbering from one to five on each side of the kelly, are being employed, with many arguments concerning their effectiveness.

Perhaps the greatest advancement since Kelly's name became a part of rotary drilling is the advent of the hexagonal kelly. Although the square kelly is still in wider usage, the hex has several distinctive advantages. Their manufacturers claim greater torque and tensional strength, greater adaptability to all drilling operations (pressure or fluid), possibility of larger bore, less tendency to bend, warp or twist and less weight per foot. An octagon-shaped kelly is now on the market, but has limited acceptance.

Although much has been done to make the kelly tougher, it is still a rig moving headache. Nothing is

For square kelly

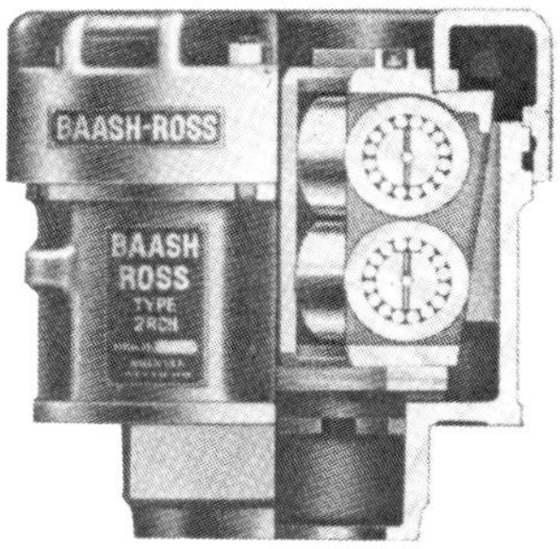

For hexagonal kelly

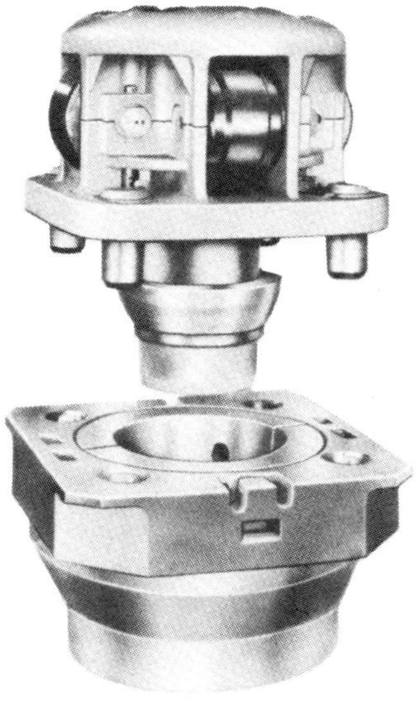

Pin drive bushing for square kelly

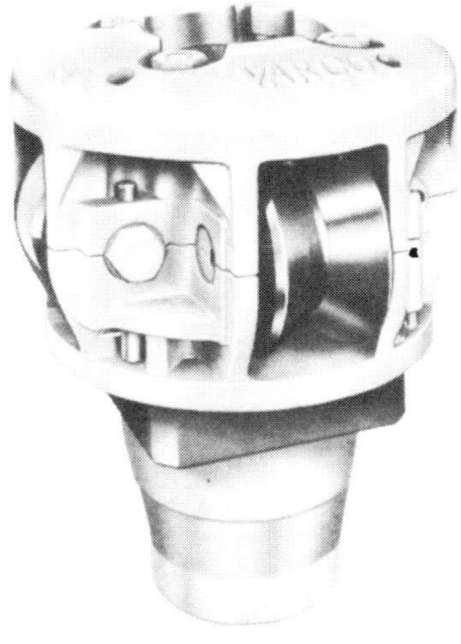

For hexagonal kelly

Fig. 3-12. Kelly roller type drive bushings.

so useless as a bent kelly. Because of its willow-like characteristics and length, a kelly scabbard has been provided for its protection. Although self-drilling scabbards are on the market, they have been deemed impractical in many areas because surface strata has required the use of a rock bit to drill the rathole. Nevertheless, a scabbard of some type has been provided as a combination rat hole and kelly protector by all thoughtful contractors. This has become common practice, just as it has to avoid breaking out the kelly with the tongs (always on the kelly saver subs).

Thinking back through all that has been said about the kelly, as well as the swivel and hooks, we must conclude that a dollar spent in the right direction toward equipment suited for your drilling needs, and time spent in protective maintenance can and will pay cash dividends. It's guaranteed to keep it turning to the right.

Kelly Cocks

Like insurance, the importance of the kelly cock is seldom felt, but is invaluable in an emergency. It is an inexpensive safety item for any size rotary rig, and should be on the same plane of importance as the blowout preventer. Actually the kelly valve serves the same purpose within the drill stem.

It stands as a positive shut-off valve above or below the kelly when drill pipe is in the hole without a float, or should the float fail when needed. It protects the rotary hose and standpipe when pressure from a formation exceeds the input. It can, in everyday use, save wear and tear on the rotary hose by opening the cock gradually when starting the mud pump, eliminating the usual jumping caused by air in the hose.

The kelly valve is actuated by a quarter turn of a special wrench on a non-protruding plug in the side of a sub. The valve is designed to permit full flow when open. Only frequent replacement in the packing around the actuating shaft and regular lubrication is required to keep the equipment in good order.

Chapter 4
Suspension Equipment

Rotary Swivel

The swivel is the changeover link which connects the rotatable drill string to the suspension system. The swivel stem (or body) attaches to the upper end of the kelly and rotates with it. The weight of the drill string is supported by the oil flooded thrust bearing within the housing and the whole assembly is suspended by the hook and the swivel bail (Fig. 4-1).

Three-fold purpose of the swivel is to suspend the drill stem, permit free rotation and to serve as the passageway for drilling fluid into the drill column. It is capable of supporting more than 650 tons and rotating attached drill column at speeds up to 400 rpm without undue wear.

Often when a piece of equipment performs its task day in and day out with little or no attention, it is completely neglected. This is sometimes true in the case of the swivel. Realizing this, and understanding the points of hidden wear, manufacturers have constructed it to permit complete overhauling at all points of wear on the derrick floor actually in minutes.

To comprehend this wear, and its correction, we must look at the principal parts involved. These are the bail, bonnet, gooseneck, body, upper and lower bearings, stem and washpipe. The body is usually of steel casting and is a grease filled protective housing for the main bearings.

These bearings, either tapered thrust, ball, straight or needle roller, are encased in oil to aid the free movement of the stem.

The gooseneck, wash pipe and its retaining cover—the bonnet—and stem comprise the core. These are constructed to receive, deflect and smooth out the high pressure, pulsating action and abrasive content of the drilling mud. The wash pipe is the heart of this focal point.

All possible wear is directed to this short heat-treated cylinder. Around it is a packing cartridge of rubber or fiber and metal rings to prevent drilling mud from reaching the bearings and to hold the protective lubricant. Manufacturers often provide complete packing box cartridges and wash pipes to be set into place in one operation. The life of these parts is prolonged by the use of extreme pressure lubricants.

Excessive replacements at this point may be an indication that the swivel is overburdened by ex-

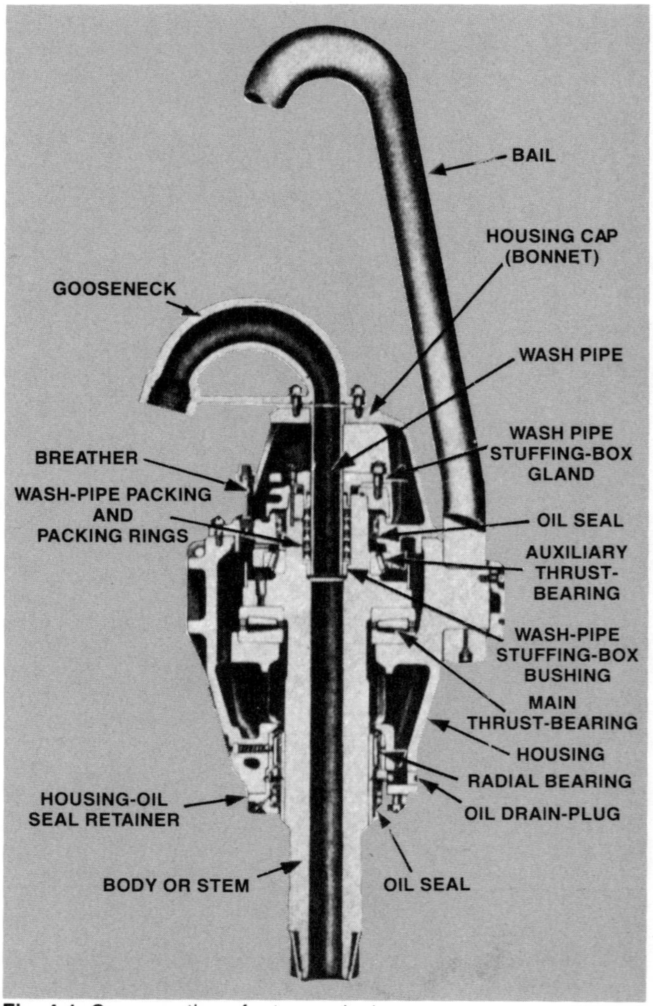

Fig. 4-1. Cross section of rotary swivel.

treme mud pump pressure. Wherever and whenever in the drilling circulation system that it is possible to increase the ID of pipe, hoses, connections or drill collars, it will decrease the drilling time and replacement costs on related parts. It would be well to check for faulty shale shaker screens, grit-filled mud pits and lumpy mud mixtures, if erosion at the wash pipe and similar connections persist.

Some manufacturers rate their swivels by the size of the stem. For example, one manufacturer puts out a selection numbering 3-S, 4-S, 6-S, the numbers representing the approximate overall dimension of each, specifically 3½-in., 4½-in., and 6⅝-in. OD stems.

Others use maximum rated depth. When consider-

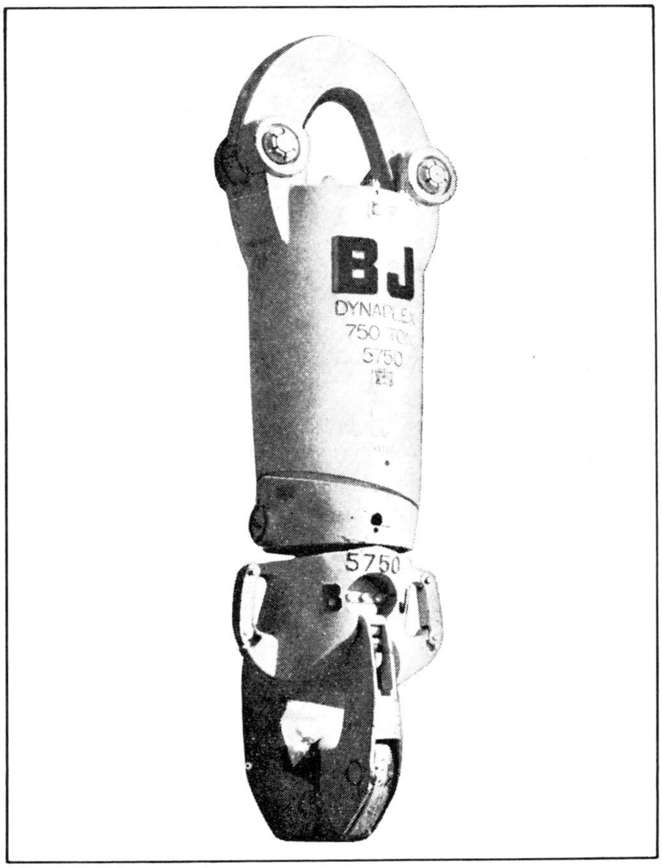

Fig. 4-2. Triplex hook.

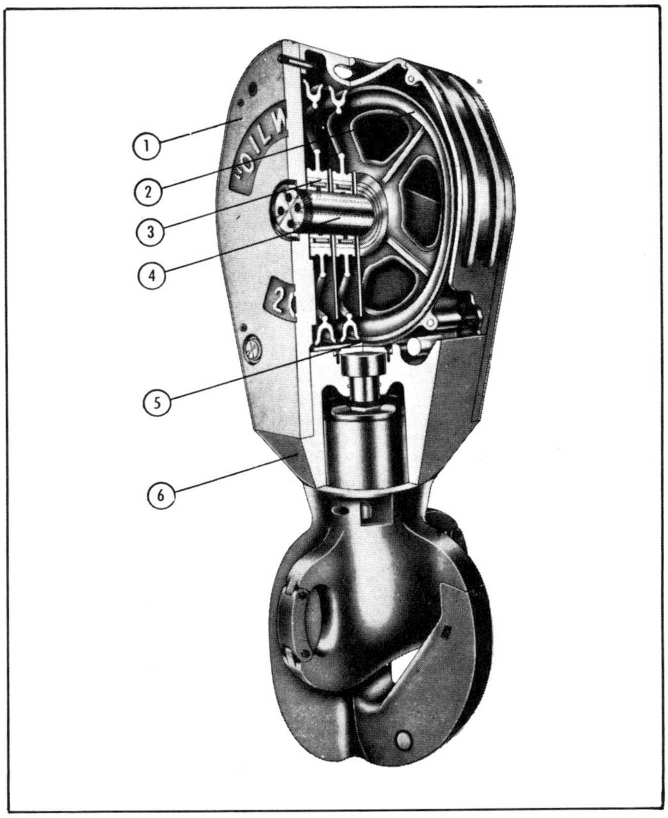

Fig. 4-3. Hook Block Combination.

1. Steel side plates are streamlined in design. There are no protruding parts. All capscrews, nuts, plugs and grease fittings are recessed.
2. Large diameter sheaves are steel. A.P.I. grooves are ground to provide maximum wireline life.
3. Heavy-duty bearings are roller type.
4. Center pin is an alloy-steel forging, drilled with internal passageways to conduct lubricant from grease fittings in either end of the pin to each bearing.
5. Steel sheave guards give full wireline protection and prevent lines from hanging-up. These circular guards also guide the wireline around each sheave, simplifying stringing operations.
6. Cast-steel adapter is furnished for coupling blocks and hooks together.

ing a swivel from this standpoint, one must not forget that these figures are based strictly on drill pipe. Knowing that 4½-in. drill pipe averages 17 lb/ft, we see that the number of drill collars, ranging from 120 to 170 lb/ft will shorten the maximum string tremendously. The buoyant effect of the drilling mud will offset this increased weight considerably.

Other swivels are rated by the dead load tonnage capacity, resulting in the 150, 250 and 300, etc., thus simplifying rating procedure substantially. Although different means of rating offer a definite safety factor over and above normal capacity, purchase of larger equipment than anticipated requirements demand often is advisable.

Manufacturers vary their product in outward appearance, safety and service features, choice in number, type and placement of bearings, grease seals, etc. Nevertheless, they are guided by API specifications in the selection of fittings relevant to corresponding pieces of equipment, primarily the threaded ends. The upper gooseneck connection, that is adjoined by the rotary hose, comes with 2½-in.,3-in. or 4-in. line pipe connection. The lower part of the stem is outfitted usually with a 6⅝ or 4½-in. API regular pin (left hand thread).

Below this is a kelly cock or box substitute which makes up into the kelly joint, a regular API thread. (Both of these connections are left hand.)

Hook And Hook-Block Combination

The conventional spring hook served the industry very well for many years. It was upgraded from time to time both in capacity and in regard to safety and convenience. The hook proper had to have a throat wide enough to accommodate elevator links and swivel bail simultaneously. The triplex hook, with one large safety hook for the swivel bail and with two smaller hooks or ears on the sides of the main hook body to accommodate the elevator bails, seemed to offer many advantages when introduced, but it cost considerably more. Conventional hooks didn't wear out very fast but as they did they usually were replaced by triplex hooks, some of which have a capacity of 500 tons or more.

In an effort to reduce both overall cost and the

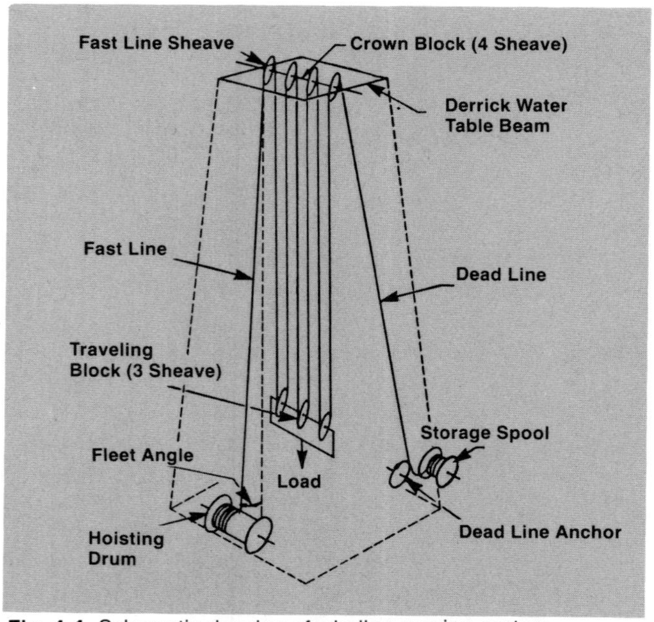

Fig. 4-4. Schematic drawing of wireline reeving system.

vertical space required, a triplex hook combined with a traveling block has proved popular. This is particularly true of sizes up to 250-300 ton capacity. (See Fig. 4-2 and 4-3.)

Wire Rope Hoisting System

Within the hoisting system, the crown and traveling blocks, when strung with the drilling line, multiply the line pull of the hoist so that the heavy weight of the drill string may be raised and lowered during drilling.

The blocks and line act as a highly engineered pulley system to provide the necessary mechanical advantage or load multiplication. This multiplication of force is accomplished, as in any plain pulley system, at the expense of motion. Thus, the hoisting drum must spool on much more line than is equivalent to the distance through which the traveling block moves.

A schematic picture of a simplified hoisting system is shown in Fig. 4-4. This figure shows a block with mechanical advantage of six. Six is found by counting the number of lines that lift the traveling block. If a total load of 120,000 lb was to be lifted 100 ft with the illustrated tackle system, the hoisting drum would have to spool 600 ft of line, but the line pull of the hoist would only be 20,000 lb, providing negligible friction was encountered in the hoisting system.

Actually, some effort would be required merely to move the line and rotate the sheaves. Due to the many variables encountered in this effort, it is the usual practice to add 2% of the applied force for each line. In our example, the application of this 2% factor

Fig. 4-5. Mast type crown block.

would add 12% or 2400 lb to the necessary line pull of the hoist. The resulting 22,400 lb, as the necessary hoist line pull to lift a load of 120,000 lb, would be accurate enough for any but the most extreme cases.

Prior to passing on to an actual description of the blocks, it should be noted that in the majority of cases the load on the crown block is appreciably greater than the load on the traveling block. If you realize that the line pull or the line tension is approximately constant, at about 20,000 lb throughout the system, it becomes apparent that the load on the crown block is 160,000 lb (8 lines by 20,000 lb) as compared to 120,000 (6 lines by 20,000) on the traveling block.

Crown Blocks

The crown block, as the stationary pulley in the hoisting system, is firmly fastened to the top main, or water table, beams of the drill structure, whether mast or derrick. Block beams are often an integral part of the mast structure rather than a separate assembly.

The placement of the crown block is usually fixed by design. This design placement is governed by the necessity of the pipe hanging over center of the well and by attempting to get the fast line sheave as close to the center (longitudinally) of the hoisting drum as possible. If the centering relationship is off enough so that a fleet angle of more than 1.5° is obtained, poor

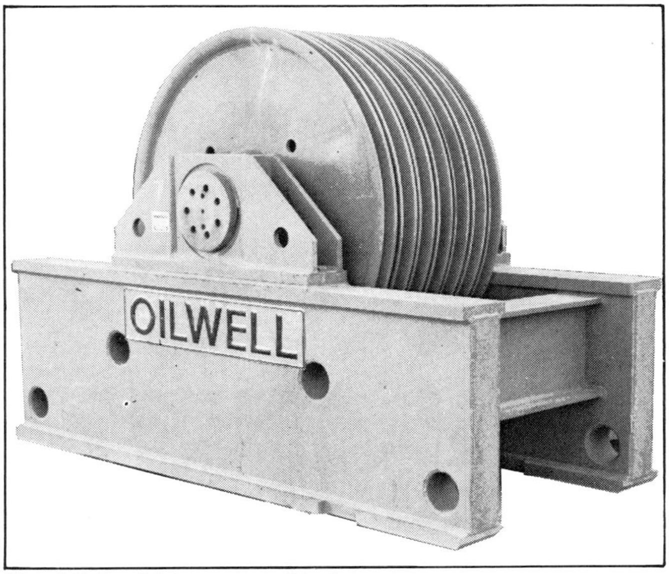

Fig. 4-6. Derrick type crown block.

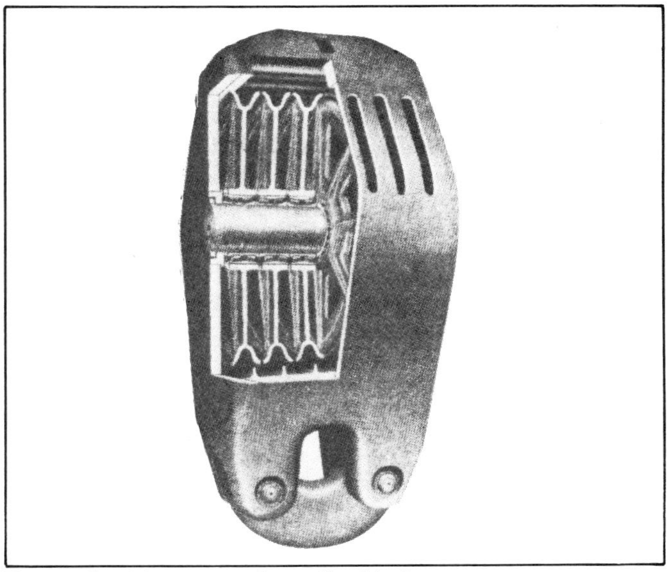

Fig. 4-7 Partial cutaway of traveling block.

drum spooling and excessive line wear will result.

Although there are a multitude of crown block designs, including double-deck assemblies with right-angle top sheaves, some of the most popular models for normal modern drilling incorporate parallel sheaves and a single-deck assembly. Other designs, particularly for mast use, utilize special design placements (Fig. 4-5). For derrick use, where there is room for the line inside the structure, an inline sheave arrangement is used (Fig. 4-6).

Although not actually a part of the crown itself, a sand line sheave is often attached to one end of the main crown block beams. In today's operations, the sand line sheave is usually hung underneath the crown block, out of the way of the drilling lines. The actual location of the sheave is determined by the location of the sand reel, but either drawworks (rear) or V-window (front) side are common locations.

Traveling Blocks

The majority of modern traveling blocks incorporate parallel, inline sheave arrangement shown in Fig. 4-7. Traveling blocks are designed and selected to satisfy a number of different, and in many cases conflicting, requirements. A primary requirement is that in order to insure stability the block should have a low center of gravity, as any tipping or canting of the block during operations makes the derrickman's job very difficult. The block should also be short so that less head room, especially in the shorter portable masts, is used at the top of the hoisting cycle. Similarly, the block should be narrow or slim so that the inside clearance space within the upper mast will not be endangered. In some mast designs this clearance space is further limited by the derrickman's platform and pipe racking structure. Both slimness and shortness combine to provide a lighter weight block, so important in these days of portable equipment.

On the other hand, a block that is too light will not overcome line friction fast enough to drop with an empty hook, so that some compromise is often necessary. Further, like all drilling equipment, both blocks and especially the moving traveling block must be designed of heavy enough material to take the handling encountered in tough oilfield use.

These requirements have recently resulted in the design and use of the combination block and hook, built as a unit. This design, although providing more rigid connection between the two units, saves a great deal of operational space. As the traveling block operates in human working areas, the points where the cables run on the sheaves must be guarded against personal contact. In practice, only very narrow entry points are left in the block guards for the lines. A typical safeguard is shown installed on the traveling block in Fig. 4-8.

Stringing of Blocks

The stringing or reeving of crown and traveling block can be accomplished in many ways to suit special conditions encountered. Main factors involved in the reeving decision are being sure that the traveling block hangs and travels straight, that the derrickman is not endangered by the fast line and that additional lines may be strung easily as the well progresses.

Block Sheaves

Both the crown and traveling block sheaves are manufactured from high-quality steel and usually

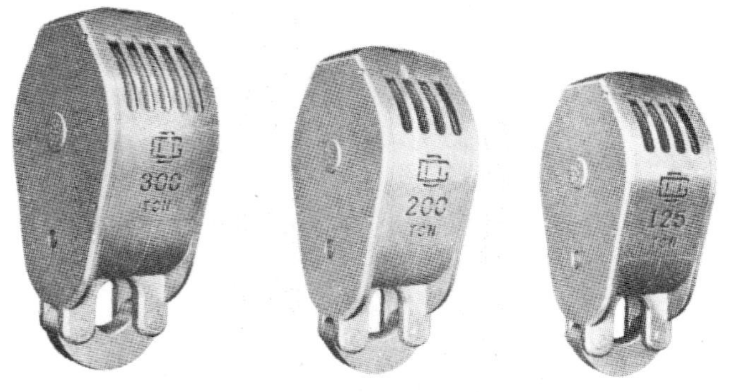

30 T-542 8,000′ to 15,000′ | 20 T-436 5,000′ to 10,000′ | 12 T-430 4,000′ to 8,000′

Fig. 4-8. Modern traveling block.

rotate on large-diameter, anti-friction bearings. The sheaves and bearings are then, in most cases, supported by the shaft. For easy service, the blocks are usually provided with grease lubrication systems so that the lubricant is channeled independently to each bearing through passageways inside the supporting shaft.

An easily reached centralized lubrication system especially for the suspended traveling block should be provided whenever possible. Such a system, optionally available from either side, is utilized in the block shown in Fig. 4-8.

Sheave diameter is a compromise between the requirement of maximum wire rope life and the overall design balance of the hoisting system. As a general rule, 6 x 19 wire rope should not be used on a sheave of less than 30 times the diameter of the line unless very special conditions are encountered. The proper factors pertaining to rope life under such special conditions have been worked out in the API Recommended Practice No. 9B on "Application, Care and Use of Wire Rope for Oil Field Service". This publication also contains a great deal of other data concerning both rope and sheaves.

Sheave grooves are made slightly larger than the size of rope to be used, providing a cradled support of the line for 150° of rope circumference. It should be remembered that sheave grooves get smaller with use, and as a result, a larger line can never be used on worn sheaves. When necessary, however, it is possible to use a slightly smaller line in a larger sheave groove.

Block Rating

Crown and traveling block rating is usually in terms of the mainbearing rating, which conforms with API, Standard 8A, "Rating of Drilling and Production Hoisting Equipment". The following data on Safety Factors and Bearing Rating is reproduced by permission from API Standard 8A.

"Safety Factors. Safety factors shall be as follows:

Calculated Rating, tons	Safety Factor
150 and less	3.00
Over 150 to 500, incl	3.00 $= \frac{0.75(R-150)}{350}$
Over 500	2.25

Wherein: R = calculated rating, in tons."

Bearing Rating ... of crown and traveling blocks shall be determined by the formula:

$$W_b = \frac{N\ W_r}{585Y}$$

Wherein:

W_b = calculated block bearing rating, in tons
N = number of sheaves in the block
W_r = individual sheaves bearing rating, in pounds at 100 rpm for 3000-hr minimum life for 90% of bearings
Y = 1.25 for stationary shaft
Y = 1.00 for rotating shaft"

Block Maintenance

Both crown and traveling blocks should be well lubricated periodically, depending upon the service and working conditions. Once a tour might be a nominal period during heavy duty operations.

As regrooving is preferable to poor line life, sheave-groove wear on both blocks should be watched and checked with API profile gauges.

Wire Rope

Wire rope is one of the most versatile tools used in the oilfields. Take a look at different field operations and you will find that there just are not many jobs in which wire rope does not play an important role. As for oilwell drilling, it is so vital that no method presently used could exist without wire rope to hoist and lower the tremendous weight of the tools used.

Engineers know that to use a tool best you must understand its characteristics and limitations. This section will reveal the characteristics of wire rope—show how it affects drilling and how drilling affects it.

Wire rope strength is a function of the cross sectional area and grade of steel used in construction. Distribution of load among wires and strands must be proportional to their respective cross sections in order that the total strength may be utilized. The line should be in a balanced condition under load.

Flexibility is defined as resistance to bending fatigue. It is obtained by the use of smaller and more numerous wires, or more strands in manufacturing the rope. The type of lay and core also affect the original flexibility of a rope. The most commonly used

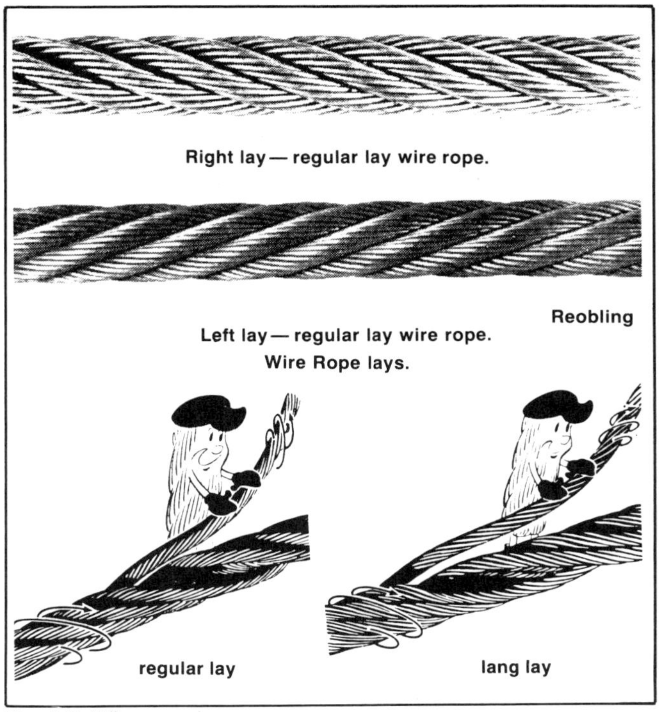

Fig. 4-9. Wire rope lays.

lays are shown in Fig. 4-9. Formation of martensite due to rapid and intense temperature changes during scrubbing on the drum and sheaves continuous and concentrated flexure as at the dead sheave or short-radius bending adversely affect flexibility.

As wire rope is a flexible tool, it always works in tension. It works best when slack is reduced to a minimum. Sudden excessive slack in a loaded line has the same potential effect as placing it in compression; that is, a kink may be thrown into it. Also, the chances of shock loading are greatly increased as the slack is removed by the load.

Elasticity affords an element of safety when shock loads are applied. Elasticity is inherent in the helically bent (not twisted) wires and strands of the rope. The length and type of lay determine the degree of elasticity. Shock loads of relatively low magnitude and/or overloading reduce elasticity.

Resistance to abrasion is a function of outer wire size. It is aided by mechanical peening of the outer wires which presents a greater surface area to abrasive forces. Ductility of the steel used is such that this peening takes place early in the service life of the rope. Lang lay wire rope has greater resistance to abrasion than regular lay ropes because it has greater exposure of the wires at the crown of the strands. Lang lay ropes are not commonly used for rotary drilling lines, however, because the total requirements of this service are better met by regular lay ropes.

In order for each wire and strand to carry its share of the load, it is necessary for each to maintain its position relative to the other wires and strands. Crushing and/or distortion tend to destroy this relationship. The primary purpose of the core is to space and support the strands, i.e., provide crushing resistance. Support is most important on the drum as the rope nears the end of its useful life.

Distortion has two different affects on strength. Temporary distortions, as over a sheave, cause uneven distribution of the load so that the wires farthest from the groove carry more than those closer to the groove. This is most important at the dead sheave because the line is very restricted as to movement at that point and the same wires and strands carry disproportionate loads over extended periods of time or service. Partial elimination of this condition is the purpose of slipping between cuts. Permanent distortions such as doglegs, kinks and high strands throw the entire load on part of the wires and strands.

Wire Rope Care

Corrosion may reduce strength by reduction of cross-section area and/or flexibility by the formation of rust between wires so that their ability to slide along each other is handicapped. Corrosion is a result of lowering the rope into the well fluids (as a coring line or sand line) or dragging it over corrosive mud materials or earth when pulling from the reel to string up or unstring.

Protection from corrosion is largely a matter of factory-applied lubrication aided by proper cleaning and field lubrication on removal of the rope from corrosive media. Corrosion is especially insidious in lines that have most of the lubricant squeezed or otherwise removed from the fiber core and wires. The core may even act as a wick, drawing the corrosive agent up into the rope where it attacks the hidden inner wires while the outer wires remain bright due to abrasion.

The materials used and methods of making and using wire rope prohibit the utilization of any of these characteristics at 100% efficiency. Flexibility is often sacrificed to attain resistance to abrasion, and vice versa, dependent on the ultimate use of the rope. Maximum resistance to crushing must be sacrificed to obtain flexibility. An independent wire rope core may be used in preference to fiber because of its greater resistance to crushing, etc. These difficulties are resolved in the design of the rope. For instance, the Seale design is the most common for rotary drilling lines. This design has large outer wires to resist abrasion and smaller inner wires to give flexibility. Some typical oilfield wire rope designs are shown in Fig. 4-10.

Regardless of the rope design used, maximum service requires that the auxiliary items used to guide and control the rope (the deadline anchor, line stabilizers, sheaves, drums, drum-line turn-backs,

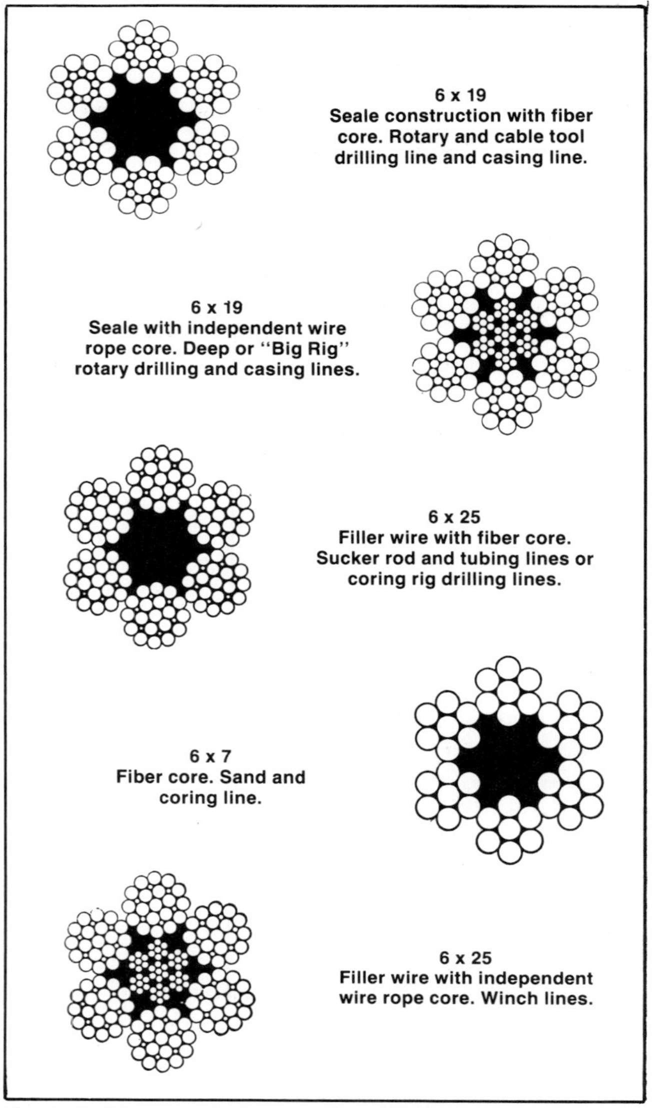

Fig. 4-10. Wire rope designs used in oilfields.

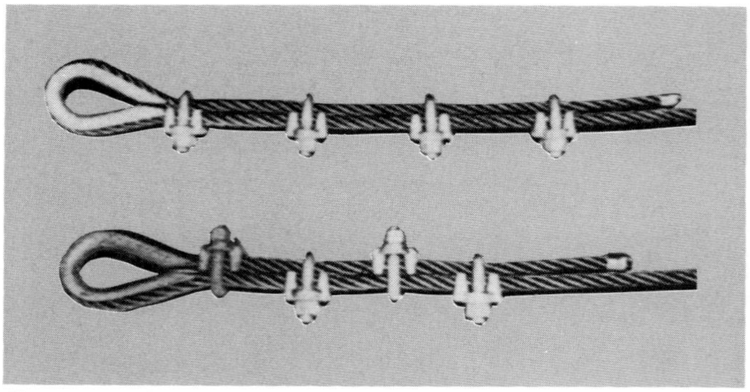

Fig. 4-11. Method of attaching wire rope clips.

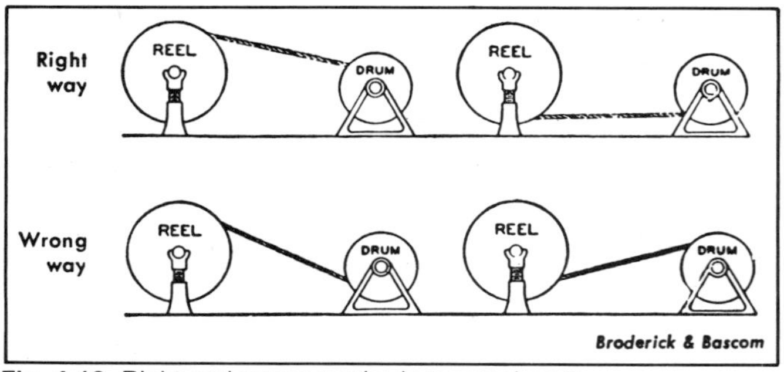

Fig. 4-12. Right and wrong methods to spool on lines.

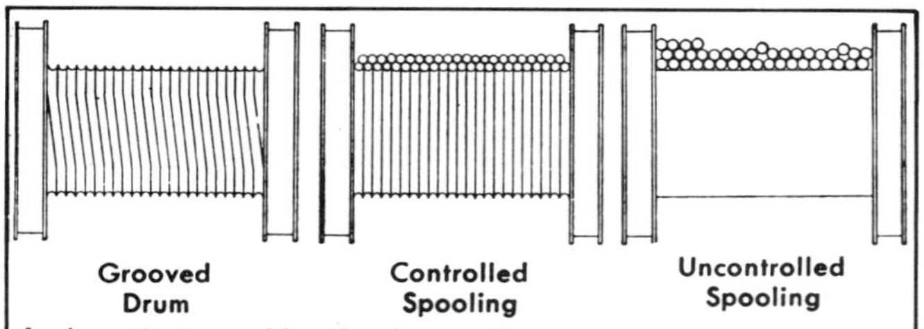

An important consideration in spooling of the wire line is to have sufficient weight on the line at all times so as to provide spooling on the drum under tension. Insufficient weight will cause loose spooling and result in unsatisfactory rope service. It is therefore imperative that the combined weight of the traveling block, hook, elevators, links and swivel be heavy enough to cause the line to spool under tension.

Wire Rope Corp. of America

Fig. 4-13. Wire rope spooling under tension.

clips, etc.) be the proper size, in good condition and properly applied. (See Figs. 4-11, 4-12 and 4-13). The characteristics have shown the ways in which wire rope can become damaged and point the way to avoidance of damage. The control mechanisms provided by engineering know-how not only aid in handling rope, they also serve to avoid damage when properly used.

Hoisting System Calculations

Examples will best show the simplicity of hoisting system calculations. In order to avoid lengthy problems for determining the exact effect of drill string buoyancy, hole friction, etc., assume that the maximum hook load is 200,000 lb and that the rig has an 8-line string-up that needs to be changed to 10 lines. Let us see how the rig will be affected. Two equations are involved. They are: $\frac{A}{B} = C$ and $(B+2)C = D$

In which:

- A = Hook load, in pounds.
- B = Number of lines in the string-up.
- C = Average tension in lines of the static loaded system, in pounds.
- D = Derrick load, in pounds.

$$\frac{200{,}000\text{ lb}}{8} = 25{,}000\text{ lb} \quad (8+2)25{,}000 = 250{,}000\text{ lb}$$

$$\frac{200{,}000\text{ lb}}{10} = 20{,}000\text{ lb} \quad (10+2)20{,}000 = 240{,}000\text{ lb}$$

First, it is evident that the tension in the lines has been reduced by 20%; or saying it another way, if the safety factor was 3, it is now 3.6. Next, the derrick load has been decreased 4%. Further, though the same weight is suspended, the engines, clutches, shafting, chains, brakes, hook, swivel, links and ele-

ROTARY DRILLING LINE SERVICE RECORD

SHEET_____ OF _____ SHEETS

COMPANY_____ WELL AND NO._____ RIG NO._____ MAKE AND TYPE DWKS._____

DRUM DIAM._____ PLAIN OR GROOVED DRUM_____ CROWN BLOCK SHEAVE DIAM._____ TRAVELING BLOCK SHEAVE DIAM._____ WT. OF TRAVELING BLOCK ASSEMBLY (FACTOR "M")_____ SIZE AND WT. DRILL PIPE_____

MAKE OF LINE_____ SIZE AND LENGTH_____ CONSTRUCTION_____ GRADE_____ REEL NO._____

DATE LINE PUT INTO SERVICE_____ DATE LINE RETIRED FROM SERVICE_____ NO. LINES STRUNG_____

INITIAL	1ST CHANGE	2ND CHANGE

WELL DEPTH WHEN STRING-UP INCREASED_____

1ST CHANGE	2ND CHANGE

1	2	3	4	5	6	7	8	9	10	11	12	13	14	15	16	17	18
						DRILL COLLARS											
DATE	Trip No.	Depth of Trip	Operation to be Performed & Remarks	Mud Weight lb. per gal.	Effective Wt. of Pipe. Fig. 4.2	O. D. and Bore	Effective Wt., Ec Fig. 4.3	Excess Wt (Col. 8 Minus Col. 6)	No. of Feet	Factor C (Col. 9 x Col. 10)	Factor M+½C	Ton-Miles Service This Operation Fig. 4.1	Cumulative Ton-Miles Since Last Slip	Length Line Slip-ped, ft.	Cumulative Ton-Miles Since Last Cutoff	Length Line Cut Off, ft.	Length Line Remain-ing, ft.

TON-MILES SERVICE — PREVIOUS WELLS_____ TON-MILES SERVICE ON TRIPS — THIS WELL_____ TON-MILES SERVICE DRILLING — THIS WELL_____ TON-MILES SERVICE CORING — THIS WELL_____ TON-MILES SERVICE SETTING CASING — THIS WELL_____

TOTAL TON-MILES SERVICE — ALL WELLS_____ TON-MILES PER FT. OF INITIAL LENGTH_____

Directions for filling out this form, including use of charts, are given in the instruction sheets included with each pad, and are also given in API RP 9B: *Recommended Practice on Application, Care and Use of Wire Rope for Oil-Field Service.*

Fig. 4-14. Sample of "Ton-Mile" record form.

vators do not have to work so hard to carry out their various functions.

There are some drawbacks to consider before adding lines to your present system. First, if the drum speed remains the same, the rate of hook ascent will be lowered. Next, more line must be pulled onto the drum to obtain the same distance of hook movement. In many cases, this will require a radical change to the moving and cutting program, which a wire rope engineer will have to analyze and prescribe.

Moving and Cutting Programs

Most drilling companies use some system for moving the line through the string-up at a uniform rate and cutting it off as it reaches the end of its safe and economic life. This presupposes having reserve line in excess of that required for string-up. Of course, if the money saved in greater wireline service is exceeded by the cost of moving a large reserve of line between locations, a practical limit is placed on the length of the reserve. The average rotary drilling line purchased is between 3500 and 4500 ft long; however, on large rigs working on long-term wells, the length may be 7500 ft or longer.

How To Find The Minimum Length Of Drilling Line Required

Multiply the height of derrick in feet by number of lines strung plus two. Round off answer to the nearest 50, then add number of feet of rope on drum.

Example: Using 176 ft for height of derrick; 8 for number of lines strung; 300 for number of feet of rope on drum at lower pick up. $176 \times (8 + 2)$, or $176 \times 10 = 1760$; round off to nearest 50. Add number of feet of rope on drum (300). Minimum length required = 2050 ft. This minimum length allows deadline to reach anchor base of substructure.

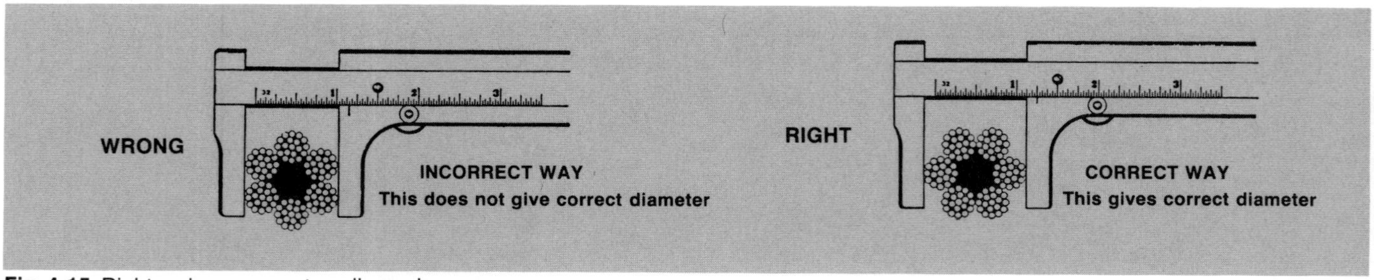

Fig. 4-15. Right and wrong way to caliper wire rope.

Evaluation of wire rope service is made in terms of the ft-lb of work done; however, due to the magnitude of work performed the common unit used is the ton-mile. The API method of calculating wire rope service, as given in API Recommended Practice 9-B, is the most accurate used (see Fig. 4-14). It is the basis for most of the slide-rule type of calculators and tables that are used in the field. Most moving and cutting programs for rotary rigs are based on accurate records of the ton-miles of work a drilling line has performed. The drum end of rotary drilling lines is the end cut off. Cable tool drilling lines are cut off on the tool end.

A close visual inspection should take precedence over the ton-mile record as a very real need for cutting may exist long before the record indicates the need. This policy must be tempered with good judgment and experience. For the first few cuts on a new drilling line, it will not have the appearance of being worn out because it has not passed through the entire system. In this instance it is better to heed the record and make the cut, than to wait and be sorry.

Always keep a close watch on the line from the dead sheave all the way to the drum. If too many ton-miles are allowed to accumulate before a move and cut, it is almost certain that a long cut will have to be made later. Upon examination, it generally turns out that the bad place was over the dead sheave at one time or another. A few scattered broken wires on the drum at lower pick-up should not cause alarm, but the line should not be allowed to run until wire breaks become numerous or localized in any short section of the system.

A wire rope caliper is useful during inspection in measuring the diameter of the rope where it appears to have "pulled down". This is usually due to core failure. The proper method of using the caliper is shown in Fig. 4-15.

Ton-mile records no longer apply if the line is kinked. Also, there is no way to evaluate an extra hard pull, such as on stuck drill pipe or casing. The record is questionable until all the line between the deadline anchor and the drum end at the time of the hard pull is removed from the system. This can be partially offset by arbitrarily adding 200 or 300 ton-miles to the record at the time of the pull, but there is no assurance that this will remedy the situation.

Possibly, the rule most universally agreed on by wire rope engineers and rotary drilling people is that the moving and cutting program must be tailored to the individual rig. Beyond this point, methods of determining the economic and safe life of a line, and the length and frequency of cut vary widely. The principal idea is to choose some goal for service in terms of ton-miles per foot of line and devise a method for reaching the goal. One such method is presented here as an example.

For purpose of analysis, assume that a 4500-ft. line has been purchased, of which 1500 ft. is required for the string-up, and that it is desired to obtain 15 ton-miles per foot of the reserve line. Further, assume that upon observation it has been determined that an accumulation of approximately 1000 ton-miles between cuts and slipping at intervals of about 500 ton-miles is practical, safe, and economical. This method of determining slipping and cutting lengths is as follows:

4500 ft — 1500 ft = 3000 ft of reserve line.

15 T-M × 3000 ft = 45,000 T-M service (in this case, 10 T-M per foot of total line equals 15 T-M per foot of reserve line).

$$\frac{500}{15} = 33.3 \text{ ft/slip} \qquad \frac{1000}{15} = 66.7 \text{ ft/cut}$$

$$\frac{3000 \text{ ft}}{66.7 \text{ ft}} = 45 \text{ cuts should be made}$$

Since ton-miles do not accumulate in lots of 500, this is how it works in service:

$$\frac{520}{15} = 34.7 \text{ ft on } \textit{this} \text{ slip.}$$

$$\frac{969}{15} = 64.6 \text{ ft on } \textit{this} \text{ cut.}$$

If the above procedure is followed and nothing occurs that requires a radically out-of-schedule or unduly long cut, it should be possible to reach the goal without difficulty. After trying this method for a while, if it is found by close wireline inspection that the goal is too high or low, it should be altered in small increments until the proper goal is found.

Whatever method of determining slipping and cut-

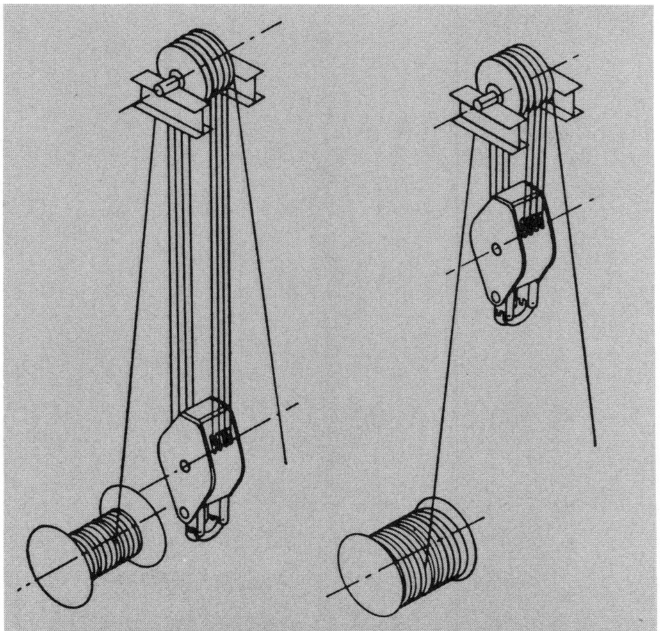

Fig. 4-16. Two spooling conditions conducive to severe wear.

ting lengths is used, it is most important to see that the length slipped does not place any part of the line in a position of severe wear a second time. These points of severe wear are the parts of the line over any sheave at either the upper or lower pick up (see Fig. 4-16) and the places where succeeding wraps cross over those beneath on the drum. It is worthy of repetition that visual inspection should be more important in determining time to cut than the ton-mile record, but when the record says cut, don't wait, do it now.

Chapter 5
Supporting Structures

Despite the fact that most of the steel associated with an oil well is below ground, the oil derrick is the symbol by which an oil well is known. To many people it is the trademark of the oil industry — in fact they often give little thought to what is going on below the ground, seeing only that which is above.

The oil derrick, however, has undergone an extensive period of transition. In the majority of cases the permanent derrick is a thing of the past. Conventional derricks are seldom left over the hole. Many manufacturers have discontinued production of conventional derricks. Also there are derrick construction contractors with a large supply of conventional derricks which they rent for drilling purposes.

Derricks in use today fall into three broad classifications: conventional, portable (usually free standing), and portable mobile types. All can be used for drilling and servicing but the first two types are predominantly used for drilling and the last type for both drilling and servicing.

Conventional Derricks

Considering the conventional derrick first, the reader is referred to API Standard No. 4A API Specifications for Steel Derricks (including standard rigs) and to Fig. 5-1 which shows derrick nomenclature.

There are nine API sizes of conventional derricks. These have the following dimensions:

Derrick size no.	Height A Ft.—In.	Nominal base square B Ft.—In.	Water table opening D Ft.—In.
10	80-0	20-0	5-6
11	87-0	20-0	5-6
12	94-0	24-0	5-6
16	122-0	24-0	5-6
18	136-0	26-0	5-6
18A	136-0	30-0	5-6
19	140-0	30-0	7-6
20	147-0	30-0	6-6
25	189-0	37-6	7-6

Tolerances: A, ± 6 in. B, ± 5 in. D, ± 2 in.

The height A of an API conventional derrick is the distance measured along the neutral axis of the derrick leg from the top of the derrick floor joists to the bottom of the water table beams or bumpers.

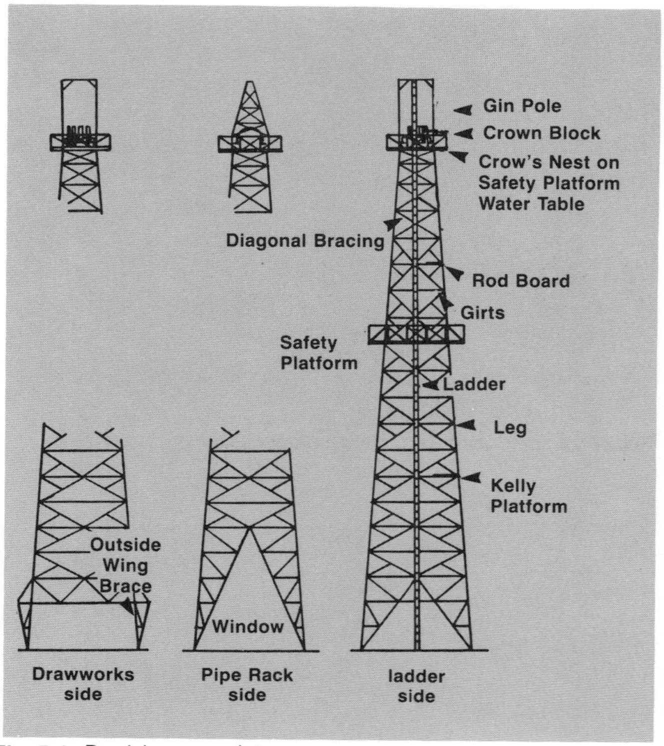

Fig. 5-1. Derrick nomenclature.

The base square dimension B is the distance between the neutral axis of adjacent legs at the top of the derrick floor joists.

The water table opening dimension D is the inside or the clear dimension of the opening in the water table.

The gin pole structure is used to lift crown components from the derrick floor to the top of the derrick and through the water table. It extends across the center of the water table opening. Minimum clearance between the horizontal header of the gin pole and the tops of the water table beams is 8 ft for derricks under 122-ft high; 10 ft for 122-ft derricks; 12 ft for 136-ft derricks and 17 ft for 140-ft and 189-ft derricks. Usually higher gin poles may be obtained on special order.

A safety platform is usually placed around the crown which conforms to the various safety codes.

Horizontal members (Fig. 5-1) are called girts and the cross bracings are called diagonals. Cross bracing comes in several forms depending on manufacturer. Typical examples are X bracing, K bracing and diamond bracing.

The V window opening used for rotary drilling permits the drawing pipe from the pipe rack into the derrick. The minimum height of the window opening measured in the clear parallel to the centerline of the derrick side from the top of the derrick floor joists is 23 ft, 8-in. for the 136-ft and smaller derricks, and 26½-ft for the 140-ft, 147-ft and 189-ft derricks.

Rotary drilling derricks have a clear opening on the drawworks side. For the 94-ft, 122-ft and 136-ft derricks this is a minimum of 10 ft to each side of the centerline of the derrick and 7½ ft above the top of the floor joists. For 140-ft, the 147-ft and 189-ft derricks no actual minimum is specified except that the area on the drawworks side of the derrick between the legs, the first girt and the floor joists shall be clear of any cross bracing or interfering members. Hence, the design of the derricks utilizes outside bracing on the drawworks side, called wing bracing.

Practically every derrick manufacturer makes some derricks which are outside the API standard dimensions.

Supporting Capacity of Conventional Derricks

Conventional derrick capacities run from 86,000 lb to 1,392,000 lb. The API safe working load capacity (not failure) is the maximum dead load the derrick is designed to support with a factor of safety of approximately 1.94. This value, given on the nameplate attached to the derrick, is the allowable column load capacity of a single derrick leg multiplied by 4. A factor of safety of approximately 1.94 is provided in the API column formula to provide for the added effect of wind loads, drill pipe setback in the derrick, dead weight of the derrick, dynamic loads and the effect of anchoring the deadline in different positions.

The safe working capacity changes when the deadline is anchored in different positions. These are as follows:

1. Anchored in corner opposite drawworks side — reduces capacity 20%.
2. Anchored at calf wheel opposite drawworks — reduces capacity 7%.
3. Anchored at corner on same side as drawworks — reduces capacity 40%.

These are approximate deductions as they will vary with hook load, single line pull and number of lines strung to the block.

Design of Conventional Derricks

Design requirements for conventional derricks is covered in API Standard 4A. Most of the complete stress analysis of a conventional derrick can be solved graphically, and a typical space and load diagram for the applied wind loads and pipe set back is shown in Fig. 5-2. The stresses in various members can be determined from these loads. For members in tension the member load is divided by the net cross-section area to give the tensile stress in pounds per square inch (psi). For members in compression the API column formula is used. This also is used to determine the allowable column load of derrick legs in compression.

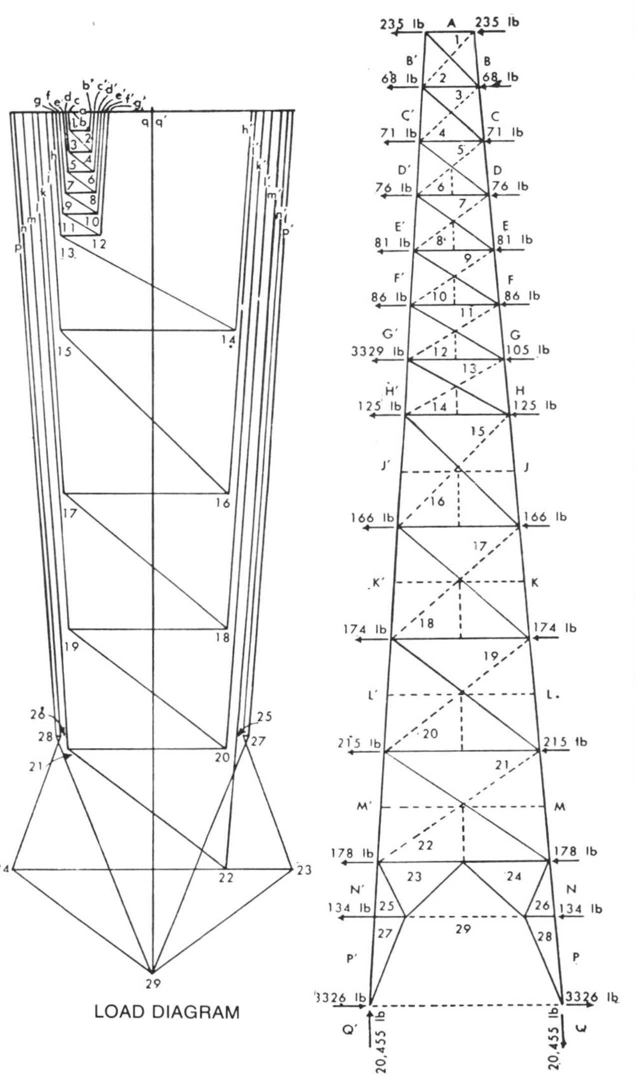

Fig. 5-2. Typical space and load diagram for graphical analysis of conventional derrick.

Because the advent of portable and mobile mast structures introduced many steels of various yield points a new API formula was adopted to fit all steels and this is given as follows (From Table 1 of the API Standard 4D Specification for Portable Masts):

Compression:

Columns gross section for $KL/_r$ less 24,000/Y allowable unit stress

$= [Y/1.941] - [Y^2 (KL/_r)^2/2.245 \times 10^9]$ for $KL/_r$

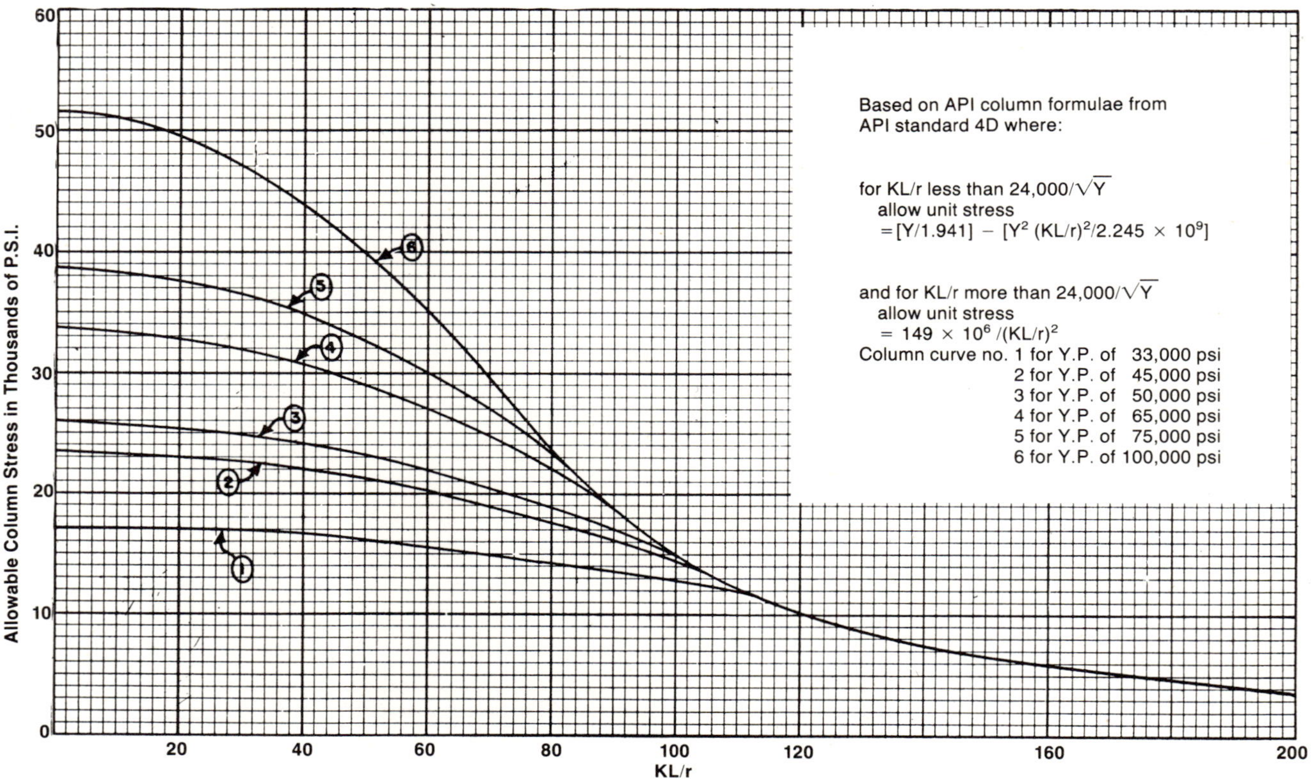

greater than 24,000/Y allowable unit stress

$= 149 \times 10^6/(KL/_r)^2$

where

K = effective column length factor (This can be left as 1.0 if the designer desires not to take advantage of a possible reduction in effective column length.) (For conventional derricks K is assumed to be 1.0.)

L = column length in inches.

r = radius of gyration

Y = minimum yield in pounds per square inch

A family of curves based on these formulas has been prepared and is shown in Fig. 5-3. Instructions for use of these curves are as follows:

1. Determine the minimum yield point of the steel being used. (For example—mild structural steel would be 33,000 psi minimum yield point; silicon structural steel would be 45,000 psi minimum yield point; Bethlehem Mayari R or U.S. Steel Corten would be 50,000 psi minimum yield point; SAE 4130 in the normalized condition would be 75,000 psi minimum yield point, etc.)
2. Determine the effective column length in inches. (Same as actual for conventional derricks)
3. Determine the radius of gyration. This can be obtained from a handbook or can be calculated.
4. Determine the cross-section area in square inches of the member or column in compression.
5. Divide the effective column length by the radius of gyration to obtain the slenderness ratio ($L/_r$). This will be $KL/_r$ if the effective column length rather than the actual column length has been used.
6. From the chart in Fig. 5-3 select the column curve corresponding to the minimum yield point of the steel used. Read up from the calculated $KL/_r$ and across to the allowable unit column stress in pounds per square inch.
7. Multiply the allowable unit stress by the cross-section area of the member to obtain the allowable column load. This is the load the member will safely carry with a factor of safety of approximately 1.94.

Wind Loading and Other Design Factors

Wind load in miles per hour is calculated from the following formula:

$P = 0.004V^2$

where

P = unit wind load in pounds per square foot

V = actual wind velocity in miles per hour

Actual wind velocities corresponding to unit wind loads of 11.76 psf, 22.50 psf and 52.90 psf are 54 mph, 75 mph and 115 mph respectively. An actual wind velocity of 54 mph is equivalent to 70 mph indicated velocity as determined by the Robinson 4 cup anemometer.

The minimum unit wind load for the 80-ft, 87-ft, 94-ft, 122-ft and 136-ft derricks with pipe set back is 11.76 lb/sq ft (54 mph actual wind velocity.) The minimum unit wind load for the 140-ft, 147-ft and 189-ft derricks is 22.50 lb/sq ft with pipe set back and 52.90 lb/sq ft without pipe set back.

The application of wind load assumes that the wind

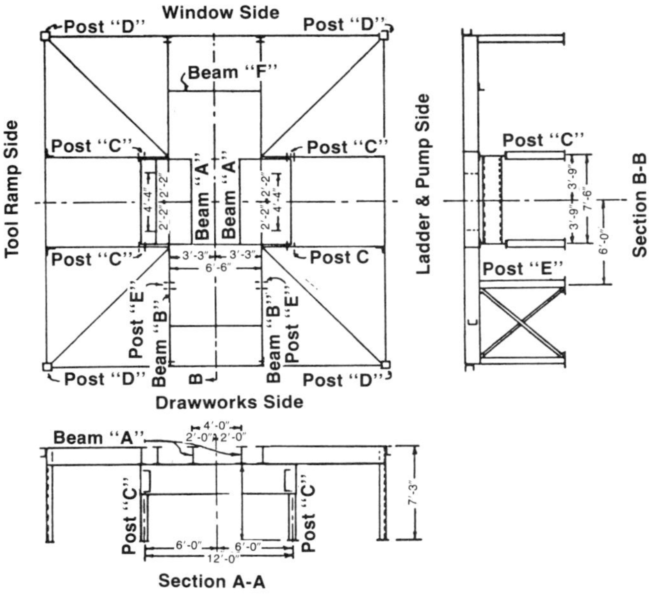

Fig. 5-4. Type A substructure.

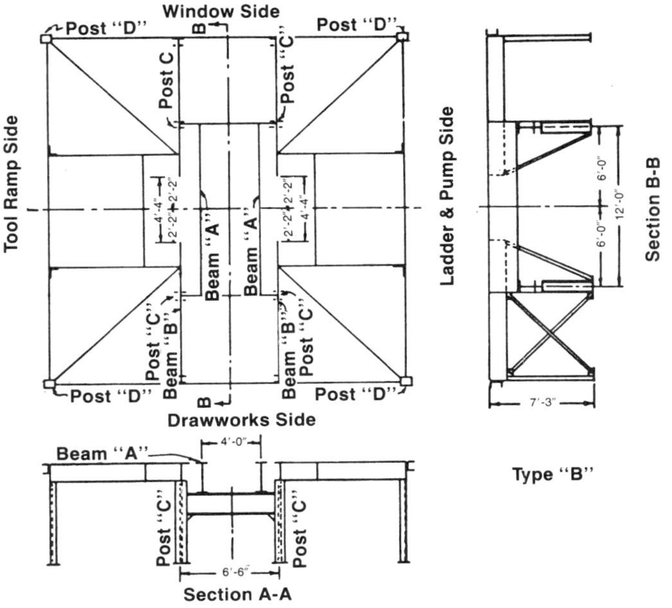

Fig. 5-5. Type B substructure.

load is applied to the exposed area of two directly opposite sides of the derrick and also to the pipe set back standing in the derrick. Additionally the horizontal component of the weight of the pipe set back is assumed acting at the pipe racking platform. For the convenience of the engineer, user and designer, Table 3 in API Standard No. 4A gives proper values for pipe set back weight and wind loads in conventional derricks.

Diagonal wind loading must be considered. To provide for this contingency, the legs, leg splices and anchorage are designed to resist loads 1.414 times those previously mentioned.

Other items which are considered in the design calculations of conventional derricks include:

For joints, girts and braces in addition to the wind loads and pipe set back loads already mentioned, a load condition consisting of three co-acting loads is as follows:

1. A concentric crown block load equal to 1.2 times the derrick maximum safe load, assumed to act concentric with the derrick.
2. A pipe set back load previously mentioned.
3. A horizontal load of 3000 lb or 0.01 times the maximum safe load of the derrick, whichever is least, applied at the top of the derrick leg at one corner of the derrick, in the plane of the derrick side and in the same direction as the horizontal component of the pipe set back weight.

For heavy girts on opposite sides of the derrick used for the support of the working platform, derrickman, pipe tong and counterweights, the girts must be designed to sustain a 1200 lb load 3 ft from the derrick leg or adjacent derrick side.

The water table beams are designed in accordance with the following formulas. For derricks 136 ft and shorter:

$$W = \frac{12M}{L} = \frac{12\,(S_m)\,(f_c)}{L} \text{ or 570,000 lb, whichever is smaller.}$$

For derricks 140 ft and higher for both 6½ and 7½-ft water table openings:

$$W_t \frac{15M}{L} = \frac{15\,(S_m)\,(f_c)}{L}$$

where

W or W_t = Maximum safe load for two water table beams in pounds.

L = Distance in inches between neutral axes of legs at the bottom of the water table beams or between the load axes if derrick is reinforced.

S_m = Section modules of one water table beam in inches.

f_c = Allowable unit stress on extreme fibers as stipulated by API Standard 4 A in pounds per square inch.

Selection of Proper Derrick Capacity

There is considerable argument concerning the proper way to select derrick capacity. One method is to match the drawworks to the derrick. Thus if a drawworks has a known maximum bare barrel line pull then this line pull times the maximum number of lines to be strung to the block plus the pull of the fast and deadline and the effect of the deadline position should not exceed the safe load capacity of the derrick.

Another method is to determine the safe load

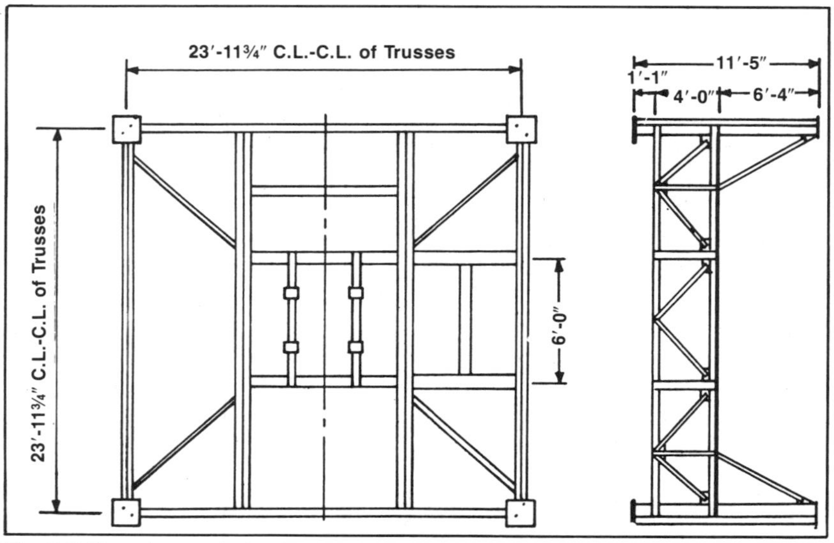

Fig. 5-6. Truss type substructure.

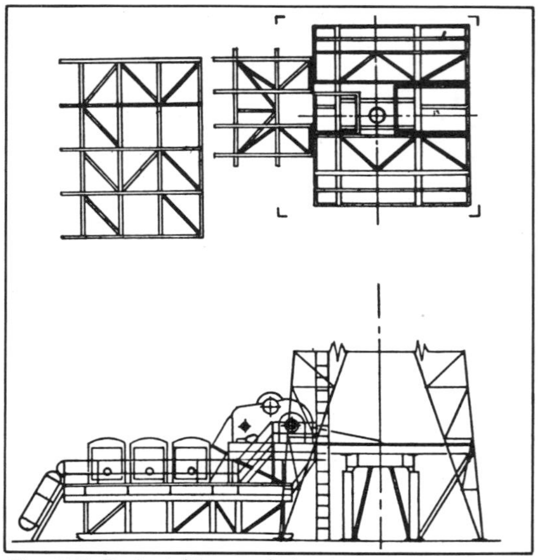

Fig. 5-7. Pedestal type substructure.

capacity from the hook load required to part a casing joint by jarring if the casing is stuck in the hole. Assuming a casing string of known size and joint strength is first necessary — which will equal the required hook load — after which the required number of lines and line pull can be determined with the effect of the fast and deadline and their position effect added to determine proper safe capacity of the derrick.

Substructures For Conventional Derricks

The API has now adopted 3 types of substructures for use with conventional derricks. These are Type A shown in Fig. 5-4, Type B shown in Fig. 5-5 and a Truss or Girder type—a typical example of which is shown in Fig. 5-6.

Pedestal type substructures to permit step down drives such as shown in Fig. 5-7 are also used. There are numerous variations in the Truss or Girder type.

The Type A substructure is designed to permit pipe set back and rotary loads to be carried separately when it is desirable because of soil conditions. This is accomplished by the use of six foundation piers. The Type B substructure uses four foundation piers to carry the pipe set back and rotary table loads. With the Type A substructure the blowout preventer equipment is lowered through the derrick floor while the Type B it may be withdrawn to the side from beneath the floor. Usually with the Truss types it may also be taken in or out from underneath.

The height of Type A and B substructures has been standardized as follows:

Base size feet	Substructure height
24	7.25
26	7.25
30	7.25, 10* or 14*
37.5	14*

*Tentative.

The substructure height is the vertical distance from the bottom of the corner post base to the top of the substructure framing sills on which the floor joists rest.

In the design of substructures the API Standard 4A advises as follows:

For truss or girder type substructures having a four point support the corner columns shall be designed for *two* loading conditions as follows:

a. Weight of casing and pipe set back acting *simultaneously*.

b. Derrick loading and weight of pipe set back acting *simultaneously*.

Assumed minimum loads for substructures are given as follows:

Substructure base Size, ft	Derrick loading lb	Casing load lb	Pipe set back load lb
24	700,000	300,000	150,000
26	950,000	450,000	200,000
30	950,000	450,000	200,000
30	950,000	600,000	350,000
37.5	950,000	750,000	450,000

The casing and pipe set back loads were determined as follows:

For the 24-ft base size, the pipe set back load of 150,000 lb is based on 8000 ft of 4½-in. 16.6-lb drill pipe with a full hole tool joint at each connection (18.5 pounds per foot). The casing load of 300,000 lb is based on 8,000 ft of 28-lb casing and an impact factor of 33⅜%.

For the 26-ft and 30-ft (136-ft height) base sizes, the pipe set back load of 200,000 is based on 11,000 ft of 4½-in. 16.6-lb drill pipe with a tool joint at each connection (18.5 pounds per foot). The casing load of 450,000 lb is based on 7500 ft of 45-lb casing or 11,000 ft of 30-lb casing and an impact factor of 33⅓%.

For the 30-ft base (140-ft and 147-ft height) size, the

Fig. 5-8. Cantilever mast being raised.

pipe set back load of 350,000 lb is based on 15,500 ft of 5-in. 19.5-lb drill pipe with tool joint at each connection (22.5 pounds per ft). The casing load of 600,000 lb is based on 10,000 ft of 45-lb casing or 15,000 ft of 30-lb casing and an impact factor of 33⅓%.

For the 37½-ft (189-ft height) the pipe set back load of 450,000 lb is based on 20,000 ft of 5-in. 19.5-lb drill pipe with tool joint at each connection (22.5 lb/ft). The casing load of 750,000 lb is based on 12,500 ft of 45-lb casing or 18,750 ft of 30-lb casing and an impact factor of 33⅓%.

For Type A and B substructures the casing load on beam A is considered as two concentrated loads, equal to one-fourth of the specified casing load and distributed as shown in Fig. 6 of API Standard 4A.

Also for Type A and B substructures the pipe set back load is considered as two-thirds the specified pipe set back load, uniformly distributed over two-thirds the length of one pipe set back beam as shown in the aforementioned Fig. 6 of API Standard 4A.

Truss or Girder type substructures are designed to support the pipe set back to the right of the driller as he faces the well.

Design of these substructures avoids any direct load carried by fastenings such as bolts and rivets in shear. Derrick legs are properly centered over the corner column.

The foregoing has dealt with standard API substructures. There are many other types. Sectional unitized bases to permit ease in skidding the derrick with uniform load distribution to a drilling mat are frequently used. Also substructures having 14 posts for more direct load transfer to the ground have been used. The main problem is to design the substructure to properly carry load with an ample factor of safety. API Standard 4A should be used as a general guide for all substructures including special or modified ones.

Fig. 5-9. Free standing mast ready to drill.

Portable Derricks And Masts

Portable derricks and mast fall into two groups — free standing and mobile. The free standing mast is usually a vertical or near-vertical mast made of several individual sections fastened together and mounted on a separate skid or base. Typical example of this type are the Ideco Full View (Fig. 5-10) or the

Fig. 5-10. Ideco full view mast.

Fig. 5-11. Lee C. Moore 136-ft helicopter cantilever mast.

Lee C. Moore Cantilever Mast (Fig. 5-8).

These derricks or masts are usually wireline raised with the hoist and traveling block. Free standing masts or derricks are self-supporting and stable without the use of external guys to the ground although wireline bracing might be used to the subbase or mat support frame.

Of historical significance are photographs taken in Tulsa, Oklahoma in 1938 (Fig. 5-8 and Fig. 5-9). They show the first raising of a Lee C. Moore 87-ft Cantilever Mast. The site was Carter Oil (Exxon) Co.'s yard.

Free standing masts have gained preeminence in rotary drilling operations in most areas. Fig. 5-10 shows an Ideco Full View drilling mast which has gained wide acceptance.

The decline in known petroleum reserves has made it necessary to explore for oil and gas in areas of the world where the environment often is hostile, such as the jungles of South America and remote Arctic regions, where there are few, if any roads. Rigs which can be moved into such locations must consist of relatively light segments so they can be transported by helicopter. Fig. 5-11 shows a Lee C. Moore 136-ft cantilever mast which breaks down into parts small enough to be flown in by helicopter.

As drilling depths increase and blow out preventer stacks become higher, it is necessary to increase substructure heights accordingly. Practical limits on such structures would seem to be 20 ft to 22 ft. A recent approach to this situation is the Raised Floor Cantilever Mast which is assembled near ground level with mast feet pinned directly into the front shoes. When raised it carries its own working floor along, including rotary support girders, rotary beams and set back deck. The drawworks and engines are installed at truck bed level with catworks and rotary at working floor level. Fig. 5-12 shows a mast of this type. Fig. 5-13 shows a similar rig winterized for work in arctic regions.

Mobile masts or open face derricks as they are sometimes called, consist of several types. First to be considered is the single and double pole telescoping pipe masts. These are made of telescoping pieces of pipe which extend to two or three times their collapsed height. Designs of this type include a racking platform and are frequently made tall enough to handle doubles. The biggest difficulty with pole masts is their extreme slenderness which makes them require multiple guy wires to several ground positions. For shallow well servicing with lightweight equipment, however, they do a good job.

Because of pole mast design limitations the single frame or lattice-type derrick was introduced about 1935. Originally it was designed to handle range 2 singles but later was made tall enough to handle doubles. Such derricks now are used on large core drills and slim hole drilling rigs. All are wheel mounted either truck or trailer complete with drilling equipment.

Following the single piece lattice or trussed-type construction mast came the development of the mod-

Fig. 5-12. Lee C. Moore 142-ft hi-floor cantilever mast.

ern telescoping derrick. Usually these derricks are 90-ft to 103-ft high and are suitable for handling doubles of range 2 for drilling and servicing. A triple telescoping mast having enough height to handle range 2 triples is also available (see Fig. 5-16).

Short telescoping derricks are also made to be mounted on smaller self-propelled vehicles for shallow well servicing. These are usually 55-ft to 65-ft high with hook load capacities of 75,000 lb to 150,000 lb. Taller telescoping derricks in the 90-ft to 103-ft range have hook load capacities of 150,000 lb to 300,000 lb. An example of a trailer mounted mobile type telescoping derrick is shown in Fig. 5-14.

Also important is the folding derrick which is competitive in both the shorter and taller heights to the telescoping derrick. These derricks are raised hydraulically as in the case of the Hopper or by a chain mechanism as in the case of the Lee C. Moore.

Many types of chassis are used for carrying mobile type derricks. These include tractors, trucks, trailers, self-propelled vehicles and the new drive-in-type self-propelled vehicle. This last type, currently being made by Hopper, Ideco, Franks, Wagner-Morehouse and Wilson, has many advantages (see Fig. 5-15).

First, the unit drives into the well location instead of backing in resulting in a faster, safer approach. Second, the weight on all wheels is more nearly balanced and most of the units more nearly comply to highway regulations. Third, the same large engine and torque converter used for driving the hoist is used for driving the vehicle over the road.

Mobile derricks or masts are used for both drilling and servicing operations. They are easily and quickly

Fig. 5-13. Lee C. Moore 142-ft hi-floor mast winterized.

spotted and erected (usually less than 30 minutes) and travel with the lines and blocks always strung so that they are ready for work as soon as they are erected in position. They eliminate the need for permanent derricks with the accompanying large capital investment.

Design Of Portable Masts

The design of portable masts or derricks is not nearly so limited by API specification as is the design of conventional derricks. The API does have a tentative specification for portable masts, however, and this is API Standard 4D. Generally speaking the same forces acting upon a conventional derrick act upon a portable derrick. These are the compressive leg loads due to hook loads, fast and deadline loads, hold back guy loads, wind loads and pipe set back loads.

Loads in members are usually determined by graphical analysis although the analysis is usually more involved than that of conventional derricks. Compressive column stresses and allowable loads are determined by the same method previously described using the API formula or the family of curves shown in Fig. 5-3. Wind loads on the mast and on the pipe set back are considered to be 11.76 lb/ft (54 mph actual wind velocity).

Practically all portable masts are of welded construction. The legs are made of either tubes or structural shapes. Some manufacturers make their own square or rectangular tubes from structural shapes by welding angles or channels together. The API formula for columns is based on a K factor of 1.0 as the value for a pin end condition. Some improvement of the end condition value can be expected for design purposes in a mast of this type. Because of the fixing properties of the adjoining members and the carry-through effect of a welded structure a better fixity

Fig. 5-14. Ideco Rambler Rig.

value (K) may be used in place of a pin end condition where K = 1.0.

The following items of general interest are also from API Standard 4D.

The height of a portable mast (other than single pole types) is the vertical distance from the horizontal plane through the foot of the mast legs (or from the lower side of the mast base if the mast is anchored to the mast base) to the lowest member which may contact, and thereby limit the rise of the traveling block.

The height of a single pole mast is the vertical distance from the centerline of the lowest crown sheave to the foot of the mast leg.

Load Guylines For Guyed Masts

Each guyed portable mast shall be provided with two load guylines attached as close to the top of the mast as possible. Each guyline shall have a breaking strength of at least 2½ times the maximum calculated guyline load. If stabilizing guys are used they also must have a breaking strength of at least 2½ times the calculated guyline load and they cannot be less than ⅜-in. diameter.

Mast Capacity

Most portable masts (at least of the mobile type) are now rated primarily by hook-load capacity with a certain number of lines to the traveling block. To obtain the gross capacity of the mast for the same number of the lines strung-up, the effect of the fast line, deadline and hold back guys must be added in.

Fig. 5-15. Hopper drive-in unit.

Fig. 5-16. Wilson back-in "Winchmobile".

It should be noted that this is not necessarily the gross column capacity of the mast (as in the case of a conventional derrick). It is merely the gross load imposed upon the mast under a given set of conditions. Changing the number of lines up or the position of the crown sheaves greatly modifies the input loads and their positions and influences and load carrying ability of the mast. Such changes should not be made without advice of the manufacturer.

Substructures For Portable Masts

As yet there is no API specification for portable mast substructures. Naturally it is expected that all manufacturers follow the material specifications and formula already laid down in API Specifications 4A and 4D.

There are about as many types and makes of substructures for portable masts as there are portable masts. Perhaps the most popular type for free standing masts is the trussed double box type. This type consists of two boxlike structures of truss frame construction approximately 4 or more feet wide by 9-ft high and about 44-ft long.

The reason these dimensions are so popular is that two of these box structures make a legal load in many states. Tying the two box structures together when in use are cross members at the top and bottom and knee braces.

The cross members are designed to carry rotary table loads, drill pipe loads, drawworks and engine loads into the substructure and thence into the

Fig. 5-17. Spud type drilling barge.

ground. Often the bottoms of these box structures are covered with metal plate to provide greater bearing to the soil. Or sometimes a location is prepared simply by leveling and laying down heavy planking.

It is common practice for manufacturers to make substructures of this type in various heights and widths ranging from 5 to 10-ft high and 4 to 6-ft wide and any length to fit the drawworks and engine combination.

For mobile types of portable masts, ramps and folding shelf type platforms are often used. When the mast is to be used for drilling, a trussed box frame structure carrying the rotary table, mud system, shakers and pipe rack all in one package is available.

The whole idea behind the mobile mast units is to provide a complete unit of extreme compactness on wheels to reduce moving and rig up time. All component assemblies including substructures are designed with this in mind.

Care And Use Of Derricks And Masts

The API Standards 4A and 4D have an exceptionally complete set of instructions on this important item. A properly set up derrick or portable mast simply will not fail if it is not overloaded or tampered with. Too often failures do occur because members have been removed, altered or because the derrick is loaded eccentrically or in a way not intended by the designer. It is usually the intention of the engineer and manufacturer to make every member a working member in one way or another and their removal or alteration to make room for other equipment is an extremely hazardous practice.

Fig. 5-18. Submersible drilling barge under tow.

The API recommendation is a good check list to follow when using either conventional derricks or portable masts as it encompasses every part of the use and maintenance of masts and derricks including welding, footings, soil bearing capacity, tie down, locking sections together, wireline and hydraulic raising, lubrication and loads.

Marine Drilling Foundations

As we move towards the end of the twentieth century it can be stated that most easy-to-find petroleum has already been found. A substantial portion of our new reserves therefore will be found in marine environments. Since much of the machinery used and the drilling techniques employed in marine operations are essentially the same as with large land rigs, only the drilling foundations will be covered in this chapter.

The first wells actually drilled within the boundaries of a body of water were those from platforms in Caddo Lake in Louisiana and at Goose Creek in Texas. The first well in heavy seas was drilled a short distance offshore of Cameron Parish, Louisiana in 1938. The platforms used to support the rigs in these instances, and for several years thereafter, were of wood construction on driven wood piles. Extensive

Fig. 5-19. Derrick barge transporting drilling derrick.

development of oil reserves in Lake Maracaibo in Venezuela has been going on for many years.

Inland water drilling can be defined as all drilling operations in the bays, lakes, rivers, canals and marshes which are at least partially protected from wind and waves associated with the ocean. Since the line of demarcation between onshore and offshore is poorly defined in many areas, the drilling method, rather than a geographical or political boundary, can best be used to separate the two areas.

At the turn of the century, as the search for oil spread to the Southwest, it became apparent that drilling operations would have to be conducted in many lakes, bays and marshes in that areas. The oil industry accepted this challenge in its customary manner and the oil and gas fields discovered under the inland waters of the Gulf Coast account for a substantial percentage of this country's reserves.

One of the first inland water areas to receive attention by the oil industry was Caddo Lake in Northwest Louisiana. Production had already been established on three sides of the lake and gas bubbles in the lake itself could be ignited. Most companies thought that the cost to drill the wells, produce the oil and deliver it to the shore would be too great to compete with oil produced on land. However, the Gulf Refining Co. thought that it could be done and purchased 8000 acres of lake bottom from the U.S. government for

Fig. 5-20. Island method of inland water drilling.

$30,000 cash and $70,000 in oil payments. After two years of planning, construction and drilling, the first well was completed on May 11, 1911.

Until the early twenties there was only a limited amount of activity connected with inland water drilling, but with the advent of modern geophysical methods in 1923, the rate of discovery of salt domes on the Gulf Coast was accelerated. Many of these salt domes were situated in the marshes and bays of South Texas and South Louisiana and in order to drill them a variety of marine structures to support drilling and production equipment was used.

The pioneers in inland water drilling often built roads and locations above the expected water height, which were flooded out, resulting in a loss of time and often in loss of drilling equipment. Not to be defeated, they built their roads and locations higher, using protective levees, and often were forced to place all equipment on piling platforms or man-made islands to escape rising water and floods.

These measures were expensive and in order to compete with land drilling costs, a constant effort was being made to reduce drilling costs and improve efficiency. To accomplish this, more and more of the drilling equipment was placed on barges, so that the rigging up and moving time could be cut to a minimum, until the submersible drilling barge was developed. This submersible drilling barge has probably done more than any other single piece of equipment to develop the large oil and gas reserves on the Gulf Coast (see Fig. 5-17 and Fig. 5-18).

One of the earliest methods employed as a foundation for the drilling rig in the low-lying marshland of the Gulf Coast was an earthen and wooden mat. In areas not subject to a large rise in water level, the road and location were raised with an earthen fill and a wooden mat built on top to distribute the drilling equipment load. Instead of raising the ground level with earthen fill, levees were often built around the

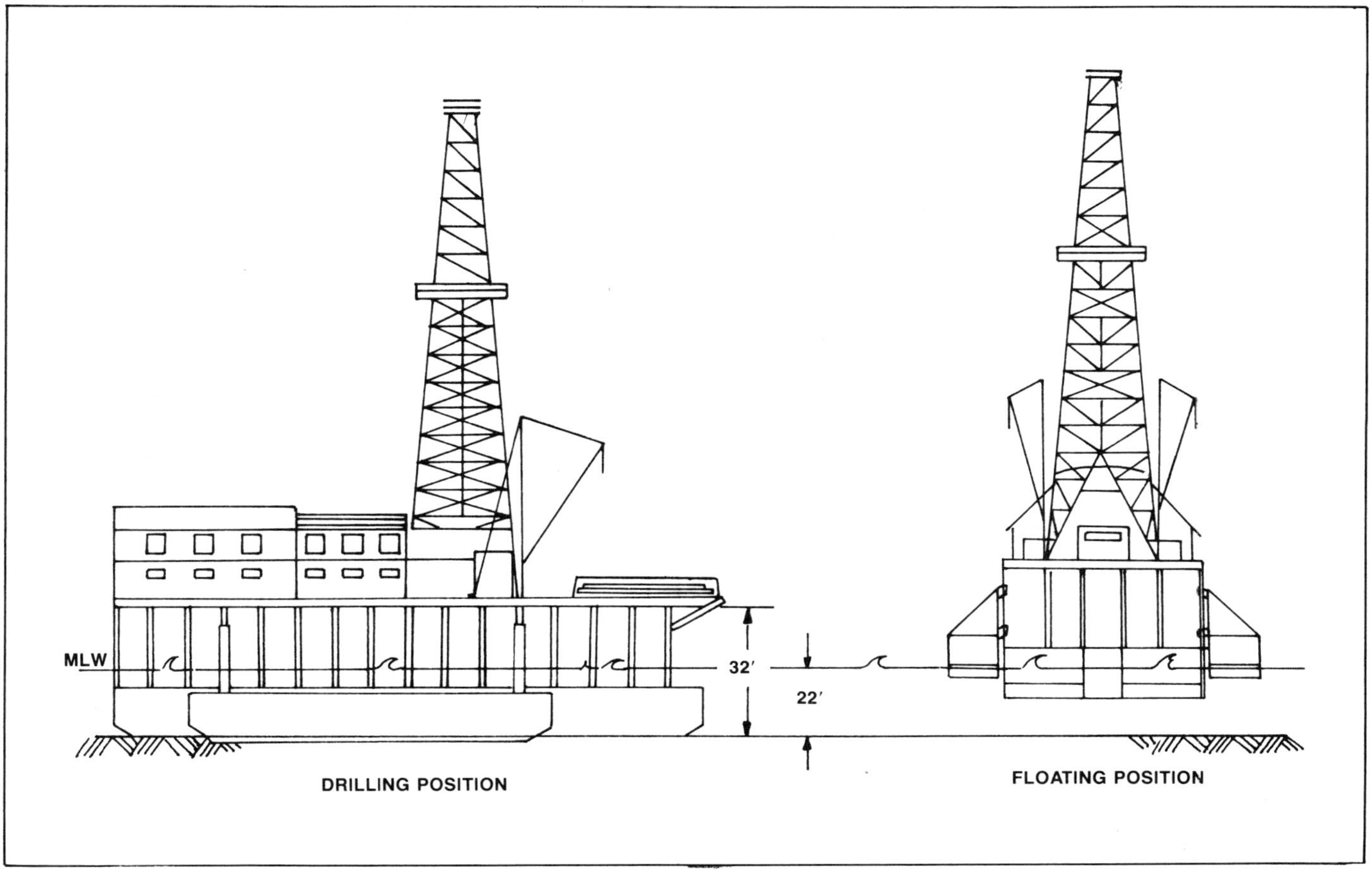

Fig. 5-21. Barnsdall-Hayward Submersible Barge

location to keep the water out. When this was done, a pump was installed to pump out seepage and rain water. When a fill or levee was exceptionally high, it was often necessary to drive wooden piling for additional support and bulkheads to prevent the earthen fill from washing out.

One of the most successful applications of this method of drilling was in the Bayou des Glaise field, in the lowlands of the Atchafalaya River Basin, Louisiana. This field is located inside the 28-ft levees that protect the surrounding country from the sudden and destructive high waters and floods that occur when the Mississippi River overflows into the Atchafalaya. After the first location was inundated by the rising Atchafalaya, an earthen island was built high enough to remain above the flood waters of the river and large enough to support the derrick, drawworks, and boilers.

The mud pumps, mud pits, mud supply, generators, tool house, etc., were installed on a floating barge. This barge was tied up to the island and allowed to rise and fall with the tide. By drilling several wells from an island, the total foundation cost was reduced and the production operations simplified.

The wells drilled in Caddo Lake in 1911 were drilled from a large piling platform. Cypress trees along the lake shore were felled and made into piling. A floating pile driver, three tugs, 10 barges and 36 small boats were brought in from the Red River and the Mississippi River to assist in fabricating the platform. It required many months of slow and backbreaking work to complete the platform and install the drilling rig. The platform was large enough to support the drilling rig, pipe rack and storage space.

Approximately 275 wells were drilled in Caddo Lake using this method during the period 1911 to 1950.

These self-contained platforms were very expensive and took considerable time to construct. To reduce this high initial cost and excessive construction time, much of the drilling equipment was placed on a floating barge and the platform reduced in size to support only the derrick and drawworks. This combination platform and floating tender method of drilling was used extensively in areas where the water was too deep for a submersible barge or where underwater obstructions, such as stumps or boulders, made the use of submersible barges impracticable.

A later development was to fabricate the small platform with tubular steel so that it could be installed as a single unit. After steel piling is driven through the jacket legs, the substructure, derrick and drawworks are installed as a single unit with a derrick barge. After the well is completed, the entire unit

Fig. 5-22. DeLong jackup drilling platform.

is moved to another platform without unstringing. The floating barge contains all other machinery and equipment necessary for drilling operations and a move can be made in a matter of hours (see Fig. 5-19).

In well protected areas and with certain limitations, floating barges have been used for drilling and workover operations. All equipment is placed on a slotted barge and held in position by spuds through the four corners of the barge and into the ground. It has been used on several occasions for core drilling, sulfur drilling and for shallow workover operations. These operations were in relatively shallow water and in areas not subject to wave actions. The time spent in moving this type of barge is less than that for any other method as the rig is ready to start operating as soon as the barge arrives on location and the spuds are dropped. Naturally, the hook load is limited to the buoyancy of the barge which is not used in drilling deep wells.

A barge was completed for drilling operations in Caddo Lake that permitted the drilling of 5500-ft wells from a partially floating barge. The depth of water, uneven bottom and presence of cypress stumps made the use of a submersible barge in this lake impracticable.

The first step to a new method was to reduce the size of the platform by placing mud pumps, mud pits, tool house and miscellaneous equipment on a floating barge, reducing the platform cost. The next step was to design and build a floating barge to support the substructure, derrick, engines and pipe racks.

As it was necessary to operate in 15 ft of water and withstand winds of 75 mph and 5-ft waves, eight 14-in. OD spuds were installed on the barge. These spuds were driven to a depth of 25 ft below the mud line and the barge clamped off to them with a special clamping arrangement. After the barge was clamped off, approximately 2 ft of water was pumped into the

Fig. 5-23A.
Penrod Drilling Co.'s Penrod 60 is shown while working on location in the Gulf of Mexico. Penrod 60 is a Marathon LeTourneau Class 116 self-elevating jackup drilline platform built in LeTourneau's Vicksburg, MS yard. The rig can drill in 340 ft of water and has 452 ft of legs. The hull is 230 ft long, 200 ft wide and 26 ft deep.

barge, which loaded the piling and resisted the uplift forces caused by wave action.

Two hydraulic jacks were later designed and used to drive and pull the spuds which eliminated the use of a pile driver. It was also found that approximately 10 ft of penetration were sufficient to hold the barge in position.

A similar barge was designed to operate in Lake Maracaibo, Venezuela, in 25 ft of water. This barge 65 ft × 135 ft × 10 ft, with ten 30-in. OD spuds, was designed to withstand the lateral forces resulting from an 8-ft wave. All equipment is installed on this barge and the two jacks on each spud will develop more than 100 tons or enough to lift the barge out of the water.

The high cost, long construction time and lack of mobility of the island structure and large piling platform were partially overcome by placing some of the equipment on a floating barge. A floating barge with all drilling equipment eliminated all of these disadvantages but was not stable enough to be used in open water or for drilling deep wells. The logical solution to this problem was the subsmersible drilling barge developed by the Texas Company (now Texaco) in 1931. The basic principle of this method of drilling in inland waters has remained unchanged.

Fig. 5-23B.
Penrod 68, a Marathon LeTourneau Class 116 self-elevating jackup drilling platform for Penrod Drilling Co. is shown undergoing final construction at Marathon LeTourneau Offshore Co.'s Brownsville, TX yard. The jackup rig can operate in 300 ft of water and has 410 ft legs. The hull is 248 ft in length, 200 ft wide and 26 ft deep.

Drilling operations in open waters of the Gulf of Mexico started in 1938 near Cameron, Louisiana on what was an extension of the onshore Creole field. The rig foundation was all wood construction and this type of structure was used almost exclusively for several years. But when water depths increased as operations moved further from shore, a large variety of marine drilling platforms evolved. Notable among these was the Barnsdall-Hayward barge developed in the 1940s by John Hayward. In many respects the basic principle of submerging the hull was similar to that of barges used on inland waters (see Fig. 5-21).

The unit consists of two upper decks, for machinery and operations, supported by steel columns which extend to a submersible hull 160 ft × 54 ft × 13 ft. Two pontoons, 100 ft × 14 ft × 12 ft, are mounted outrigger fashion on either side of the hull. The upper two decks house the machinery, auxiliary equipment, supplies and quarters for the crews. The hull

Fig. 5-24. Sedco semi-submersible barge

consists of numerous ballast and storage compartments and a room for ballast and transfer pumps. One end of the barge includes a well slot to permit moving off location after the well is completed, much in the same manner as provided for in inland drilling barges.

Sinking the unit on location is accomplished by flooding and sinking the slot end of the barge to bottom. When firmly situated, the other end of the barge is flooded and sunk. The pontoons provide stability while the hull is being sunk. The pontoons are thereafter flooded and jacked down to bottom to provide additional foundation. The submersible barge is used for both exploratory and development drilling in water depths of 40 to 50 ft.

Fig. 5-25. Western Oceanic Pacesetter No. 1

Fig. 5-26. Trans-World drillship Rig #61

Another popular type of mobile marine drilling platform is the jack-up such as the Delong (see Fig. 5-22) or the LeTourneau (see Fig. 5-23). The jack-up generally is suitable for use in deeper water than the submersible barge. The jack-up is used primarily for exploratory drilling and if worthwhile reserve is indicated the development drilling is done with a self-contained platform from which a large number of wells may be drilled Figs. 5-22 and 5-23.

A third style of platform used extensively is the tender type in which the derrick and rig floor machinery rests on a steel pile platform while the engines, pumps, mud tanks, crew quarters and supplies are carried in the tender or floating vessel which is tied up to the platform and adequately anchored.

For exploratory drilling in deeper water, the semi-submersible drilling barge has proved most popular. Depending on water depth these barges may be anchored to the sea bottom or dynamically positioned by use of thrustors which are controlled by a variety of sensors. Typical of this type of barge is the Sedco 135, shown in Fig. 5-24, with Lee C. Moore 142 ft x 40 ft dynamic mast.

Another example of the semi-submersible is the Western Oceanic pacesetter shown in Fig. 5-25.

For exploratory drilling in extremely deep water the floating platform or drillship, as shown in Fig. 5-26,is often used.

Chapter 6
Rotary Drive Mechanisms

In the 1800s various primitive type rotary machines were used to drill both water and oil wells. Birth date of the modern rotary table, however, is generally taken as 1901 at Spindletop, Texas. Although the rotary table has been modified and improved since its conception, it is still essentially the same piece of machinery. As in early models, the modern rotary table consists of a ring gear which carries a bushing to drive the drill pipe and a pinion to drive the ring gear (Fig. 6-1).

The present-day ring gear is spiral beveled, as compared to the earlier straight bevel gear. The overlapping action of the teeth tends to eliminate shock and provides a smoother running machine. The gear is machined from alloy steel and the teeth cut before heat treating. The teeth are then flame-hardened to provide maximum wear resistance.

The gear table is an alloy steel casting on which the ring gear and upper race of the main bearing are shrunk. The table skirt is provided with studs, which secure a ring to lock positively the hold-down thrust bearing. In addition to carrying up thrust, the hold-down bearing, together with the main bearing, provides accurate table alignment.

The master bushing is a split steel casting with API standard driving square for the kelly bushing and slip seat. Recesses are provided in the bushing for a lifting hook to facilitate handling on the rig floor. The bushing is seated in the table casting. Early tables had, in place of the square for the kelly bushing, a grip ring attached to the turntable (Fig. 6-2). The rollers clamped directly on the drill pipe. This drive often resulted in crushed drill pipe.

A natural modification was the introduction of a heavier-than-drill-pipe griefstem to ride in the grip ring. The familiar kelly soon followed. Much later a table ring guard, with a non-skid tread on its upper surface, was introduced to provide greater safety for the men working around the rotary. It also provided a convenient place to set the slips on trips.

The pinion shaft assembly consists of an alloy steel shaft with the pinion gear shrunk on it. The shaft is supported by two roller bearings mounted near the respective ends. Shaft lock collars are mounted on the shaft between the bearings. Two pawls manually actuated by handles mounted on the outside of the table engage the lock collar and may be used to lock the table against rotation in either direction.

The assembly is factory-aligned, or equipped with self-aligning bearings, to prevent required adjustment of the contact with the main gear in the field. The teeth of the pinion gear also are hardened to resist wear.

Most of the moving parts in the rotary table are flood lubricated, the remainder lubricated through conveniently located grease fittings. The main base acts as a sump which holds the oil in which the ring gear and main bearing run. Although the pinion is flood lubricated, it is not always lubricated from the same sump as the ring gear. Some manufacturers provide a separate reservoir to carry the lubricating oil for the pinion gear and inner pinion bearing. The outer pinion bearing and hold-down bearing are usually grease lubricated. The particular advantage in a machine completely flood lubricated is that greasing up every tour can be eliminated. Lubrication oil need be added to the rotary table infrequently.

The main base, in addition to serving as an oil sump, provides support for the table casting, through the main bearing, and for the pinion shaft assembly. It is frequently an alloy steel casting, but may be welded from preformed steel plate. The top of the base is provided with a non-skid surface. A tongue-and-groove labyrinth seal between table and the base make the lubrication system slush proof.

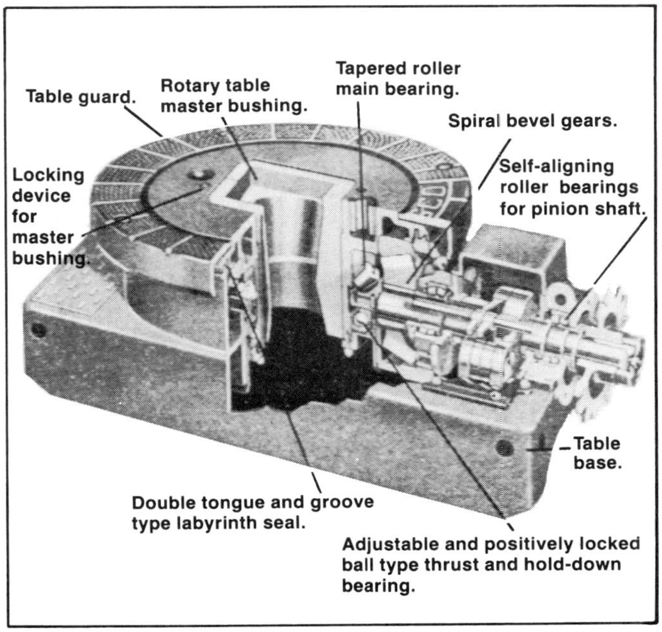

Fig. 6-1. Cutaway of modern rotary table, showing various parts of a modern unit.

Fig. 6-2. Early model rotary table had a grip ring attached to turntable.

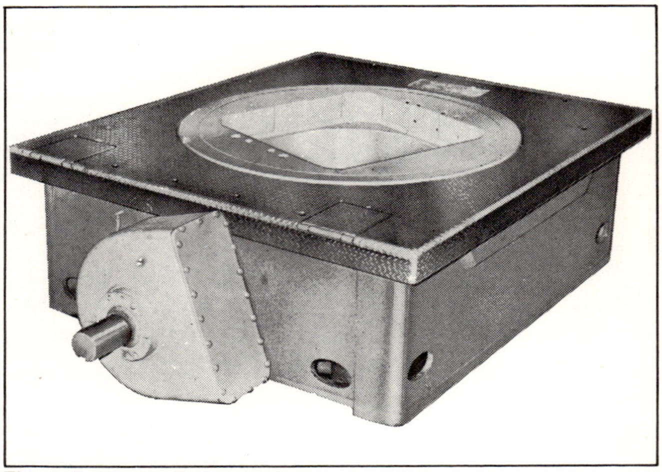

Fig. 6-3. 49½-in. Rotary table.

The size of the rotary table is taken from the diameter of the opening in the table through which the bit passes. The practical limit to size of opening is governed by both ability of moving parts to withstand the increased loads imposed upon them by virtue of their larger size, and the logical upper limit of hole diameter. The largest tables today have an opening of 49½ in. (Fig. 6-3). The 27½-in. table, however, suffices for most drilling purposes. With new drilling techniques, rotary speeds greater than 300 rpm sometimes are necessary.

Most rotaries take their power through the drawworks, by means of chains and sprockets, from the primary engines. This results in an obstruction, in the form of a chain and guard, extending halfway across the derrick floor. In recent years, with the trend toward deeper drilling, there has been a move toward using a separate engine to drive the rotary table.

The first truly independent rotary drive was developed during the steam era. It consisted of a small 2-cylinder steam engine mounted on skids with the rotary table and speed increaser gears in line with the pinion shaft. Next came drilling units with internal combustion engines mounted in like manner but with a gear box, having two speeds forward and one reverse, between engine and rotary pinion shaft (Fig. 6-4).

The large diesel electric rigs, currently in use, employ a separate d.c. motor of 500 hp to 800 hp capacity, connected to the pinion shaft of the rotary table through a 2-speed gear box (Fig. 6-5).

The principal advantage of the auxiliary power is that, with the proper layout, it may be used in an emergency in place of the primary power plant. The elimination of the obstruction caused by the chain drive reaching from the drawworks to the rotary table is also a great convenience.

One of the more significant advances in rotary table design in recent years has been the inclusion of power slips as an integral part of the table (Fig. 6-6). The disadvantage of the overhang associated with power slips that are mounted externally on the table is eliminated. The slips are actuated by air pressure, controlled through a pedal at the driller's position. Through the use of slip bowls the conventional master brushing is no longer required. Separate bowls are used for drill pipe and casing, except for casing 5½ in. OD or smaller which can be run through the drill pipe slip bowl. Drill pipe slips can be changed to casing slips in only a matter of minutes.

The kelly is driven directly from the table. This arrangement avoids the trouble that is sometimes experienced by the drilling crew in getting the kelly bushing to slip into place. The operating levers for the pinion shaft lock assembly are on top of the pinion shaft housing, as contrasted with the usual side location. This arrangement permits the table to be mounted flush with the derrick floor if desired.

Turbodrill

The turbodrill is primarily a Russian development. A prototype of their tool, a 1924 Russian patent, was exhibited and briefly demonstrated at the 1928 Tulsa oil show. The design of the tool was a well guarded secret but it is believed that it was a single stage high speed mud turbine with reduction gears to produce a normal rotary bit speed of about 30 rpm. Also in 1924, a U.S. patent was granted to an American on a multi-stage turbodrill which is the forerunner of current designs. The Russians, a few years later, designed a multi-stage tool which they have refined from time to time and are now using in drilling nearly all their wells. The French have been active in turbodrill development and although their tool compares favorably with the Russian tool, it has not replaced conventional rotary drive in French operations.

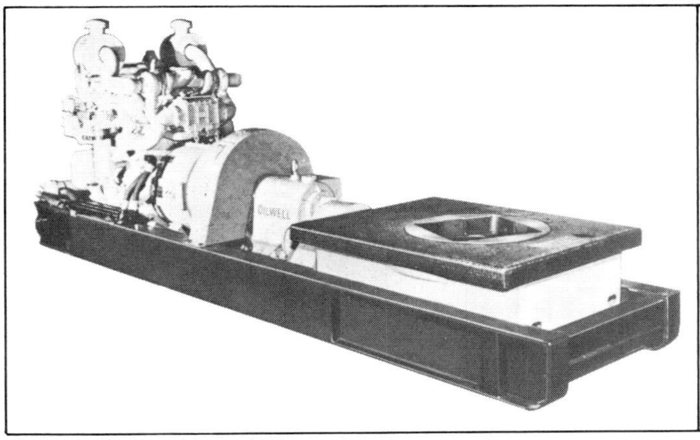

Fig. 6-4. Engine driven drilling unit.

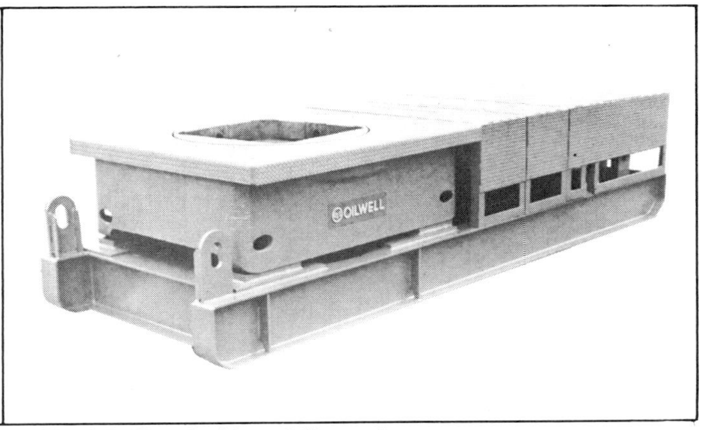

Fig. 6-5. Electric driven drilling unit.

The modern turbodrill is a multi-stage hydraulic turbine, usually positioned immediately above the bit and driven by the conventional mud stream. While drilling, the drill pipe is rotated very slowly or not at all and the downward thrust is provided by weight of the tool. Drill collars are not used since additional weight, if required, can be supplied by the drill pipe. Wear on tool joints and damage to drill pipe in compression is alleviated by low (or no) rotation. Bit rpm is especially high, hence low bit weight must be used. Normally the rate of penetration is greater than in conventional drilling but cone-type bits do not stand up too well. Diamond bits may be better suited for light weight, high speed operations. Turbodrilling has not proved economic for normal drilling operations in the U.S. However, it is used for special services such as directional drilling. (see Fig. 6-7).

It is puzzling why, with the reported success of turbodrilling in Russia, it has not been used more widely in the U.S. Several reasons have been suggested, first the economy of the two countries is entirely different. Russians give priority to greatest number of wells drilled with a limited number of rigs, cost being secondary. Considering the relative drilling technology of the two countries, rigs in the U.S. outperform turbodrilling in Russia. It has been suggested that our drill pipe and tool joints have been improved continuously while the Russians concentrated on improving the turbodrill and gave less attention to improving conventional drilling methods. Therefore, higher penetration rates and fewer fishing jobs favor turbodrilling in Russia.

Considering the interest developed in the U.S. by reported turbodrilling success in Russia, Dresser Industries, about 1955-56, decided to import both Russian and French turbodrills for evaluation. Dresser's Mr. O'Connor visited Russia and France and purchased several tools from each. After these tools arrived, arrangements were made to rent them to various operating companies and drilling contractors for thoroughly testing the efficiency of turbodrilling compared to our conventional system. Exhaustive field tests at 77 wells were made during 1957, 1958 and 1959. Careful records were kept and a summary of test results was reported on by Mr. W. E. Bingman of Shell Oil Co. in his API paper titled "The Status of Turbodrilling in the U.S.A." (Page 15, API Drilling and Production Practice, 1960).

Mr. Bingman's paper covers his subject very well and is recommended reading. His conclusions were as follows:

"As a result of extensive field testing of current turbodrilling methods and equipment, including both diamond bits and rock bits, the following conclusions were reached:

"1. Turbodrilling cannot, as yet, compete economically with good rotary practices. Where effective bit weight cannot be used with rotary (crooked hole, directional drilling, small hole size and large surface hole drilling), turbodrilling should eventually prove competitive.

"2. Rock bits have generally performed poorly on the turbodrill. Penetration-rate improvements have not offset the drastic reduction in bit footage (often less than 30% that of rotary). The type of bit wear indicates that the cutting structure, as well as the bearings, will require modification for turbodrilling.

"3. A few economical turbodrill diamond bit runs have been made where the conditions were favorable. However, inconsistent turbodrill and bit performances have cancelled such savings on subsequent runs. More basic research should be conducted to determine the factors affecting diamond bit performance. Surprisingly, little is known about the effects of crown shape, waterway design, diamond size, quality, concentration and exposure, and drilling variables on the penetration rate and bit life.

"4. The turbodrills, as well as the bits, need modification before turbodrilling can be used effectively. The characteristics should be altered to provide optimum drilling conditions at the bit (subject to the findings from bit research). The

Fig. 6-6. Modern rotary table with integral slips.

tool should also be redesigned for field servicing and provided with a better shaft seal.

5. A reliable turbine tachometer has been developed and field-tested satisfactorily below 12,000 ft.
6. Current turbodrill equipment, bits and techniques have been adequately tested under our drilling and economic conditions. Before further field testing can be recommended, basic research in bit design (both rock bits and diamond bits) is needed, and significant modifications in the turbodrill should be made.

"It has been decided to discontinue field testing until such time as bit development justifies a new program. The turbodrill purchaser has announced that the turbodrill equipment has been 'moth-balled' and research emphasis is being shifted to the development of a suitable cutting tool for the turbodrill.

"Certainly the industry recognizes the need for research in developing a better drilling method using some down-hole power unit. Such a method should ultimately offer the best opportunity for drastically reducing drilling costs of our deeper, more difficult directionally controlled wells."

Dynadrill

The dynadrill tool is an outgrowth of the Moyno pump which has been used extensively in various industries for liquid transfer. Dynadrill uses a specially designed Moyno pump running backwards as a power source (Fig. 6-8). The three stage motor assembly, which occupies about one-half total length of tool, is lined with rubber that contains a spiralled cavity or passage of elliptical cross section throughout its full length. Running through the length of this cavity is a solid steel shaft, which is mounted eccentrically at both ends and is shaped in a regularly recurring wave

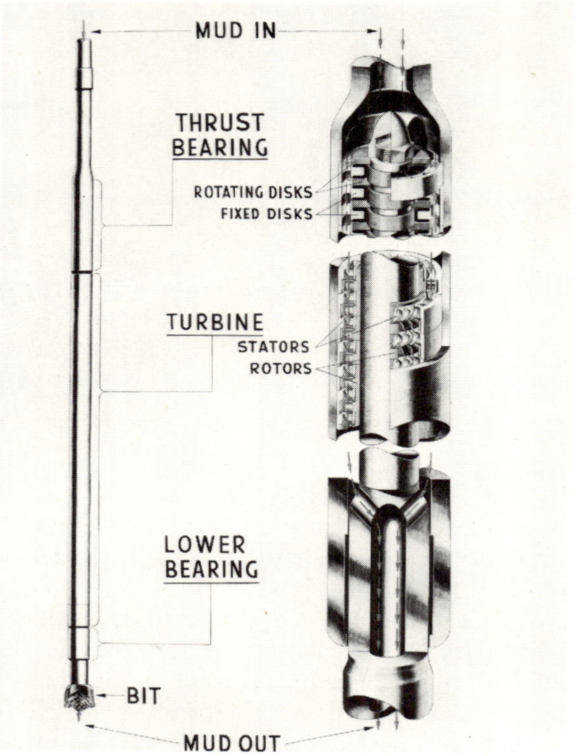

Fig. 6-7. Turbodrill.

form. When mud is pumped under pressure from above, it seeks an exit through the annulus between the eccentric shaft and the spiralled passageway in the surrounding rubber. Since this annulus can be cleared all the way to the bottom of the passageway only through the displacement of the eccentric shaft, the mud pressure causes the shaft to rotate, thus actuating the motor assembly.

Below the motor assembly is a connecting rod assembly with universal joints at each end to accommodate the eccentric rotation of the motor shaft and transfer this rotation to the bearing and drive shaft assembly which is fitted with thrust bearings to carry vertical loads in either direction. The rotating bit sub below the large thrust bearing is the only external moving part of the tool. Various bit types with API shanks can be accommodated.

The dynadrill motor is a positive displacement device which operates at relatively low rpm, comparable to conventional rotary speeds. It can be used with either mud or air as the circulating fluid. Since drill pipe rotation is not required, dynadrill is adapted easily to controlled directional drilling. A bent deflecting-oriented sub immediately above the tool takes the place of a whipstock as used in conventional drilling. It has proved to be a useful tool in offshore directional drilling. (See Fig. 6-9)

There will always be considerable interest in the development of a good downhole motor to turn the bit. It seems a little strange that we must actuate the bit

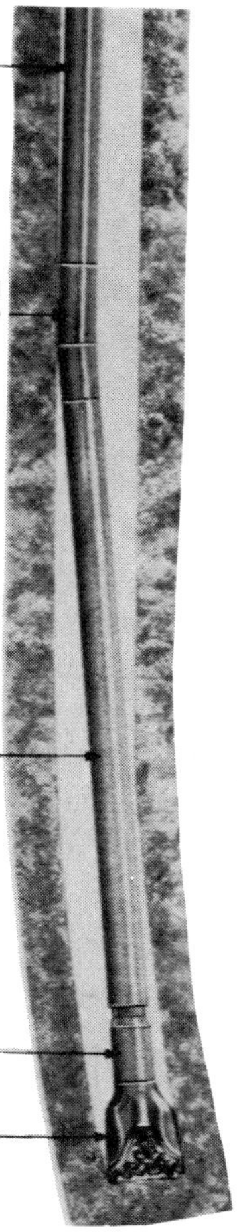

Fig. 6-8. Cutaway view of Dynadrill.

Fig. 6-9. Dynadrill deviation hook up.

with a power source in some instances as much as five miles away.

The challenge is quite great to develop a dependable power unit closer to the bit.

Efforts in this direction include vane-type air motors, where air drilling is used and numerous rotary mud motors. Downhole electric motor experiments are being conducted and offer possibilities.

We should not overlook the possibility that the rotary bit might be displaced by some exotic means of destroying rock such as pellet drills, explosive drills, electric air drills, jet piercing drills, plasma drills and a host of others.

Fig. 7-1. Two-shaft, 3-speed drawworks on three wood jack posts. Typical of the late 1920s.

Chapter 7
Hoisting Equipment

Drawworks

The drawworks is the kingpin of rotary rig machinery. Its brand name, model or nominal capacity are often used to designate or identify the particular rig size of which it is a part. The drawworks is a hoist of special design, which performs a number of essential functions in rotary drilling operations. The primary function: to exert a pull on a line or cable, not unlike hoists used in mining and other industries or the winch used aboard ships. All of these machines developed from the windlass, a simple apparatus consisting of a drum or cylinder revolving on a horizontal axis, upon which is wound a rope, cable or chain attached to the object to be lifted. In the simplest form, a windlass with crank is hand powered to draw buckets of water from a well or to remove debris from a manually dug mine shaft.

From the beginning, as at Spindletop, the drawworks performs five essential functions: (1) exerts a pull on the drilling line, which when reeved through blocks and with other suspension equipment, withdraws the drill string from the hole; (2) through its braking mechanism retards and controls the rate of lowering drill string or casing into the hole; (3) also through its braking system it limits and controls the amount of weight or downward thrust being applied to the bit; (4) provides a power takeoff for a chain driven rotary table; (5) cat shaft is equipped with catheads which, by exerting pulls on rope or cable, perform numerous rig floor services.

The drawworks used drilling the Lucas wells at

Fig. 7-2.
Three-shaft, 4-speed drawworks on two steel jack posts. Rear view with guards removed. Typical of the 1930s.

Fig. 7-3.
Three-shaft, 4-speed drawworks unitized on steel skids. Guards removed.

Fig. 7-4.
Three-shaft, 4-speed bath drawworks.

Spindletop in 1901 was chain driven from a single cylinder steam engine. It had two shafts with bearings on wooden jack posts, one brake and a single hoisting speed. Since that time we have come a long way, step by step, in upgrading this piece of machinery to what now seems to be a very sophisticated machine, which later on will be illustrated and described fully. Although improvements were more or less continuous, with a few busts now and then, there have been some milestones along the way worth noting. Improvements generally have been related to well depth: two or three drum speeds and two brakes increased rig capacity (Fig. 7-1); addition of a countershaft, or jack shaft, made four speeds on the drum shaft possible (Fig. 7-2); unitization of the drawworks, whereby wooden jack posts were replaced by steel beams mounted on steel skids (Fig. 7-3). This improved portability of the machine and reduced rig up and rig down time. The early unitized drawworks were field shop conversions of existing machines and were semi-enclosed. Fully enclosed drawworks with oil bath lubrication is illustrated in Fig. 7-4. Auxiliaries, such as automatic catheads and auxiliary brakes are shown in Figs. 7-5, 7-6, 7-7, 7-8 and 7-9. Fig. 7-10 illustrates a medium to large drawworks embodying nearly all of the improvements made up to the end of the Steam Era.

Spark plug rigs, as they were first called, came into the picture in the early 1930s and for the most part were designed for light duty. The reasons being: (1) There were few, if any, highly portable internal combustion engines of sufficient size to compete with large steam rigs; (2) they were pioneered by manufacturers of well servicing equipment, like Wilson, or by personnel formerly active in such equipment design. Friction clutches, were of course, essential parts of an internal combustion engine driven drawworks. Torque converters, fluid couplings and air controlled heavy duty clutches were incorporated in drawworks design. Friction clutches became so successful that they replaced positive jaw-type clutches on some steam drawworks for rotary drive and for both high and low drum clutches (Figs. 7-11, 7-12, 7-13).

Modern drawworks, regardless of manufacturer, are very much alike, size for size. They usually are rated according to the input horsepower capacity and range in capacity from 500 hp to 3000 hp. The larger machines naturally are more sophisticated and therefore, in order to cover the various features available at this time, only a large unit will be described in detail. One such unit is illustrated in Fig. 7-14. This machine is adaptable to either electric or mechanical drive. The mechanical or engine driven shaft speed layout is shown in Fig. 7-15, whereas

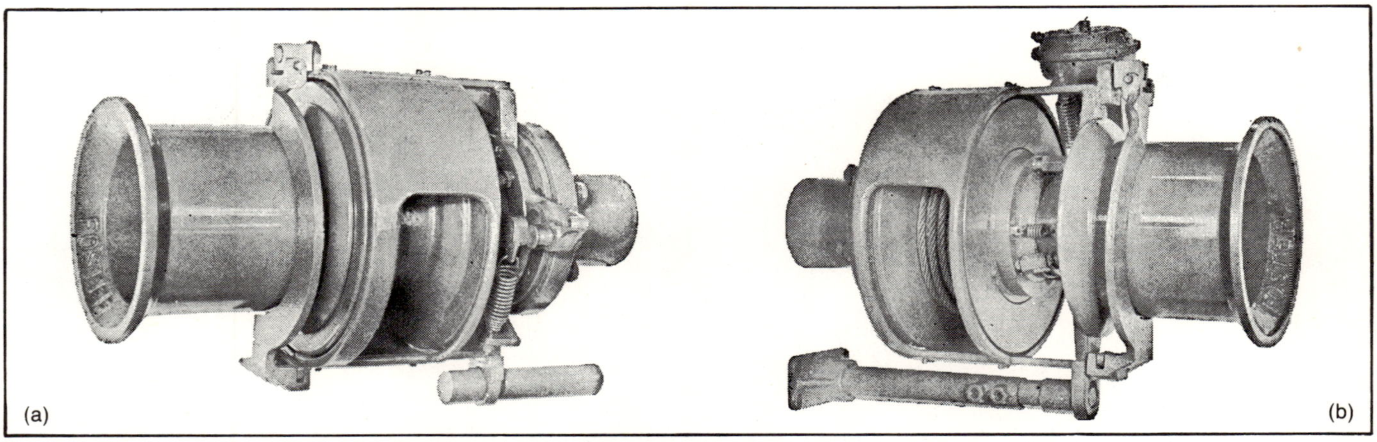

Fig. 7-5 a&b. Automatic catheads. At left is the spinning or makeup cathead. At right is the breakout cathead.

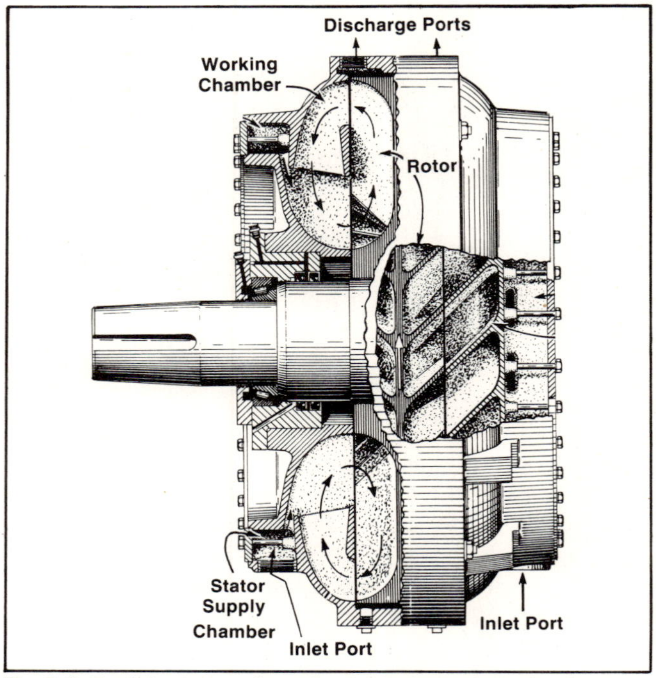

Fig. 7-6. Partial cutaway view of a hydrodynamic brake. Common trade name in use is hydromatic.

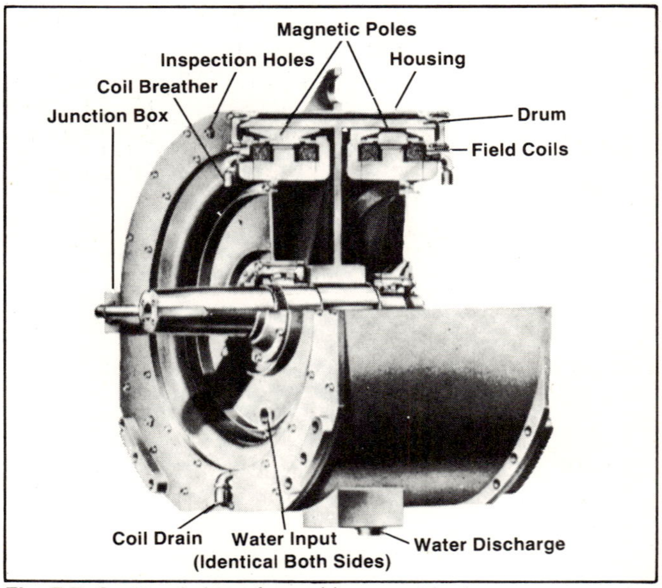

Fig. 7-7. Sectional view of an eddy-current brake.

Fig. 7-8. Overrunning and disconnect clutch.

the electric-driven shaft speed layout is shown in Fig. 7-16.

It will be noted that multiple strand roller chain, flood lubricated, is used extensively. Air clutches are used for both high and low drum drives and for rotary table drive. On the engine driven adaptation, an air actuated master clutch on the engine compound connects power to the drawworks.

All clutches, whether air actuated or positive, are air controlled from the driller's console. Fig. 7-17 illustrates a console for an electric-driven drawworks. The console for the engine driven unit is basically the same except the top section, which concerns only the electrical power input, is omitted and the engine drive group of controls and gauges sits alongside the main control cabinet.

Auxiliary Brakes

The auxiliary brake may be either the hydromatic or eddy-current type. An overrunning clutch is recommended when the hydromatic brake is used.

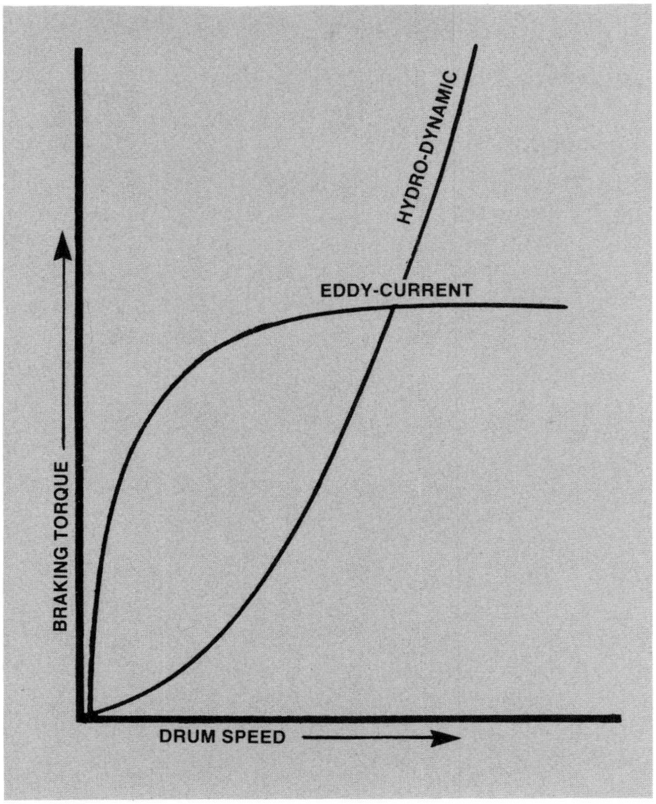

Fig. 7-9.
Performance curves of the two major types of auxiliary brakes.

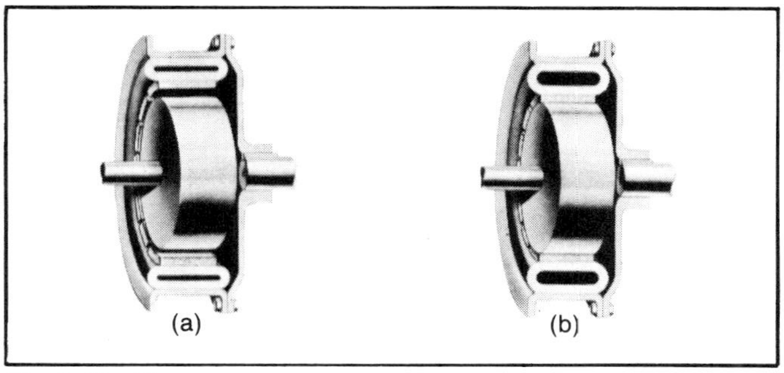

Fig. 7-11 a&b.
Two views of a drum type air clutch. (a) engaged, (b) disengaged.

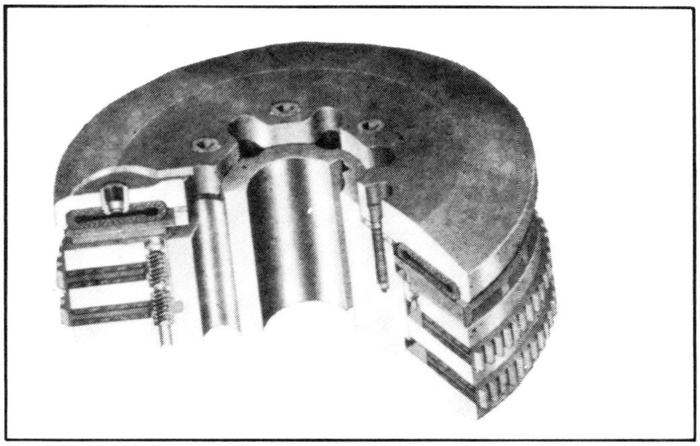

Fig. 7-12. Air actuated 2-plate clutch.

Serving with the brake disengaging coupling, between the hydromatic brake shaft and drawworks drum shaft, the overrunning clutch engages automatically when lowering the load and free-wheels while raising the empty blocks. When the eddy-current brake is used a simple disengaging coupling will suffice since little or no drag exists when the field coils are not electrically energized.

The hydromatic brake, shown in Fig. 7-6, is a hydraulic device that absorbs power by converting mechanical energy into heat. Resistance is created exclusively by agitation of water circulated between the veined pockets of the rotor and stator elements with the conversion of mechanical energy into heat taking place directly within the water itself. The amount of mechanical energy that can be absorbed in this manner depends on the quantity and velocity of the water in the working chamber.

With any specific quantity of water in the working chamber, the velocity of the water circulated is increased with increased speed of the rotor, with resulting increase in fluid friction. In this manner the torque capacity of a hydromatic brake increases approximately in proportion to the square of the speed.

Fig. 7-10. Large unitized flood lubricated steam driven drawworks. Typical of the 1940s.

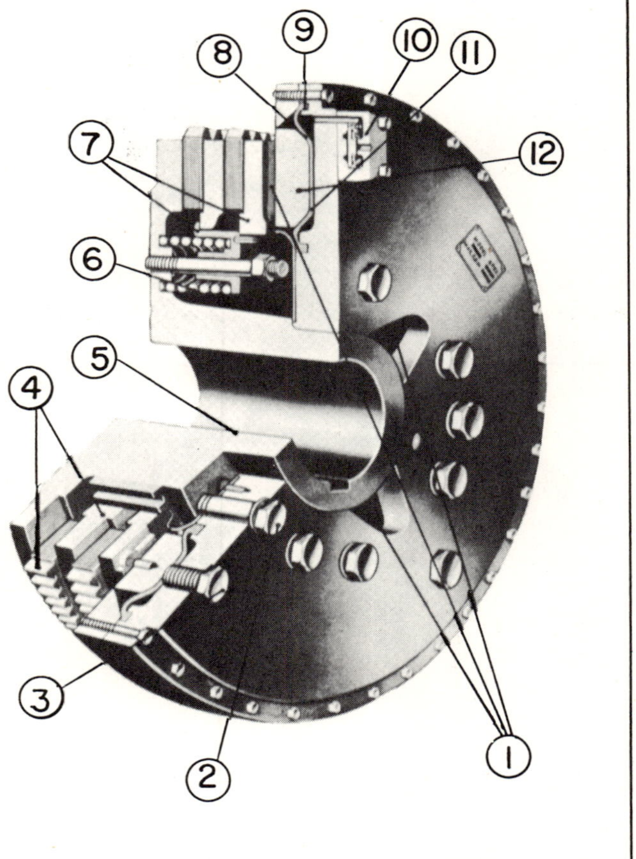

1. Air passages
2. Cap screws
3. "Come home" to provide for manual operation
4. Teeth on driving plate
5. Narrow width for compactness
6. Release springs
7. Positive positioning of plates
8. Diaphragm
9. Air seal
10. Quick release valve
11. Diaphragm
12. Insulation plate

Fig. 7-13. Piston-type air clutch utilizing a diaphragm.

The heat is removed from the brake by circulating cool water into the chamber to displace the warm water. The amount of water remaining in the chamber is regulated by adjusting the water level in the standpipe set alongside the brake unit. The warm water, which discharges into top of vented standpipe, goes out the adjustable overflow pipe and into rig storage by gravity. Brake capacity at any speed is dependent on level maintained in the standpipe and this level is adjusted readily by raising or lowering overflow outlets.

The eddy-current brake also generates a lot of heat which must be removed by circulating cool water through it to cool the revolving drum which is heated by eddy-currents. The cool water usually is pumped into the brake and overflows by gravity into a cooling tank set at a lower level. Auxiliary brakes normally remain disengaged except while tripping into the hole and then only after the load becomes appreciable. Operating characteristics of the two types are quite different as shown by performance curves in Fig. 7-9.

Fig. 7-14. Large modern drawworks for either electric or mechanical drive. Electric version shown.

Practically all modern drawworks, regardless of size if used for drilling, have water-cooled brake rims. Water under pressure from the rig supply is circulated through the rim jackets and discharges into storage tanks where it cools. Circulation rate is regulated by a simple globe valve.

Hoisting drums generally are designed for maximum barrel diameter, length between flanges, and adequate braking surface compatible with power, space and fleet angle limitations. It is very important that each wrap of wireline lay as close to the preceding wrap as possible. Present-day practice is for the manufacturer to machine a spiral groove in the parent metal of the drum barrel to assure proper spooling of the first layer so that succeeding layers will also be tightly coiled. The grooving must be spaced to accommodate wireline of specified size. Originally, to accomplish this grooving effect, grooved jackets were available which could be bolted or welded onto the barrel.

On the earliest models of drawworks, catheads were mounted outboard on both ends of the uppermost revolving shaft which then served other functions but is now referred to as the cat shaft.

Automatic Catheads

On modern drawworks we usually find automatic catheads on both ends of the cat shaft. The automatic cathead on the driller's side, usually referred to as the

SHAFT SPEED DIAGRAM—1625-M Drawworks

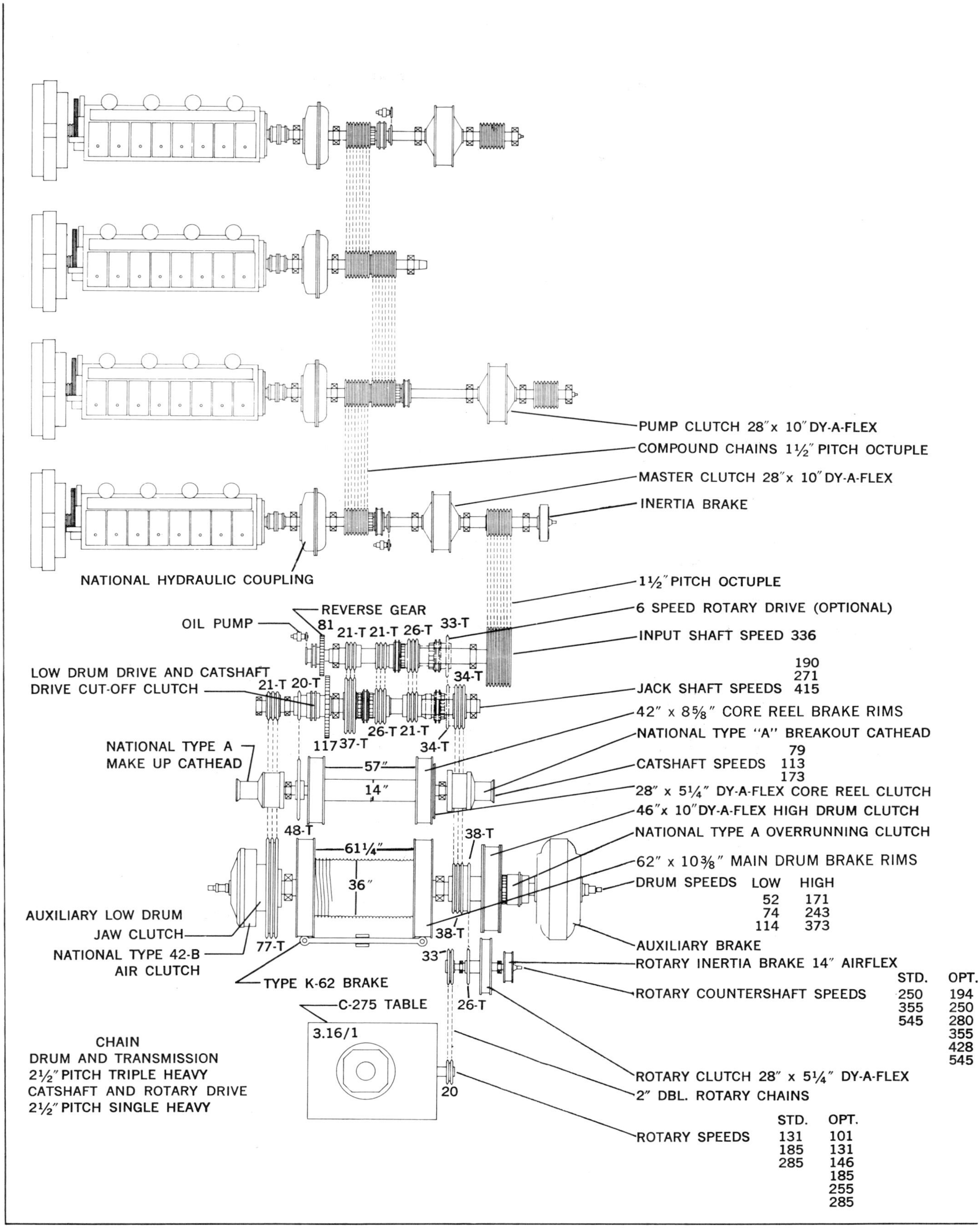

Fig. 7-15. Large modern drawworks. Shaft speed layout for engine drive.

SHAFT SPEED DIAGRAM—1625-DE Drawworks

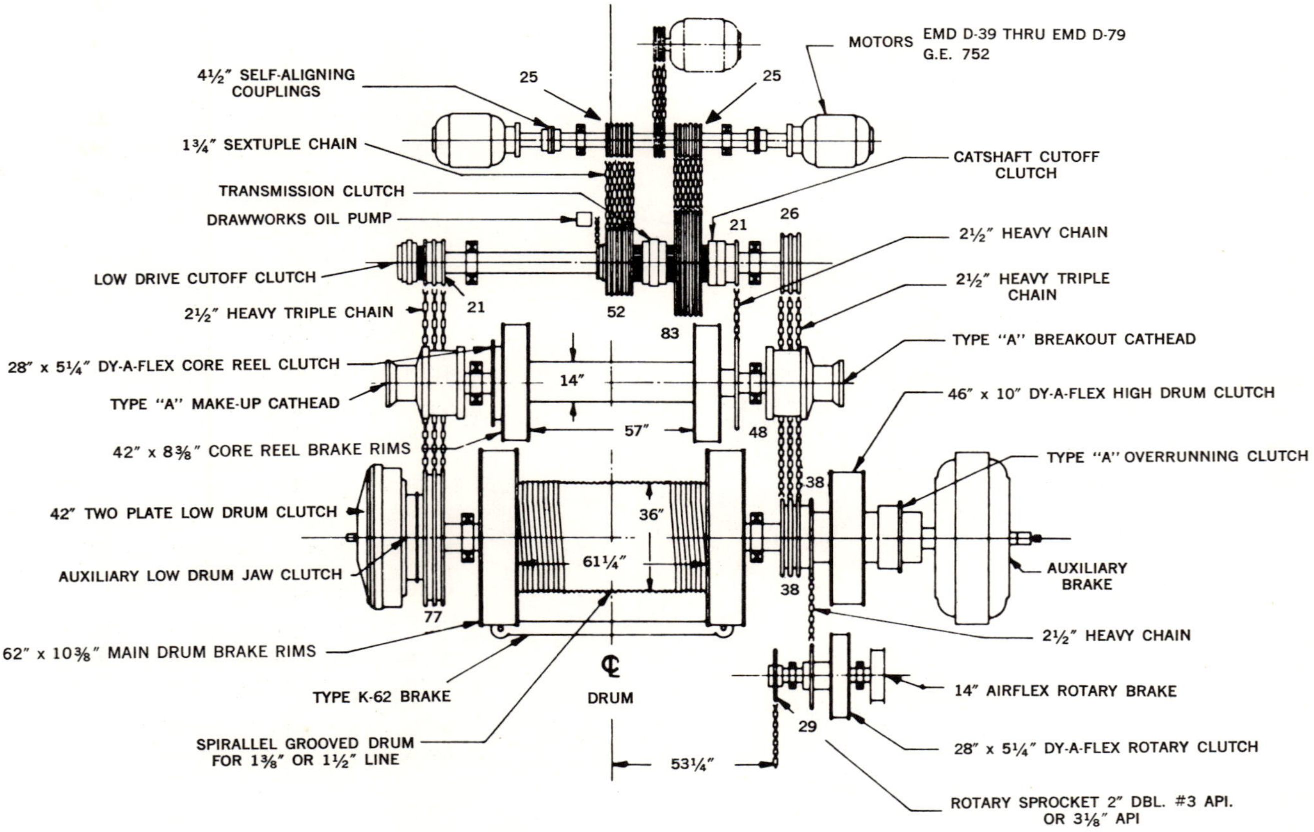

Fig. 7-16. Shaft speed layout for large modern drawworks — electric drive.

1625-M

Main Control Cabinet

1. Make-up cathead control
2. Master clutch and inertia brake control
3. Group 2 engine throttle
4. Group 1 engine throttle
5. Pressure gauge — main water supply
6. Pressure gauge — main air supply
7. Pressure gauge — low drum clutch
8. Pressure gauge — high drum clutch
9. Coring reel clutch control
10. Rotary drive clutch and rotary inertia brake control
11. Breakout cathead control
12. High and low drum drive clutch control
13. Main brake lever
14. Coring reel brake lever
15. Operating specification plate
16. Reverse and high transmission shifter
17. Low and second transmission shifter
18. Low drum and coring reel drive shifter

Auxiliary Control Cabinet (Typical 4-engine control system illustrated)

19. Group 3 engine throttle
20. No. 1 pump clutch control
21. No. 2 pump clutch control
22. No. 3 pump clutch control
23. Pressure gauge — No. 1 pump clutch
24. Pressure gauge — No. 2 pump clutch
25. Pressure gauge — No. 3 pump clutch
26. Water control valve for auxiliary brake
27. Water control valve for drum brake rims
28. Water supply data plate
29. Pressure gauge for drawworks spray lubrication system
30. No. 4 engine clutch
31. No. 3 engine clutch
32. No. 2 engine clutch
33. No. 1 engine clutch
34. Pressure gauge — No. 4 engine clutch
35. Pressure gauge — No. 3 engine clutch
36. Pressure gauge — No. 2 engine clutch
37. Pressure gauge — No. 1 engine clutch
38. Engine selection instruction plate
39. Engine selector cocks (4) for grouping of engines and throttle controls
40. Foot throttle cutoff cock
41. Foot pedal for auxiliary low drum jaw clutch
42. Foot pedal engine throttle

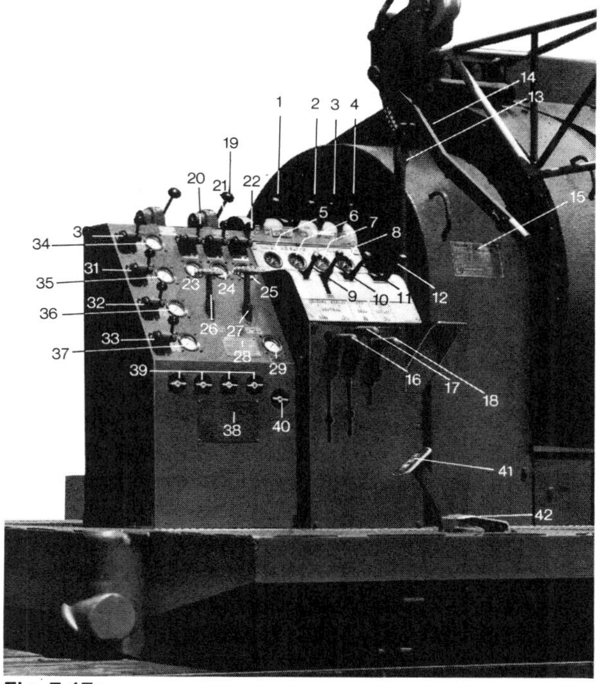

Fig. 7-17a.
Driller's console for large electric powered drawworks.

Fig. 7-17b.
Main control cabinet for large N.S.Co. mechanical rig.

1625-DE

The control arrangement for the Type 1625-DE drawworks is basically the same as that of the 1625-M control arrangement. There are main and auxiliary control cabinets in the same positions. Drive group controls and gauges are omitted in these cabinets, and an electrical control cabinet (included with electrical system) sits atop the standard control unit.

Fig. 7-18. Driller's protective guard.

spinning cathead, (Fig. 7-5a) is used for spinning in and tonging up drill string connections, while tripping in or making connections. The automatic cathead opposite the driller's position, usually referred to as the break-out cathead (Fig. 7-5b) is used to break tool joint connections while coming out of the hole. After breaking or loosening the connection, they are spun out with the rotary.

Break-out catheads normally have heavy duty clutches and also stronger jerk lines than do spinning catheads. For this reason they are often used in combination with tong block, to tighten large drill collar connections. Use of the tong block doubles the pull on tongs and cuts effective jerk line speed in half.

The cat line spools of both catheads are connected rigidly to the cat shaft and revolve with it. The drums, which wind in the jerk lines, are mounted on roller bearings and provided with built-in friction clutches to govern their operation. Clutch control may be manual, hydraulic or pneumatic. Manual control is still used on some spinning catheads so as to provide a feel to the operator during a tightening operation. Air controls are used on the more remotely located breakout catheads and are gaining favor for control of the spinning cathead.

Catheads and cat lines have been the source of many personal injuries in years gone by but automatic catheads, sometimes referred to as safety

Fig. 7-20. Catworks for use on high-floor masts, with coring reel. Model shown is for either electric drive or for mechanical drive from drawworks below.

catheads, have alleviated that hazard. Also a protective guard rail (see Fig. 7-18) between console and drawworks, projecting onto rig floor, effectively protects the driller from injury in event the jerk line or spinning line should break.

A recent innovation in mast design, the high-floor mast, has made possible the placing of drawworks, engines, pumps etc, at the desired truck-bed height but only if rotating and cathead functions are provided for at high-floor height, the working level. Rig manufacturers have met that challenge by developing a new piece of equipment known as catworks. It may include, optionally, a sand reel, chain drive or shaft drive to rotary, for either electric or engine drive. The catworks shown in Fig. 7-19 has no sand reel, is electric motor driven with rotary and its drive train is flush with the floor. The catworks illustrated in Fig. 7-20 has a coring reel, is electric driven, with chain drive to rotary.

Another recent innovation is the helicopter rig which can be broken down into packages weighing no more than 4000 lb. each thus making them portable by helicopters of moderate size. Such rigs have proved successful for exploration and development in remote jungle areas and in the Arctic. A drawworks for one such rig is shown in Fig. 7-21.

Fig. 7-19. Electric driven universal drilling unit with catworks.

Fig. 7-21. Diesel electric helicopter rig.

Chapter 8
Transmission Systems

A drilling rig requires power for hoisting, rotating, pumping and service auxiliaries. This power is supplied by the prime mover, such as an internal combustion engine, a turbine, a steam engine, or an electric motor. Power of the prime mover, however, must be transported or flow to its point of final use. It is also necessary in many cases to adapt or change the original form of the power in some degree.

Means by which the power is transmitted to its point of use, adapted or converted for application, and controlled is called power transmission. In a drilling rig the transmission of power is usually accomplished by mechanical means such as chains, belts or gears.

Conversion or adaptation of power utilizes both hydraulic and electrical devices in the form of special couplings and converters. Power control during such transmission or adaptation is provided by, in addition to the foregoing, such equipment as air or mechanical clutches.

The design of a power transmission system is a difficult compromise between a number of conflicting requirements. On one hand heavy duty design is best served by high rigidity and great weight, while on the other hand quick and easy portability are the backbone of present drilling operations. Cost, field serviceability and design features plus simplicity serve to further complicate the production of any perfect design.

Mechanical Transmission

In medium and large rigs it is necessary to combine the power of two or more prime movers for pumping or hoisting operations. This is accomplished by a compounding transmission. Although formerly other means such as V-belts were used, all present day compounding transmissions use chain to combine the power of the engines before transmitting it to its point of application. In addition to combining power, most modern compounding transmissions incorporate various multi-clutching means so that the power can be combined or directed in a number of combinations to satisfy the multitude of power requirements for all drilling phases.

Jet bit techniques changed the power requirements of drilling so that now pumping power requirements often overshadow hoisting power requirements. Further, pumping requirements are semi-continuous in nature as opposed to the intermittent nature of hoisting. A large multi-engine chain compounding transmission with the inspection covers removed is shown in Fig. 8-1. Two single lever controls are used to mechanically actuate splined clutches so that a variety of engine power flow selections is provided. This allows the necessary amount of power (number of engines) to be directed to the output shafts. Actual provision for pump drive is made in this transmission by one long and one short pump pedestal at the rear of the compound.

Power is usually transmitted to the slush pumps by V-belts from the integrated rig compounding transmission, or in the case of an individual pump package from the pump engines or special compounding case.

Where power requirements are particularly heavy, or where relative location of driver and driven sprockets can be fixed, chain drives (Fig. 8-1) are sometimes used. Two examples, using Poly-V and V-belts

Fig. 8-1. Compounding transmission with covers removed.

Fig. 8-2. Poly-V drive operating mud pump.

Fig. 8-3. V-belts on two mud-pump drives.

respectively between mud pumps and engine compound, are shown in Fig. 8-2 and Fig. 8-3.

Power flows from the compounding transmission to the drawworks where it is used to hoist and rotate the drilling string and for such important auxiliaries as catheads, integrated sandreels, etc.

Most mechanical transmission systems include, in addition to the engine compound and pump drive assembly, some form of selective speed change unit. In small to medium size rigs this item normally is incorporated as an integral part of the drawworks assembly. In most large rigs it is not possible to consolidate the whole system, due to road width and weight limitations. Therefore it becomes necessary either to split the drawworks into two sections, consolidate the selective speed transmission with the engine compound or to make it an independent unit between engine compound and drawworks.

Regardless of where situated the selective transmission unit usually provides three speeds forward and one reverse which, with high and low drum drives, gives the drawworks six forward and two reverse speeds.

Both rotary drive and cat shaft drive are taken from the jack shaft, hence they each have three forward speeds and one reverse speed. The reverse speed usually is accomplished by using two spur gears on same working centers as are the sprockets (See Fig. 8-4). Figure 8-5 illustrates the layout of an entire mechanical transmission system.

On earlier mechanical transmission systems it was common practice to use the compound as a power source for auxiliary equipment such as lighting generators, oil pumps, air compressors and water pumps, each of which might be either direct connected or driven by chain or V-belt, from any convenient power take-off. Current practice is to provide an independent self-contained utility unit for all of these services or to have an engine driven electric generating unit (with 100% standby capacity) for lighting and for electrically driven pumps, compressor, shale shakers, etc.

All engine driven mechanical transmission systems must incorporate means to (a) connect and disconnect engines to the compound; (b) split or divide engine power between drawworks and pump loads; (c) connect and disconnect power from compound to pumps and drawworks; (d) halt motion of units when disconnected.

To accomplish these purposes several different pieces of auxiliary equipment are used, such as (a) clutches of various types; (b) hydraulic couplings (c) hydraulic torque converters, (d) electro-magnetic couplings and (e) inertia brakes. Each will be discussed on following pages.

Mechanical Clutches

The two general types of mechanical clutches used are jaw, and gear or splined clutches. The once-popular jaw clutch has been replaced largely by air controlled friction clutches where it was used for multiple connections and disconnection during operations. In many cases, however, jaw clutches are still used. Perhaps the most common of present day uses is on the steam or electric rigs where positive mechanical action can be utilized more easily due to the cushioning and wide speed range of steam or electric power. The same thing can also be noted concerning steam rigs as they are adapted for electric motor drive.

Where clutching action is fairly non-operational for extended periods splined or gear type clutches are used generally. These clutches are controlled by shifting rings which in turn are manually or air actuated.

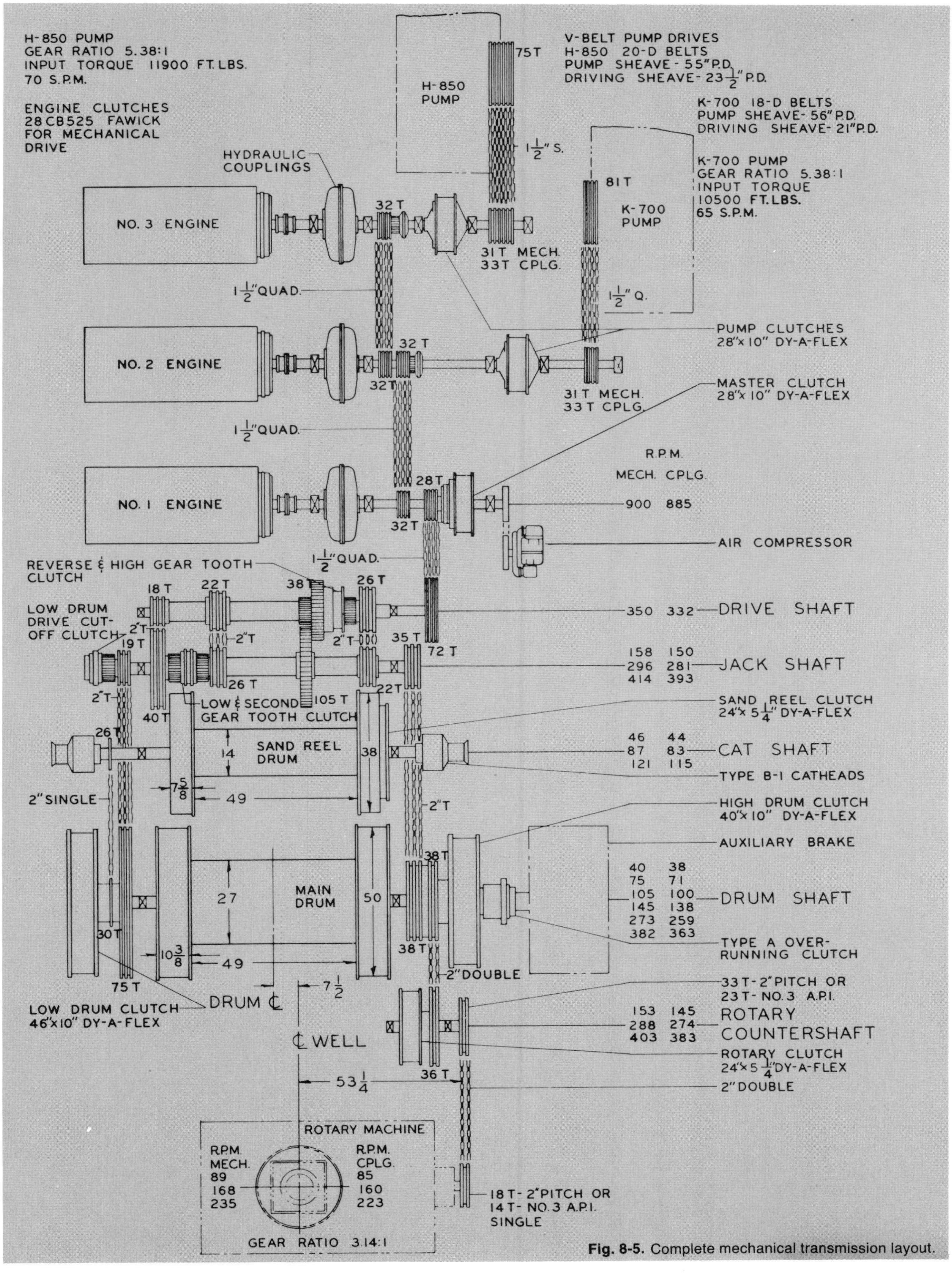

Fig. 8-5. Complete mechanical transmission layout.

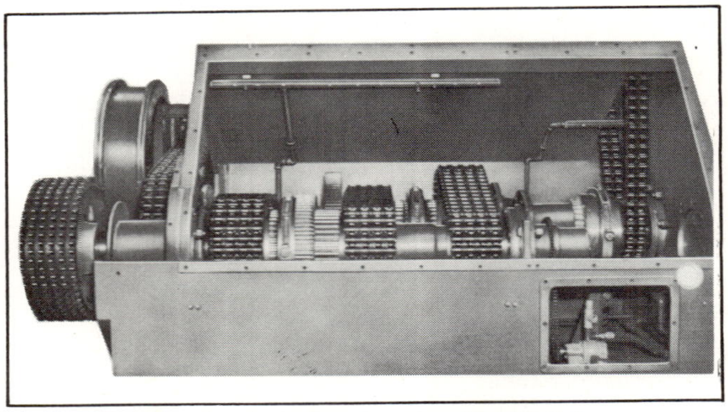

Fig. 8-4. Selective transmission of engine driven hoist.

It is extremely important that these clutches only be operated where there is no relative motion between the clutched parts, otherwise the teeth or splines will become easily burred and inoperable. A multiple position splined clutch used operationally is shown in Fig. 8-5. This clutch is controlled by a shifting ring. This system is equipped with an interlock control so that the clutch cannot be engaged until a neutral brake has stopped relative motion between the parts to be clutched.

Friction-Air Clutches

Practically all friction clutches on today's drilling rigs are the direct air type, where the air is conducted directly into the clutch. Some mechanical or external air actuated friction clutches may still be used but adaptation to new drilling equipment is rare.

Of the direct air friction type there are two basic designs, the piston-disc type and the circumferential tube type. A multi-plate diaphragm or bag-type disc clutch is shown in Fig. 8-6. In the disc-type friction clutch, air is introduced into the clutch so that by piston or bag action it compresses the surfaces of one or more flat moving plates or discs against the surfaces of an equal number of other flat stationary discs. Friction of the moving discs against the stationary discs thereby drags the stationary discs after the moving discs. As the pressure is increased both sets of discs become locked together and the clutch is engaged. When the air is exhausted from the clutch internal springs push the discs apart and disengage the clutch.

In the circumferential tube-type clutch, air is introduced into a bag or endless tube which is located along the outer circumference of a circle. An illustration of this type clutch is shown in Fig. 8-7. As air is introduced into the tube it expands, either externally or internally, until any circumferential clearance is taken up. As the bag is equipped with friction shoes these shoes grip or drag on their contacting surface so that the surface is brought into accompanying rotating motion with the bag section. Circumferential tube-type clutches (Fig. 8-7) are used in compounding engines and controlling pump drives because of ease of assembly and a flexible alignment feature. A more rigid version of this type clutch is being used in drawworks service applications.

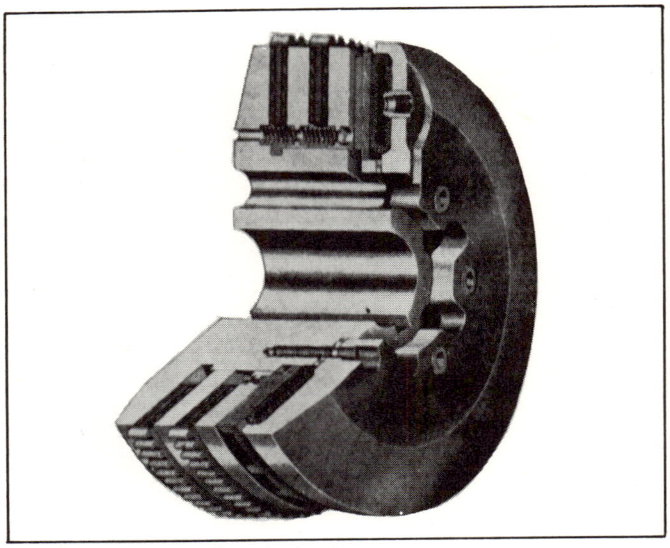

Fig. 8-6. Multi-plate disc air clutch.

The circumferential-type clutch provides an easy and apparent means of a complete disconnect. The disc-type clutch, however, has been used in oilfields for a longer period than the drum or circumferential tube-type and has been accepted as an operational control clutch. Both types are in wide use but trends in air clutches seem to be toward the drum type.

The maintenance of both disc and circumferential tube-type clutches must be determined by the specific manufacturer. For example, many disc-type clutches require special lubrication for their internal seals, while other disc types cannot stand any lubrication, and practically no circumferential tube clutches require it.

There are, however, one or two items in common in regard to clutches that must be watched. First of these is for the operator to get as positive an engagement as quickly as possible. Do not allow the clutch to slip any more than necessary as even a little continuously repeated slipping will glaze the friction surfaces and ruin the efficiency of the clutch. Continual gross slippage on the other hand will not only glaze the surfaces but is apt to so overheat the clutch that the plates will become warped and tubes or bags will be ruined by the heat. Slippage can be caused by a number of factors, nearly all of which can be prevented by maintenance or proper supervision. Factors that cause slippage are: (1) improper operation of the controls — the clutch control should be engaged firmly and quickly and no overlapping of controls should be used; (2) dirt, especially oil, on clutch friction surfaces — these should be cleaned with solvent

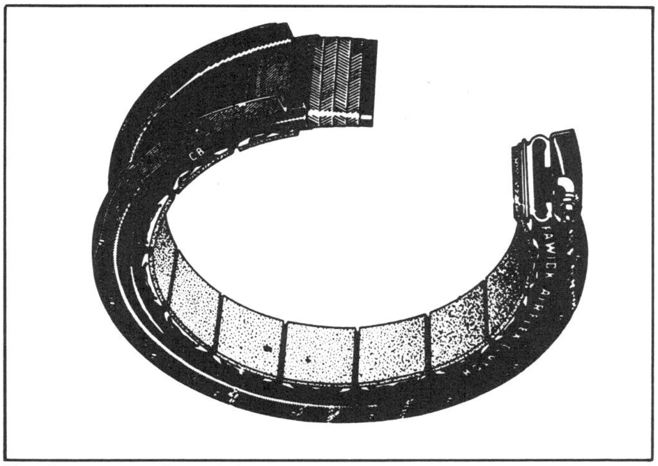

Fig. 8-7. Circumferential air tube clutch.

and kept guarded against entry of oil, grease or dirt; (3) overloading – the right gear should be used whenever possible — "frogging" the pipe out of the slips with a high gear when going in the hole should be avoided; (4) wet clutches, or low air pressure — clutches and air lines should be drained of condensate periodically and the minimum recommended air pressure should be maintained at all times.

Hydraulic Couplings

The simple hydraulic coupling is used as a shock reducing means of power transmission (Fig. 8-8). By putting a fluid cushion between the driver and driven equipment it serves to isolate each from any detrimental power shocks or vibrations that might otherwise be transmitted between the two. The fluid drive can be obtained in either a constant speed or an adjustable speed design. With a full complement of oil in the fluid drive, and when the output shaft is operating at top speed, the device has a high efficiency. In all cases, for either the constant speed or the adjustable speed fluid drive the loss in efficiency is, for all practical purposes, proportional to the percent slip. The principle of the coupling is illustrated by the special turbine-type fans shown in Fig. 8-9. In this illustration one fan serves to transmit the input power from its blades or impeller through a fluid to the output shaft of the second fan or runner. Any shocks, vibrations, etc., are thus isolated to driver or driven system as there is no mechanical connection between the impeller and runner.

The operational characteristics of the fluid coupling are shown in graph form in Fig. 8-10. There are a number of commercial adaptations of the basic fluid coupling principle. Two popular forms allow the fluid within the coupling, which acts as the power transmission means, to be removed either by degree or wholly. In this manner the partially filled coupling has its basic speed-torque characteristics reduced or

Fig. 8-8. Cutaway view of hydraulic coupling.

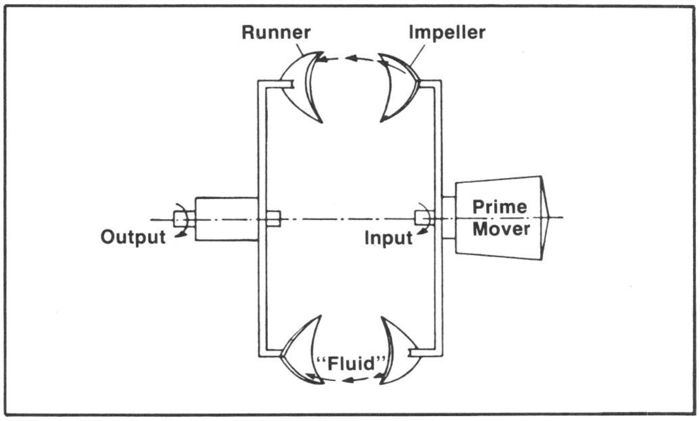

Fig. 8-9. Sketch of hydraulic coupling principle.

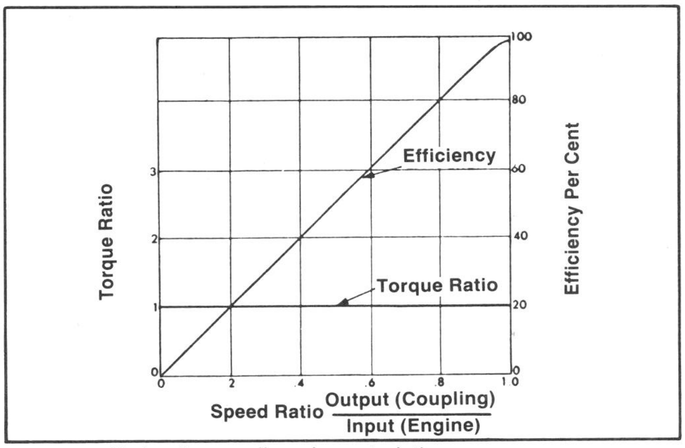

Fig. 8-10. Hydraulic coupling characteristics.

Fig. 8-11. Gyrol drive installation on rig compound.

Fig. 8-12. Compound showing torque converters.

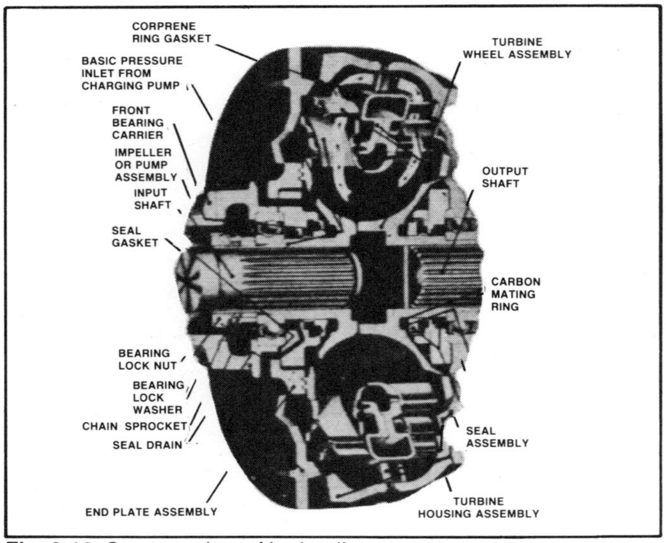

Fig. 8-13. Cutaway view of hydraulic torque converter.

controlled by the amount of fluid kept in the coupling. Either the adjustable filling coupling or the dumping coupling can be used as a power disconnecting or clutching means, in that the slip of the coupling can be varied from nearly zero when full to 100% when completely empty. Applications of medium or large adjustable speed type couplings are shown in Fig. 8-11 installed on the compounding transmission of a deep drilling rig.

Hydraulic Torque Converter

The hydraulic torque converter is a basic modification of the simple hydraulic drive or fluid coupling, as shown in Fig. 8-12. The converter provides the same shock reducing action as the coupling but it is so designed that the input torque of the engine or motor is converted and multiplied, at the sacrifice of speed, when made available at the output shaft. This is accomplished at greatly reduced efficiency compared to the very high efficiency available within the coupling. A cutaway of the torque converter is shown in Fig. 8-13.

It is possible to approximate the operational principles of the converter by again using the illustration of the two turbine-type fans. In the converter a stationary reaction member is added that takes the fluid that leaves the runner and directs it back onto the impeller again. In this manner the rebound energy of the fluid serves to precharge the fluid into the impeller. Due to this added reaction, conditions are such that the efficiency is materially reduced in the converter. As a result of the pre-charging the torque or twisting moment of the runner increases. As the runner speed slows in relation to the impeller speed the fluid rebounds from the runner harder and is further pre-charged into the impeller so that maximum output torque is achieved at a multiplication of input torque when the runner is stalled. The turbine or bucket bladed fans with the reaction member added to illustrate the torque converter principle are shown in Fig. 8-14.

The operational characteristics of the torque converter are shown in graph form in Fig. 8-15. As shown in the figure, the converter's top operational speed ratio is in the neighborhood of 2 to 3 and in this area the efficiency reaches a maximum of approximately 80%. Notice that any stopping, or enforced slowing, of the output shaft quickly increases the torque multiplication and attempts to overcome the resistance. In practical application terms the increased ability to turn the hoist, rotate the table, or operate the pumps without additional mechanical speed changes is somewhat paid for by decreased efficiency and resulting increased fuel costs. A balance of these two factors, plus the basic features of shock reduction and infinite intermediate torque multiplication must be obtained in the proper utilization of torque converters to drilling equipment.

Converter-Coupling Combination

It is possible to combine the operational characteristics of the converter and fluid coupling so that a portion of the advantages of each can be utilized. A graph of the operational characteristics of such a combination is shown in Fig. 8-16. This combination starts out like a standard hydraulic torque converter. However, as the output shaft comes up to speed and converter efficiency starts to fall off, the reaction member within the converter is unlocked by pressure on its rear side and free wheels or rotates so as to be nearly inoperative. As a result of this shift the converter approaches the operation of a coupling and con-

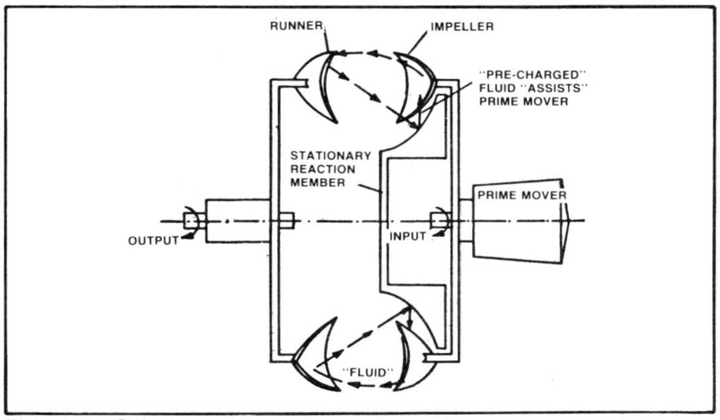

Fig. 8-14. Schematic illustration of hydraulic torque converter principle.

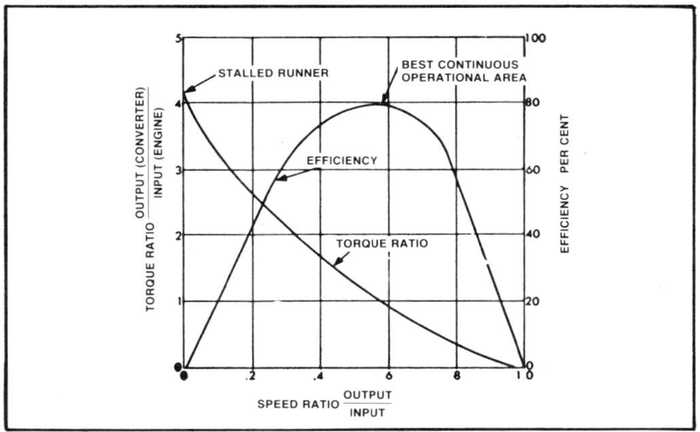

Fig. 8-15. Hydraulic torque converter characteristics.

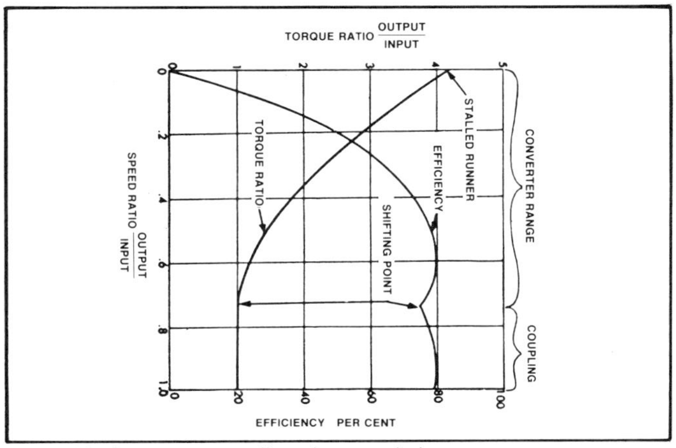

Fig. 8-16. Converter-coupling characteristics.

tinues in this manner until the speed of the runner is reduced enough in relation to the speed of the impeller to re-lock the reaction member and turn the coupling back into a converter.

Electromagnetic Coupling

The electromagnetic drive or electric eddy-current coupling has operational characteristics very similar to those of the hydraulic coupling. Control of the electric couplings, however, is obtained by controlling the input excitation current rather than by controlling the amount of fluid. By cutting off the external excitation current it is possible to de-clutch the coupling completely.

It is possible with eddy-current couplings to dissipate full engine power continuously. It is also possible to drive drawworks pumps or rotary table at any desired speed with the engine or electric motor running at governed speed.

As a result of its similar operating characteristics the electric coupling is used for the same purposes as the hydraulic coupling. The main difference between the two has to do with the practical application features of control, cost, service and availability. A photo of electric couplings is shown in Fig. 8-17.

Due to the fairly complicated and specialized nature of all special drives very few generalized maintenance rules can be stated. It is, of course, best to follow the manufacturer's instructions as closely as possible. In hydraulic drives probably the most trouble is provided by (1) dirty oil as a result of improper filling or filter maintenance; (2) overheating as a result of over-loading or inefficient operation from running in the wrong gear; (3) air in the system, sometimes resulting in reduced pressure and cavitation, as a result of improper filling and system bleeding.

Roller Chain

Roller chain is by far the most popular means of power transmission used on a drilling rig. Two typical applications of roller chain are seen in Fig. 8-5 where a double strand of 2-in. pitch chain is used as a cat shaft drive within a selective transmission and a single strand of motorcycle-size chain is used to drive a lubrication pump. Some advantages of roller chain drives are:

1. Efficient operation — 98%.
2. Positive drive, without slip.
3. Compact drive, requires relatively little space.
4. Requires no initial tension.
5. Can provide great speed reduction (10-1).
6. Needs slightly less perfect shaft alignment than gears.
7. Is easy to maintain and replace.
8. Is not affected by reasonable changes in climate or surroundings.
9. Is durable.
10. Can be run in either direction.

As the speed of chain drive has been continually raised during recent years the majority of chain drives on most modern internal combustion powered rigs now use chains of 2½-in. pitch or less rather than the long pitch API chain.

The roller chain used in drilling rig main power transmission drives is usually manufactured in ac-

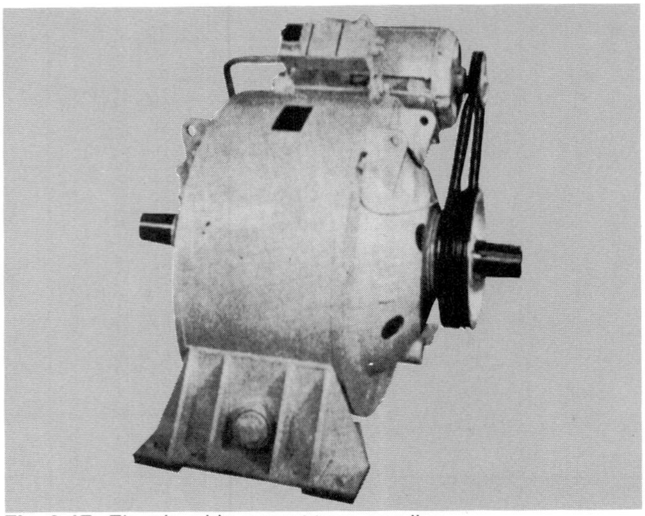

Fig. 8-17. Electric eddy current type coupling.

Fig. 8-18. Three speed and reverse selective tranmission.

cord with the requirements of the American Standards Association. Normally used sizes are 2½, 2¼, 2, 1¾, 1¼, and 1-in. pitch. With some exceptions, chain nomenclature when divided by 80 results in the chain pitch. For example, chain number 140H is (140/80) 1¾-in. pitch chain with heavy (H) strength side bars.

Fig. 8-18 shows a precision steel roller chain installed on drawworks transmission. These are manufactured in single, double and multiple widths and a variety of chain pitches.

Lubrication of roller chain, as shown on the API graphs, becomes more critical as speed increases. On most high speed drilling rig drives it is advisable to provide force-feed lubrication to the chain from the inside of the drive so that centrifugal force will throw the lubricant through the chain and not off the chain. Some designs now introduce the lubricant into the sprocket through the shaft and thence out the root diameter of the sprocket into the chain. Actual lubrication is extremely important to good chain life and the recommendations of the manufacturer should be followed as closely as possible. Particularly troublesome have been such items as the incorrect oil; lack of oil, or clogged lines resulting from dirty filters, and sludge or rags and soft rope getting into the system.

During installation the alignment of roller chain drives should be rechecked. Also some slack should be allowed on one side of the chain, preferably the lower side. A pretensioned drive is detrimental to both chain and shaft bearings. However, excessive slack is also harmful due to chain whip and possible tooth disengagement. On long drives it is sometimes possible to remove a link to shorten the drive if stretch-wear provides excessive slack. The chain should, of course, be kept as clean as possible and washed with solvent if it should be subjected to grit or dust; after washing, the chain needs to be relubed.

Worn sprockets should be replaced, if possible, or reversed so that the chain is not hooked by work teeth. If it becomes necessary to use less strands of chain on a drive designed for a greater number of strands it will be necessary to replace or reverse the sprockets when the full multi-strand chain can be obtained and installed. This requirement results from uneven tooth wear and the resulting unequal load on the multi-strand chain.

Silent Chain

Silent chain types of drives combine some of the features of both roller chain and gear drives. In comparison to roller chain the silent chain-type drive utilizes gear teeth-type sprockets and a semi-articulating mesh action.

Along with their quieter operation, silent chains may be considered to be adaptable to higher speeds than roller chain drives. However, silent chains require very good shaft alignment and very rigid installations. As they also require a tighter drive with less slack than roller chain it is often necessary to utilize take up type installations on long drives.

The general maintenance lubrication procedures mentioned in regard to roller chains are applicable to silent chains. However, one trouble with such chain is its lack of standardization and, as a consequence, each chain manufacturer should be consulted in detail for the lubrication and maintenance instructions concerning his specific product.

V-Belts

In the drilling system, V-belt drives are primarily used for the mud pumps; however, smaller installations are used for a multitude of auxiliary drives such as generators, washdown pumps, oil pumps, etc. V-belts are generally considered for fairly long drives where a softer, more flexible connection is desired than would be obtained with chain. Among their advantages are lack of required maintenance, shock and vibration absorption, and their overall rugged and dependable action.

Due to the light weight of the belts and the light open guard commonly used, V-belts are often used on portable packages because of the weight factor as well as their other advantages.

V-belts are driven by the action of the sheave grooves upon the sides of the belt. In fact, the belts are always selected to fit the sheaves so that they do not bottom in the grooves and touch the sheave root diameter. Belts are usually ordered, stored and installed in matched sets so that all the belts will be of equal enough length to allow the load to be equally divided among them.

As a result of the important usage of V-belts in the oil fields, the API has developed considerable valuable data concerning their selection, application and use. This information is published in API Standard 1, and the following information on installation, maintenance and operating practices is reproduced with the permission of the API:

"21. INSTALLATION. Proper attention should be given to the following items during the installation of V-belts:

a. Sheaves should be aligned and shafts should be parallel and free to rotate.
b. V-belts should never be run on or forced over sheaves. Centers should be slacked off until belts can be placed in the grooves by hand. Before initially tightening any set of belts, care should be taken to trace the slack in each belt to the same side of the drive. Tightening the drive before this is done can result in damage to the belts. Tension in V-belts should be such that when the drive is idle the belts appear snug, and when drive is under full speed and load, a slight sag is noticeable on the slack side. Vertical drives, extremely short center high ratio drives, and drives carrying pulsating loads need additional tension.

"22. MAINTENANCE AND OPERATING PRACTICES. V-belts require very little maintenance, but lack of attention to the following items will result in reduced service life:

a. Lubricate the bearings of sheave and idler shafts to prevent freezing.
b. Maintain operating tension by periodic adjustment of centers of idlers. Changes take place more rapidly when new belts are first installed, and these should be checked after 24 and 48 hours of operation.
c. Keep the shafts parallel, sheaves in alignment and running true. Excessive wobble or eccentricity may result in vibration and overload and cause damage to the belts.

Fig. 8-19. Two speed selective transmission for electric driven hoist.

d. Do not permit the belts to rub or strike splinters, spikes, sharp edges or belt guard while operating.
e. Keep belts and sheaves reasonably free of lubricating oil and other foreign material.
f. Belt dressing should not be used with V-belts. If slippage occurs due to an oily or dirty condition of belts or sheaves, both the belts and the sheaves should be thoroughly cleaned with a cloth and volatile cleaning fluid.
g. The practice of using a pipe wrench to turn a sheave by hand should be avoided, as the groove ribs can be severely damaged.
h. Extremely rough, broken or chipped grooves result in reduced belt life. Worn sheaves should be regrooved.
i. Dampen impulse loads as much as possible by balancing the equipment. Belt vibration may often be reduced or eliminated by slight changes in speed or tension.
j. If the tension section of a belt is ruptured by installation, accident or use, the condition is evidenced by a necked down section, and such a belt should be removed from the drive.
k. Belts should not be allowed to run turned over in the sheave grooves. The tension section of a turned belt is severely stressed and therefore subject to rupture or permanent damage.
l. Only matched belts should be used on multiple drives. New and used belts should not be used in the same drive. Used belts which are still serviceable should be matched to make full sets.
m. Belts of different makes should not be mixed in the same drive.
n. V-belts should be kept in position and properly ventilated and drained.
o. Guards should be kept in position and properly ventilated and drained.
p. Provision should be made for applying tensions, either by adjusting the center distance or by the use of idlers. If idlers are necessary, it is best to use a grooved sheave on the inside of the slack side of the drive, with a diameter not less than the

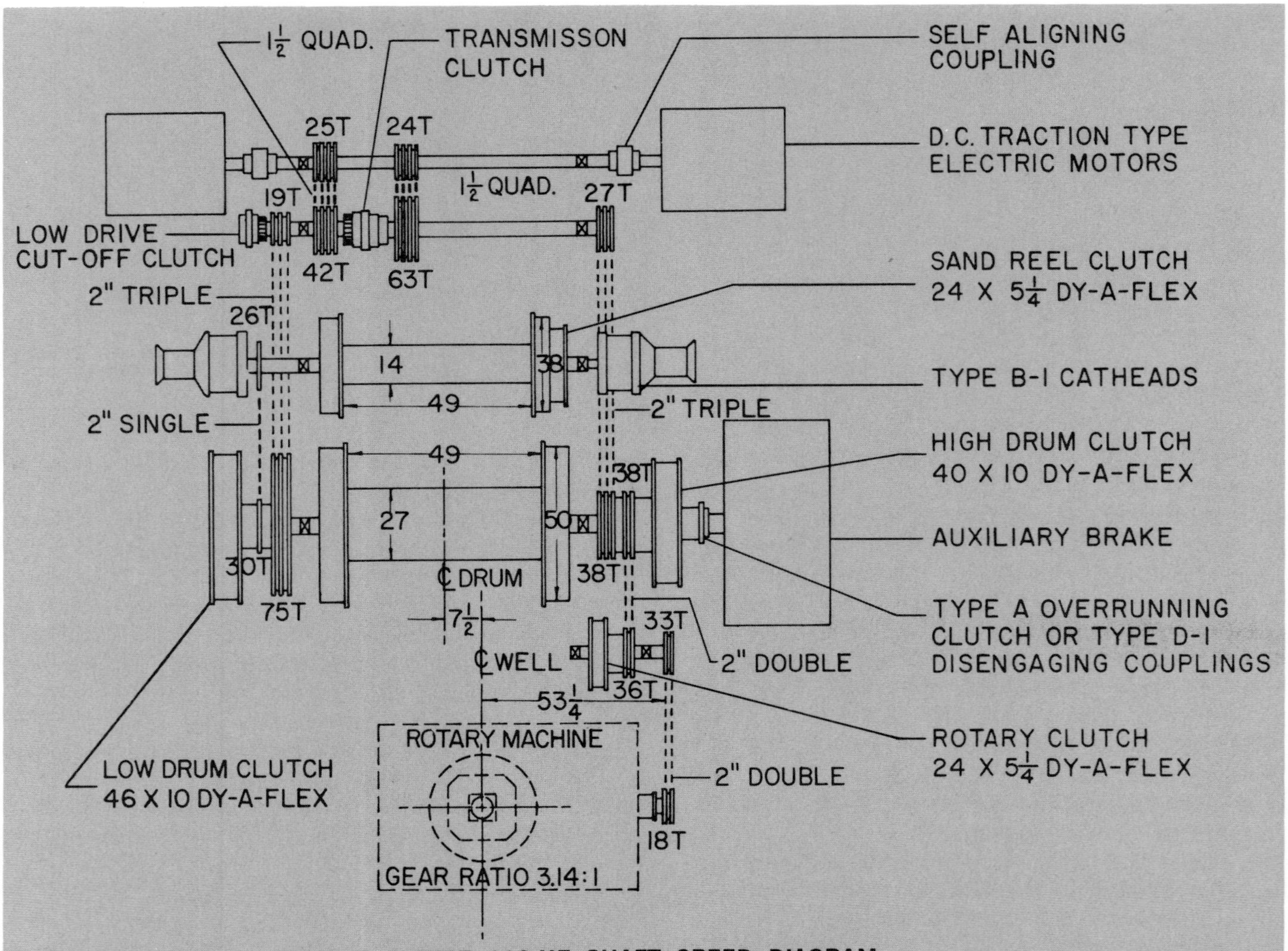

TYPE 110-UE SHAFT SPEED DIAGRAM

Rotary Table Speeds Driven From 110-UE Drawworks. 33T. to 18T. 2" Dbl. Sprocket Ratio. Table Gear Ratio 3.14:1.

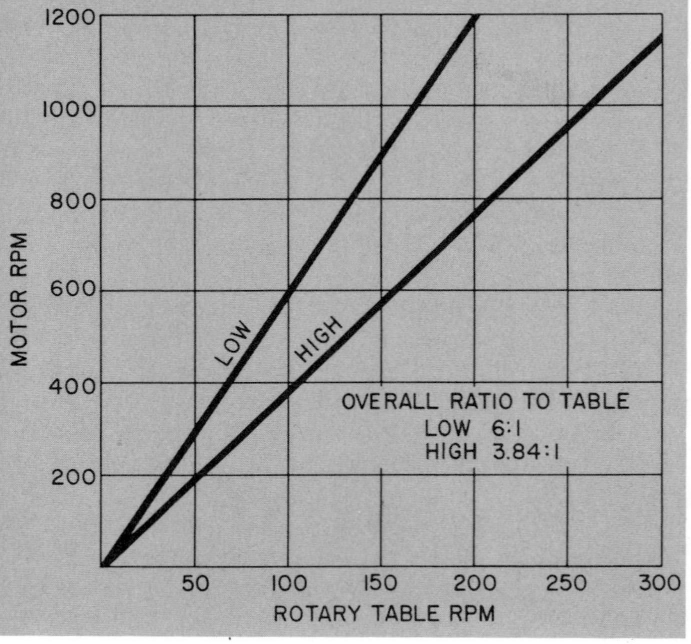

Fig. 8-20a. N.S.Co. Type 110-UE Shaft Speed Diagram.

minimum recommended for drive sheaves.

q. For pump drives with separate skid mounting, it is recommended that the pump skid be mounted in a cradle equipped with screws or turnbuckles to provide ample take-up for center adjustment. The cradle should be anchored to the substructure base and arranged to float on a pad of such size and design as to resist shifting and settling out of alignment.
r. Wide and narrow grooves resulting from poor machining or uneven wear in service will result in "differential driving" and reduced belt life. Such sheaves should be regrooved or replaced.
s. Some V-belt drives on slush pumps exhibit strong vibration of the belts, thus contributing to premature belt failure, particularly on drives employing flat (back-bending) idlers on the slack side. On many such drives the difficulty has been overcome by the use of devices which dampen the vibration of the belts on the slack side."

Poly-V drive is a single unit belt in which the strength member covers the entire width of sheave and eliminates V-belt matching and turnover.

Gears

Gears are used most where compactness of design, quiet operation and large speed variations and ratios are required. There are seven types of gears that are used in the oil fields. These are:

1. Spur gears
2. Helical gears
3. Herringbone gears (double helical)
4. Bevel gears
5. Spiral gears (including Hypoid)
6. Worm gears
7. Rack and pinion

Of these the helical and herringbone, and the bevel and spiral gears are probably the most commonly used in modern drilling equipment for power transmission and speed changing. The spiral and bevel gears and their many derivations and combinations are familiar in their application to rotary tables. The herringbone and helical gears are used in many rig selective transmissions and mud pumps, and more probably are used in pumping units than anywhere else.

The herringbone gear is actually a double helical gear cut together so that the end thrust obtained with a helical gear will be cancelled out.

Alignment is extremely important and any gear installation must be very rigid to forestall any working that might allow the gears to get out of line. Dirty rough or operations involving severe shocks are also quite bad for gear installations.

Gears usually are considered long-life installations if properly engineered and maintained. Along with general cleanliness and alignment, lubrication of the proper type and amount should be watched carefully. The recommendations of the manufacturer in regard to all maintenance and especially lubrication should be followed closely.

Electric Transmission Systems

Electricity is the most widely used medium for transmitting power from prime mover to points of utilization. Power plants of considerable capacity may be situated wherever favorable conditions exist and their output distributed either by an independent high line, or over a network grid system, to serve numerous loads. Commercially, such systems generate and distribute high voltage three-phase alternating current, thus minimizing transmission energy losses. At some point near ultimate use, transformers are used to step down the voltage to suit the specific requirements of the units to be served.

In areas where suitable boiler water was not available and where commercial electric power already was available, electric rotary drilling rigs were frequently used during the so-called steam era. On these rigs the drawworks was powered by two 100 hp to 200 hp variable speed, three-phase, 440, wound-rotor induction motors connected through reduction gears to the jack shaft. (The mud pump was powered by one large motor.) Speed control of these a.c. motors was effected by varying resistance in the rotor circuit through use of external grids connected to slip rings on the rotor. In this type of motor, efficiency, at reduced speed, also was reduced proportionally because no appreciable increase in torque resulted from the reduced speed. This low efficiency made power costs excessive and the system unattractive.

Direct current electric motors were more suitable for the variable loads encountered in hoisting operations. The speed-torque characteristics of a shunt-wound motor are such that, for a given load and field excitation, the speed is proportional to the impressed voltage whereas the torque is proportional to the armature current. Rotary rig power requirements were similar to those of the railroads which had developed d.c. traction motors to a high degree of proficiency. With the development of diesel-electric railway locomotives, using the Ward Leonard system of speed control, the switch to diesel-electric power for rotary drilling rigs was brought about with a minimum of research and development.

Ward Leonard DC/DC System

In the 1890s Mr. H. Ward Leonard developed and advocated for use on long distance railways a system of d.c. motor controls that included long distance

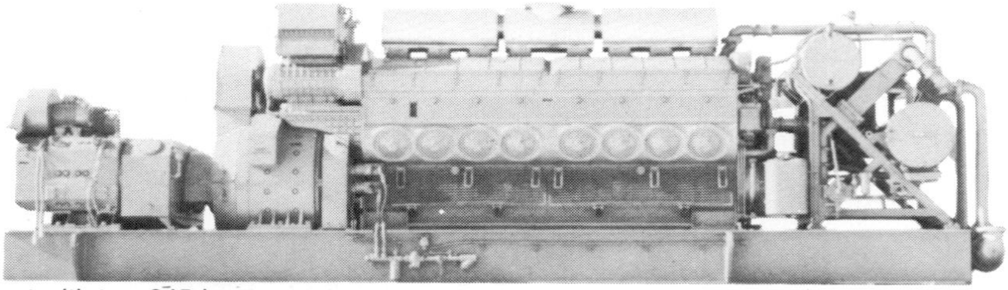

Fig. 8-21. Engine-generator set with two 645-kw generators.

transmission of power by alternating current. With motor generator sets aboard the locomotive, controlled voltage direct current could be made available for the d.c. traction motors.

The Ward Leonard system in general is a system for varying the speed of the motor for all loads carried by the motor without incurring any rheostatic losses in the main circuit. This was accomplished by interposing a motor-generator set between the variable speed motor in question and the system from which it is supplied. In this system, the excitation of the d.c. generator, and hence the voltage, may be varied across a wide range by a rheostat in the field circuit. The generator's armature and the motor's armature constitute a circuit by themselves which is not interrupted in service. By simple and efficient manipulation of field rheostats and switch gear in the separately excited circuits of the generator and the motor, the speed and direction of rotation of the motor are under perfect control at all times.

The Ward Leonard system of speed control has been applied in many industrial operations where variable speeds are required. In such instances the generator usually is powered by a synchronous motor. Among these applications was operating railway trains. In this instance Ward Leonard employed single-phase transmission to a motor-generator set on the locomotive. The generator delivered direct current to traction motors geared to the axles and wheels. Speed was controlled exclusively by manipulation of field excitation.

Along about 1930, d.c. motors, driven by diesel engines, in combination with d.c. motors and Ward Leonard controls, commenced replacing steam locomotives on all trains throughout the nation. The first diesel-electric rotary drilling rig, with which this writer is familiar, was put together by Standard Oil of California at about that same time. What use, if any, was made of the rig in California is not known but the rig and crews were moved to Rangely, Colorado in the early 1930s to drill a wildcat well. Drilling progress was very slow due to lack of sufficient power and inadequate experience in drilling hard rocks. Cable tools as of that date were doing much better in that area. Subsequently the rig was acquired by Noble Drilling Co. and was used in the drilling of Amerada's discovery well at Monument, Lea County, NM.

Since that early beginning of diesel-electric drive for rotary drilling rigs, progress has been continuous and substantial. Most offshore rigs and many of the larger land rigs are now diesel-electric driven. Fig. 8-19 shows a selective transmission which provides four hoisting speeds and two speeds to rotary and cat shaft. Fig. 8-20a shows a shaft speed diagram of an electric driven drawworks. Fig. 8-21 shows an 8-cylinder 1100 hp diesel engine connected through speed increaser gears to two 645 kw direct current generators. Fig. 8-22 shows a 16-cylinder 2200 hp diesel engine direct connected to a 1500 kw direct current generator and through increaser gears to two 645 kw direct current generators. Two such power units can supply all the power required by the largest rotary rigs with adequate standby capacity.

SCR AC/DC System

A significant recent innovation in electric transmission on rotary rigs is the use of silicon controlled rectifiers (SCR) instead of voltage controlled generators. In this system, an SCR unit is placed between the d.c. motor and a three-phase a.c. bus. The a.c. power can be supplied by a utility company, but on drilling rigs, it usually is supplied by several diesel engine driven alternators connected to a common a.c. bus. In this system the prime movers and their alternators, of whatever size, can be pooled much in the same manner as boilers were pooled to supply power to the various loads on a steam rig. Any of the traction motors of uniform size can be served by any SCR unit. Figure 8-23 shows an a.c. generating plant. Figures 8-24 and 8-25 show two views of three SCR units and one auxiliary unit.

In March, 1977, Mr. Glenn W. Webb of General Electric presented, at the IADC Drilling Technology Conference, a most comprehensive paper discussing the two electrical transmission systems, each compared with the other and with mechanical transmission systems. By the courtesy of IADC and of Mr. Webb, the following excerpts are quoted:

"Electrical manufacturers began offering factory-built drive systems in the mid-1950s when drilling went offshore in the Gulf of Mexico. This was a d.c.

Fig. 8-22. Engine-generator set with one 1500-kw generator and two 645-kw generators.

generator-d.c. motor system adapted from diesel-electric locomotives. Direct current motors are used because they provide variable and controllable speed operation, as opposed to a.c. motors which inherently operate at a fixed speed. The mass-produced domestic locomotive traction motor was modified to be used interchangeably as a drilling generator or drilling motor whose rating is now in the range of 1000 hp.

"A typical system included three or four engine-generator sets, one or two motors unitized on each pump and drawworks depending on horsepower rating, and often an independent motor for the rotary. Generators and motors were all connected by flexible electric cables to a central control panel which contained switches by which the driller could remotely assign the generators to specific motors, depending on the drilling in progress. The driller's control station contained throttles for each rig function which controlled motor speeds from zero to full speed while permitting engines to run at constant rated speed. Besides power assignments and throttles, this console could also provide torque and speed indicators and warning lights to monitor ventilation and other functions.

"Such a system obviously offered unrestricted flexibility in locating and arranging equipment on offshore structures. Also drilling motors or similar d.c. motors of appropriate sizes could be used for propulsion, thrusters, and mooring equipment, permitting the use of a common power plant for all functions. Additional a.c. generators were required for rig auxiliaries, ship's services and for most jacking systems.

"A smaller traction motor has also been adapted to serve interchangeably as a drilling generator or motor for smaller workover or helicopter rigs. This machine is rated 600 hp and weighs less than 4000 lb.

"To properly control the d.c. motor, one or more d.c. generators must be specifically dedicated to it. Speed control of the motor is then obtained by controlling the output voltage of the generator. The d.c. systems are generally arranged so that each motor can receive power from two or three different engine-generator sets to provide redundancy in case any unit is down for maintenance.

"Sometimes the load on a motor (for instance, the rotary) may not be sufficient to fully load a generator and its driving engine, or the load on a pump may be more than the capacity of one generator but less than two. Nevertheless, the d.c. generators and their driving engines must be assigned to these loads. In addition, one a.c. generator must always run to furnish auxiliary power. This could result in operating more engine capacity than the total power required by the rig would justify.

"The a.c./SCR system was developed to solve this problem. By about 1970, SCR technology had progressed in other industries to the point where power supplies large enough and reliable enough for drilling and propulsion functions were available. Briefly, the SCR power converter consists of six hockey puck sized cells in a particular arrangement which has the property of passing current in only one direction. With its cooling, protective and control equipment, this 'SCR bridge' converts 3-phase alternating current to direct current whose voltage can be controlled to regulate the speed of the d.c. motor to which it is assigned.

"This converter takes the place of a d.c. generator. Instead of being driven by an engine, it is a static device located in a control panel and receives power from an a.c. bus which is common to several converters and is fed by several generators. The driller assigns the SCR power converters to whatever d.c. motors are needed at the time and the power is supplied through the common a.c. bus by whatever engine generator sets are running and connected to it.

"In addition to all the d.c. motor loads, the a.c. auxiliary and ship's service loads can also be served from this common a.c. bus, either at 600 volts or stepped down by transformers to 460 volts which is standard for this equipment and generally preferred. Main power is usually generated at 600 volts a.c. because this converts through SCR's to the 750 volt d.c. required to get full speed from the motors.

"This system retains the unrestricted flexibility of location and arrangement, first obtained with the d.c. system, and offers the following additional advantages:

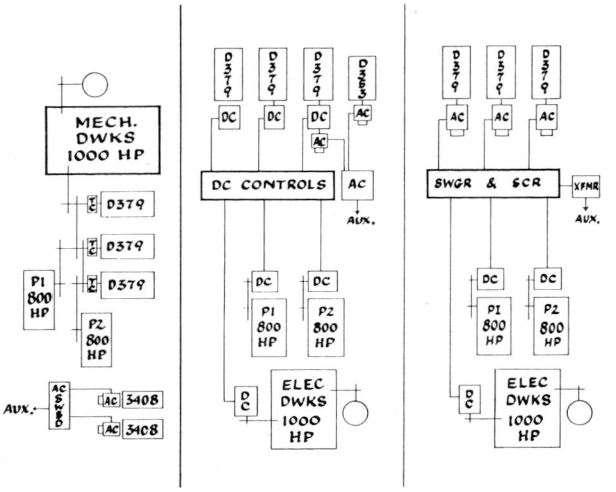

Fig. 8-26

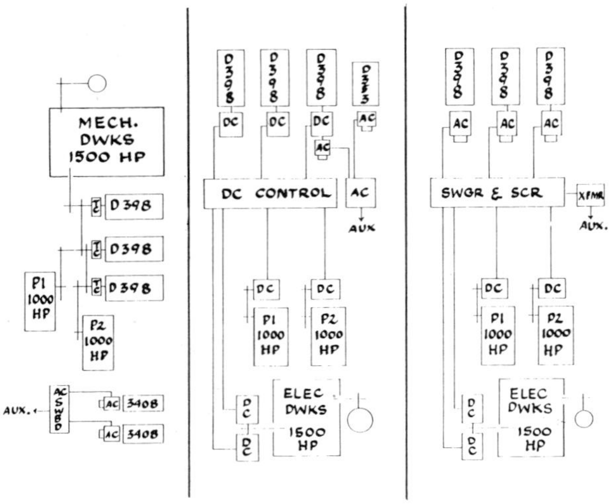

Fig. 8-27

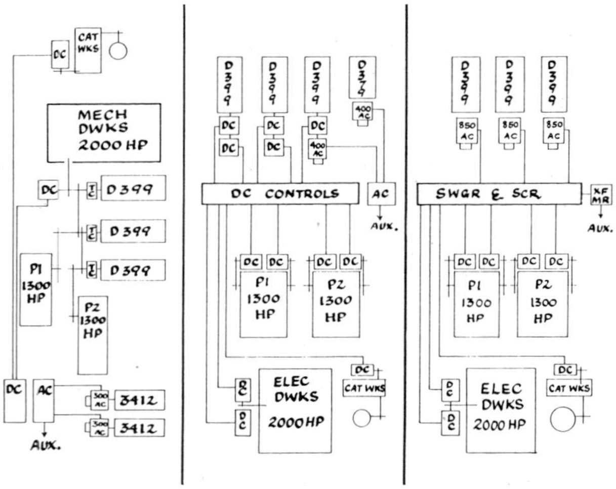

Fig. 8-28

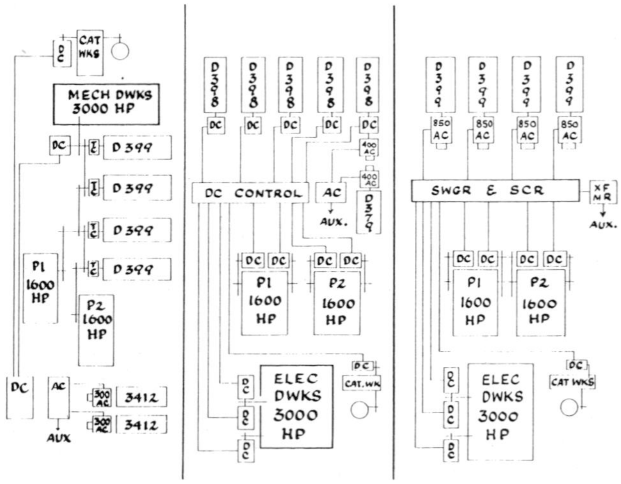

Fig. 8-29

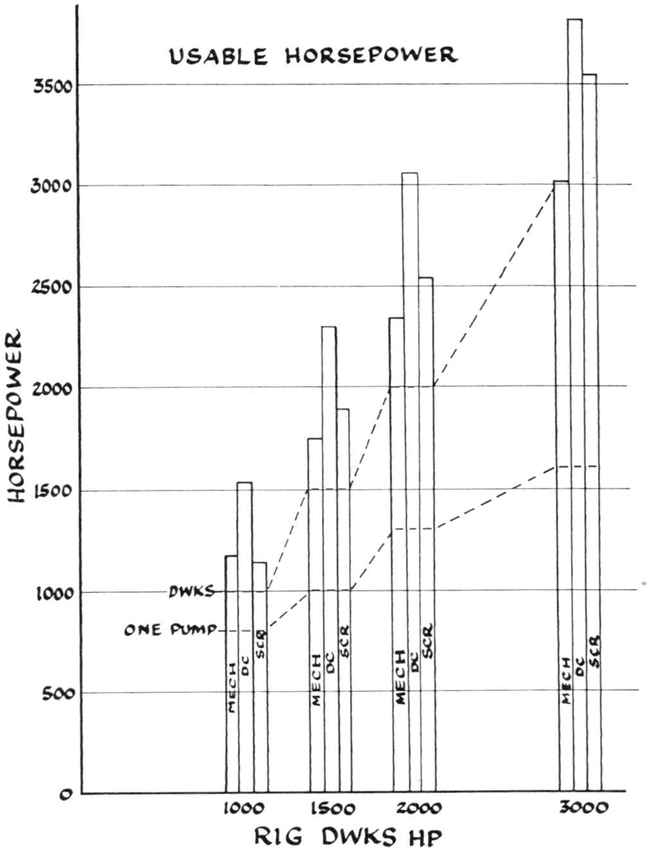

Fig. 8-30

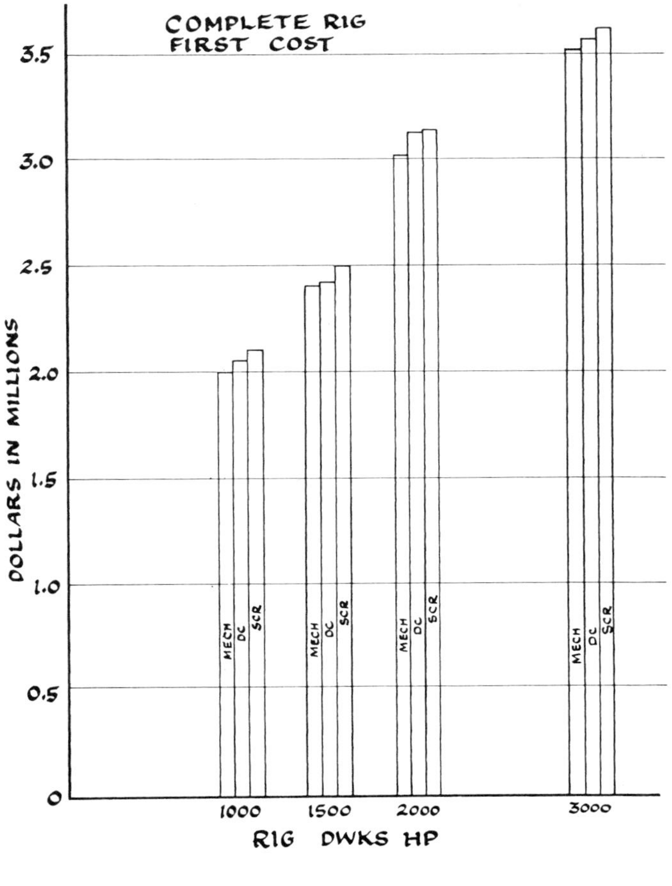

Fig. 8-31

Fig. 8-23. Diesel engine a.c. generator set for SCR drilling service.

"1). A wider choice of prime movers because a.c. generators are available to match the horsepower and speed a wide range of available engines.

"2). Power can alternately be derived from an a.c. utility supply, for example, in the Los Angeles area where engine noise is a problem.

"3). A variety of SCR ratings can be made available to match standard drilling motors singly or in pairs, or for other sizes of engines and generators.

"4). Since any engine can supply any load, or combination of loads, only enough sets need run to furnish the total drilling load, including auxiliaries. The driller can monitor total load with instruments and run minimum number of engines.

"The versatility of the SCR system is such that almost all mobile offshore rigs during the recent building boom used this type of equipment.

"But this market has become temporarily saturated and the emphasis on new rigs in the near future must be on *land* rigs. They use the same drilling machinery as offshore rigs and could be powered much the same way. Let us now examine the three available drive systems — mechanical, d.c. and SCR — and compare them for land rig application on the basis of investment cost, operating costs, performance and safety.

"Recognizing that we cannot possibly consider all variations in rig composition and arrangement that individual contractors may specify, we have selected hopefully typical rig layouts for initial cost analysis. These are in 1000, 1500, 2000 and 3000 hp with mud pumps of 800, 1000, 1300 and 1600 hp respectively. Three Caterpillar D379s, with torque converters, drive a chain compound with belt drives to the pumps and chain drive to the drawworks and rotary on the rig floor. Two 250 kw light plant sets are included.

"The d.c./d.c. system uses the same engines for main power with a d.c. generator on each engine providing power to three d.c. motors. In this system any generator or any two generators can be assigned to any motor. A fourth engine is provided for a 300 kw a.c. generator with the a.c. standby on one of the main engines in tandem with the d.c. generator. These auxiliary generators are 50 kw larger than the ones on the mechanical drive to furnish the extra auxiliary power required for the fields and blowers of the d.c. machines.

"The SCR drive system uses the same three engines driving a.c. generators feeding a common a.c. bus and SCR units and with d.c. motors on each pump and one on the drawworks. Auxiliary 460 V power is taken from the common 600 V bus through a transformer (See Fig. 8-26).

"The layout for the 1500 hp drawworks with two 1000 hp pumps in Fig. 8-27 is identical except for larger main engines (D398 s) and a second motor on the drawworks.

"Fig. 8-28 is the 200 hp rig with 1300 hp mud pumps. The three main engines are now D399s and the auxiliary sets provide 300 kw for the a.c./d.c. rig. Also a rig of this depth capacity is presumed to require a high rig floor to clear the BOP stack, so a d.c. motor is provided to drive a catworks and rotary on the rig floor in each case. The d.c. system provides five generators and seven motors including two on each pump. The SCR system, of course, includes the same seven motors.

"Fig. 8-29 illustrates a 3000 hp drawworks with two 1600 hp pumps. The mechanical drive is a 4-engine compound with D399s and, again, with an electric catworks drive for the high floor.

"Five D398s are used for d.c. generators so that each generator can get full power, and a third motor is mounted on the drawworks to match its rating. This SCR drive uses the four D399s and the same d.c. motors.

"Fig. 8-30 is a chart of available horsepower for drilling machinery which is an addition to the 300 to 400 hp provided for auxiliaries. Engine ratings for pumping and drilling are taken from Caterpillar's 1976-77 Engine Specification Bulletin LECW-1214. Mechanical ratings are 520, 777 and 1035 hp at 1200 rpm with fan for the D379, D398 and E399 respectively. Corresponding d.c. and SCR ratings are 578, 864 and 1150 hp respectively.

"The manufacturer's engine ratings are about 11% higher for electric drilling generators than for the variable speed mechanical compound drive. This is explained by the fact that the electric drives run at constant rated speed with optimized engine performance as compared to variable speed mechanical drives. The electric controls also continually monitor the torque loading on the engines and do not permit shock loading, overloading or lugging.

"Torque converters operate from 70% to 86% efficiency and compound shafts and chains lose approximately 2% each. Average efficiency of the mechanical drive from engine shaft to drilling machinery input shaft is a complicated variable, but for purposes of this discussion is estimated to be 75%. The d.c. generators, d.c. motors and a.c. generators are 95% efficient, and each d.c. machine requires 12 hp from the auxiliary supply for the blower and field excitation. SCR converters are 99% efficient, and cable losses are less than 0.5% at full load. Overall SCR system efficiency is estimated at 87% and the d.c. system at 88% with 12 hp per d.c. machine required from the auxiliary supply.

"It will be noted that the usable power on the mechanical rigs is adequate to fully power the drawworks or to fully power one pump plus the rotary, but it is not enough to fully power both pumps in parallel. For each size rig, the electric drive systems essentially equal or exceed the mechanical systems in usable horsepower for all drilling and hoisting functions as well as auxiliaries. The mechanical and SCR systems can share total power in any ratio between pumps and rotary, but in the d.c. system a full generator must be allocated to the rotary, and pumping power is obtained from the remaining generators.

"*First cost.* The four typical layouts have been priced as complete rigs including fabrication and rig-up by a major supply company as of Jan. 1977. Total list prices for the twelve selected rig layouts are plotted in Fig. 8-31. The premium for SCR over the mechanical rig is about $100,000 for the selected layouts but these include air conditioned power control rooms with plug in cable systems. The d.c. system prices are between the mechanical and SCR prices. They include plug-in cable systems and control rooms similar to those used for a.c. auxiliaries on a mechanical rig.

"*Fuel.* Probably the most significant operating cost difference between torque converter and SCR rigs is fuel. A majority of contractors concede that all engines on a compound, whether needed or not, normally run continuously, plus one light plant engine. There may even be additional smaller auxiliary engines running much of the time. In contrast, an SCR rig would normally be equipped with three or four engine alternator sets, but only two would be required for normal drilling or hoisting operations including auxiliary loads. These two would run at constant rated speed at a relatively higher percentage of rated horsepower.

"Summarizing the relative advantages of the three drive systems, the mechanical drive still represents 3% to 5% saving in first cost and enjoys the advantages of being familiar to most rig personnel.

"The d.c. system is slightly more expensive than mechanical and is also familiar to a great many people. It is next to the mechanical rig in simplicity and it is the logical answer for a partial electric rig where the rotary or drawworks is electrified.

"The SCR system currently costs around $100,000 more than a comparable mechanical rig, but it offers

Fig. 8-24.
Three SCR modules and switch gear unit.

the greatest flexibility and savings in operating costs as explained above."

Electrical Mechanical Transmission

Flexibility of variable speed d.c. motors in hoisting operations is the dominant advantage of electric drive on a rotary drilling rig. Drives for the mud pumps do not require the same flexibility of control, therefore interest has been shown in the feasibility of providing mechanical drive to the pumps while having controlled voltage d.c. generators to provide electric power for hoisting and rotating.

Several such rigs have been assembled and have proved to be quite satisfactory. The usual arrangement is to have two or three internal combustion engines connected in compound to drive the pumps and two d.c. generators. While drilling operations are in progress, the load on the engines results mainly from pump operation. The d.c. generators are lightly loaded. While tripping, the generators absorb nearly all of the engine power, even though the pump may be operated to keep the hole filled while tripping out or for whatever purpose needed.

The overall efficiency of this hybrid system is at least equal to full electric drive and costs are comparable.

Hydraulic Transmission Systems

The hydraulic principle has been used extensively for many years to transmit mechanical forces from one place to another and to modify the magnitude of the force. Hydraulic pressure can be produced mechanically by pumps of various sorts, among which are centrifugal pumps, positive variable displacement pumps and positive fixed displacement pumps. Hydraulic pressure has been most successfully utilized in producing liner motion such as in jacks, lifts etc., wherein a piston or plunger moves

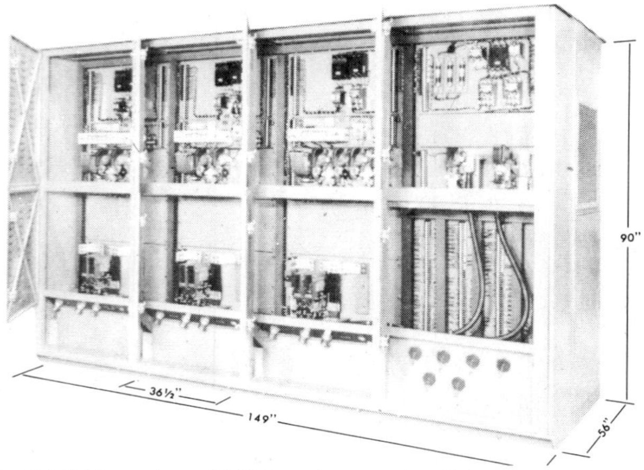

Fig. 8-25. Rear view of SCR modules and switch gear.

within a cylinder to produce a mechanical force.

The hydraulic system has been far less successful, economically, in producing angular motion. Hydraulic rotary devices commonly used as motors are really positive displacement pumps run backward. To successfully adapt hydraulic transmission systems to all phases of rotary drilling will require considerable ingenuity. But it likely will be done.

In the early 1960's British Petroleum put together a well servicing rig for use in the Persian Gulf using a hydraulic transmission system to drive the hoist and rotary. The pump is driven mechanically by V-belts. The power unit consists of two 400 hp diesel engines driving through hydraulic couplings to a compound with an air clutched V-belt sheave to drive the pump and with chain reduction drive to two variable displacement hydraulic pumps of the axial piston type.

The hydraulic motors used to drive the drawworks are fixed displacement type and also are of axial piston design. These motors are connected directly to the drawworks through flexible couplings on each end of the input shaft. Speed of the drawworks and/or rotary is varied by changing the displacement (and output volume) of the hydraulic pump. The displacement of the pump is altered by changing the swash angle from 25° positive (forward) to 25° negative (reverse). Stroke of the pistons is nil when cylinders are parallel to center line of input shaft. This type of hydraulic pump (and motor) is common in Europe and is reported to give an overall system efficiency of better than 85%.

An outstanding technical development is the automatic drilling machine developed by Ray Bromell & Assoc. during the 1960s. This machine is a radical departure from anything previously used as a rotary drilling rig, except the mud pumps and mud tanks. Instead of a derrick or mast, the rig has twin towers 46 ft high, spaced about 10 or 12 ft apart resting on a 12 ft substructure and topped out with steel beams spanning the interval between and over both towers. Hoisting is done with four hydraulic cylinders and lift rods on which rests a hydraulic drill head guided by tracks in the towers. The drill head is a massive power swivel that rotates the drill string while drilling and spins-in or spins-out drill pipe when tripping or making connections. A hydraulic make-up chuck makes and breaks all joints.

The 1500 hp hydraulic system is mounted on one trailer. It consists of five 300 hp V-8 gas engines, each of which drives a Vickers PVA-120, 325 hp hydraulic pump. A main and stand-by electric generating plant are mounted on the same trailer. Hydraulic power fluid is routed from the power center to a manifold within the tower base. From this manifold, hydraulic fluid is routed to equipment required to perform any of the several rig functions.

All phases of the operating are automated and subject to push-button control by the operator in a station some distance off the rig floor. Closed circuit TV and other electronic gear is used extensively and ultimately it is intended that it all be computer controlled. The second crew member is an onsite observer and helper in whatever capacity required. Special tasks such as fishing, running casing, etc. require additional help.

Chapter 9 Prime Movers

The term prime mover, as defined by Webster, is "the original force in a series of transmissions of force". Perhaps a more descriptive definition for our purpose is "that unit within a power system which first converts an energy source into a mechanical force".

Typical examples are windmills, water wheels, water turbines, steam engines, steam turbines, internal combustion piston-type engines, gas turbines and jet engines. Electric motors, sometimes referred to as prime movers, are considered in this discussion as part of a transmission system. The prime mover in this instance is the unit that drives the electric generator.

In study of drilling engines, the first and at one time the most important source of power was steam. This was true for two special reasons, the first being historical. In the early days steam was the only power used.

A second reason was: steam was the best and most flexible drilling power available. It was the simplest and most easily maintained; any roughneck could repair a steam drilling engine or pump.

The steam rig and its boiler plant have undergone radical changes. Once three pots, usually of less than 50 hp each, furnished steam at 100 to 125 psig. Then finally, there was a large high speed engine operating at 300 to 500 psig steam pressure.

Steam Power

Lack of space prevents anything like a comprehensive study of a steam power plant. Highlights with only a brief explanation may be mentioned. One of the more popular drilling engines was a twin with 14-in. x 14-in. cylinders. Maximum crankshaft speeds were 400 rpm for intermittent service and 250 rpm for constant speed work. Rated capacity was 1690 bhp at 400 rpm and 400 psig steam in the intake manifold. To make fast time coming out of the hole, it was necessary that the engine be capable of fast acceleration and deceleration. To obtain this, the engine was frequently built without a fly wheel. This results in extreme vibration at high speed. Because of the above, not only the shaft and bearings, but the entire engine must be very rugged.

The valve cutoff is a late development. Most low pressure industrial steam engines have been designed for a 5/8 valve cutoff. When steam pressure was increased it was customary to shorten the cutoff still more in order to take advantage of a higher steam expansion within the cylinder. This, of course, decreased the maximum power output but increased the overall thermal efficiency.

This drilling engine reverses the above idea of efficiency by being designed for a 7/8 cutoff. High power output and high steam consumption is the result. It has the old time Stephenson-type valve linkage familiar to all steam men. Piston-type valves are used.

Since the Stephenson linkage combines a reversing mechanism with the ability to shorten the cutoff, some radical part of the design is cancelled. Because taking advantage of this increase in steam efficiency would mean lowering the engine's power, the driller rarely took advantage of it. Speed and more speed is what he wants in his drawworks.

At the present, steam-powered drilling rigs are a

Fig. 9-1. One of the last steam rigs in use in Ventura Avenue field, Calif. Rough terrain calls for costly excavating.

Fig. 9-2. Gas-fired insulated boilers power steam rig at wildcat drilling site in South Texas.

Fig. 9-3. Gardner-Denver twin-cylinder steam drilling engines.

Fig. 9-4. Gardner-Denver steam pumps for slush service.

thing of the past. Among the causes would be the high price of fuel gas. Fuel costs become quite an item in drilling with steam. Nature of the terrain also enters into the picture.

Another factor is that at present the bulk of oilwell drilling is done by contractors. Since a contractor may be in an established field today and in the sticks on a wildcat tomorrow, he definitely leans toward internal combustion engines.

On steam-powered drilling rigs the power end of the steam slush pump is a prime mover. The large duplex slush pumps of latest design were effective drilling to moderate depths and, if compounded, two pumps in series were suitable for relatively deep drilling. The simplicity of this piece of equipment was one of its most valuable features. Pistons in the steam cylinders are directly connected by rods to pistons in the mud end of the pump.

The action is simple linear motion with one side starting its movement just as the other side is completing its movement thus producing a steady flow with little or no variation in rate of discharge. The thermal efficiency of these pumps is poor since the steam cylinder operates on a positive displacement cycle and thus fails to realize any benefits of steam expansion.

Internal Combustion Gas Engines

While steam is still the yardstick of all power, advent of torque converters brought the internal combustion engine to a par with steam if not better. With the torque converter, the driller feels less bounce when operating the throttle than with steam.

Also when feeling out the load, a cracked throttle on a torque converter-equipped combustion engine will produce a steady pull. The same cracked throttle on a steam engine may allow a pressure buildup in the engine cylinders, causing a constantly increasing pull on the load. Under certain conditions, this might be important on the negative side.

More new changes in the concept of engine design have appeared in the recent past than any time in history. These advances are true of all types of engines, the Otto or 4-cycle, the 2-cycle and the diesel cycle. Probably the first noticeable change was in cylinder pressures.

The gas/gasoline engines of recent time past were all low pressure naturally-aspirated engines. Today they are being supercharged and cylinder pressures are approaching that of diesels. This supercharging first made its appearance on 2-cycle gas engines and soon became universal. These high pressure designs now have reached the point where further increase in pressure will produce pre-ignition and detonation.

To overcome this and get higher cylinder pressures and horsepowers, engine manufacturers now are chilling combustion air between the supercharger and the cylinder. This work was first started on the diesel cycle. Excellent results are being reported.

Engine cooling is also undergoing radical changes. A few years ago, low cooling water volumes and temperatures were general. Temperature rises across the engine of 20°F to 30°F were not uncommon. Today higher water volumes and temperatures are a must. A temperature rise across the engine of more than 10°F is generally considered prohibitive.

FUELS. The Otto cycle or 4-cycle gas engine is probably the most versatile drilling engine today. By ordering an engine fitted with a natural gas carburetor one is able to use natural gas. If this fuel is not available either butane or propane can be used. Natural gas, butane and propane are all efficient under the same compression ratios. This means that the same cylinder heads would be used for all these fuels. If one wished to use gasoline in an engine it would be necessary to change to cylinder heads designed especially for gasoline to get maximum engine efficiency.

TEMPERATURES. The combustion engine being a heat machine, temperatures are with a doubt one of the most, if not the most important feature governing its life, reliability and cost of operation.

A range of water temperatures leaving the engine is rather narrow. As temperature increases, the

Fig. 9-5. Climax power units. Model VO 420 for natural gas, butane and gasoline.

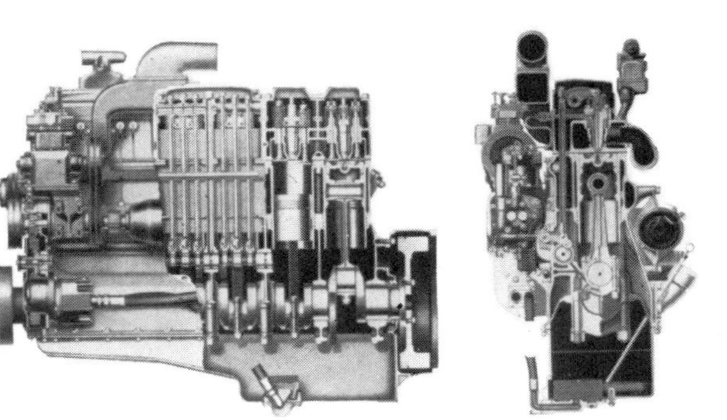

Fig. 9-6. Cummins diesel engines.

thermal efficiency of the engine increases up to about 200°F plus. When the boiling point is exceeded, an entirely new concept of cooling is put into operation. Boiling heat transfer then cools the engine. Power output increases again to a point greater than obtainable with direct liquid phase cooling.

As it is impossible to keep a liquid-cooled engine above the dewpoint, water is being constantly deposited on the cylinder walls and wiped down into the crankcase. If this water is allowed to remain, acids are formed that can seriously injure the engine.

The only way to prevent this is to keep the crankcase temperature sufficiently high to drive water out of the crankcase oil. On an engine operating indoors and protected from weather changes an oil temperature of 150°F is sufficient. A drilling engine, being out of doors and subject to weather changes, requires a minimum crankcase temperature of 165°F. The top temperature is not important so long as you stay within reason. A temperature of 225°F to 250°F is permissible, 180°F to 190°F is ideal.

In checking the crankcase temperature be sure the thermometer is well down in the oil. If it is held above the oil level, a false, high reading will be obtained due to hot oil falling off the pistons and cylinder walls onto the thermometer bulb.

LUBRICATION. The fact that lubricating oils are constantly being changed and improved indicate that lubrication engineers are not satisfied with their own knowledge of the subject.

One change in the thinking regarding lubricating oils is desirable viscosity. Twenty-five or 30 years ago an SAE 30 oil was spoken of as a light oil. Today an oil of the same viscosity is called a heavy oil.

Probably the most outstanding contribution to lube oils has been the advent of the additive or heavy duty oils. Although they are definitely an improvement, they also add certain problems, principal of which is the problem of knowing when to change oil in your engines.

Many operators use a laboratory test. Some have their own laboratories. Others make use of commercial testing laboratories that are available at reasonable costs. No lab report is infallible but it is an indication.

A visual inspection will show only some oil conditions. Probably the most important item and the hardest to detect in time is the neutralization point of the additives. Carbon entering the oil is dissolved forming a colloid. If the oil is left in the engine until the additives have been used up or neutralized the carbon starts to re-deposit on the walls and bottom of the crankcase.

This deposit is a hard flinty granular material. Its effect is similar to throwing a handfull of sand into the crankcase. Anytime such a deposit is found, the oil should be drained, the crankcase washed and new oil and filter put in, regardless of lab reports or length of time the oil has been used.

When oil changes are made on basis of visual inspection and length of time operated, play safe and change oil if in doubt. The old saying that oil is cheaper than machinery is still true.

PREVENTIVE MAINTENANCE. This is an important although simple item. It is mainly a case of using common sense. Think of an engine as a living animal. As a race horse is limbered up before, then walked until cooled down after the race, so should the engine be warmed up and cooled off slowly.

There should be no prolonged periods of overload. Keep only clean oil in the crankcase and keep temperatures up. Watch fuel mixtures. Watch for and correct minor ailments before they become major ones.

Diesel Engines

There are two distinct types of diesel engines being used in oilwell drilling today. The older is a large heavy-duty engine with high weight/bhp. The second and newer is the smaller, high compression, high speed diesel such as that of General Motors. The principal argument in favor of this type of engine is that it is a lot of power wrapped up in a small package.

The large heavy type is preferred on heavy rigs

Fig. 9-7. LeRoi Drilling engines. 600hp L3460 with radiator.

working on deep holes. In this kind of drilling, the job lasts a considerable time, frequently months. Moving the heavier machine is not a serious problem as moves are not frequent. Being built heavy and operating at low rpm and low bhp these engines are among the most economical engines today. The only engine comparable from the standpoint of economy is the spark plug engine fueled with low priced natural gas. And this type fuel is not always available, especially in wildcat drilling.

The second type, because of its low unit weight and low weight/bhp is well suited to rigs working in shallow fields where moves are frequent. Their operational cost/bhp day is not so good as heavier engines but they have increased flexibility and maneuverability.

DIESEL LUBRICATION. Lubrication poses a more difficult problem than lubrication of spark plug jobs. The compression ratio being higher, blowby is harder to prevent. Using a heavy fuel there is greater formation of carbon and dilution of crankcase because of fuel oil leaking past the piston rings. Frequently diesel fuel is found to contain sulfur. This creates sulfuric acid in the crankcase unless the crankcase temperature is kept high.

Diesel lube oil today is an additive or detergent oil. It is usually manufactured with three levels of detergency. For use with a good fuel on normal loads an oil with 5% additives might be satisfactory. An oil containing 10% additives is generally considered better. If bad fuel is used in conjunction with heavy loading, an additive content of 20% may be advisable. If there is a question of doubt, get a good lubrication engineer's advice.

As can be seen from the above, oil changes must be quite frequent. Some operators change oil as often as every three days. Few go longer than a week between changes.

DUAL FUEL CARBURETION. This type of fuel control allows a diesel engine to operate on 100% diesel fuel or part diesel oil and part natural gas. Theoretically a mixture containing 10% diesel fuel oil will give combustion within the cylinder, that is, will produce the diesel cycle. Actually a diesel explosion may be produced with as low as a 5% diesel oil content.

The selection of such a fuel arrangement would be on a drilling rig that might be drilling one month in a field where natural gas was available and the next month out on a wildcat where there was no gas. If it is desired to change from diesel fuel to natural gas, the high compression ratio cylinder heads are removed and replaced with low compression ratio heads. Larger engines usually have a lower ratio than the smaller ones. Also fuel pumps and injectors are removed. An intake manifold, natural gas carburetor, and magneto are then bolted in place.

One mechanic with helper should be able to make

Fig. 9-8. Waukesha power units. Left: Model 6-LROU, bore 8½-in. stroke 8½-in. Right: Gas or gasoline displacement 2894 cu in.

Fig. 9-9a. Caterpillar mechanical rig power. Engine: V12, D398, jacket water aftercooling with radiator. Torque converter: National series C-300 customer installed. Engine rating: 825 hp without fan.

Fig. 9-9b. Caterpillar DC power module. Engine: V12, D398, jacket water after-cooling with radiator. DC generator: GE model 752. Engine rating: 912 hp without fan. Base, packaging by Caterpillar.

this conversion on one engine in about one 8-hour day. To make this possible, some engines come with a kit containing all accessories necessary to change over to natural gas.

The kit box is designed to handle either diesel or natural gas accessories. In this manner, the parts not in use are stored and safe from loss or breakage until a conversion is necessary.

Successful operation of a diesel engine depends on fuel cleanliness more than any other one thing. Tanks and drums must be clean. Hoses used in transferring fuel especially must be watched. Temperatures must be kept the same as on spark plug engines. Torque converters are also helpful on the diesel for smooth operations.

Turbo-Electric Power

Gas turbines, like A.C. motors, are essentially constant-speed machines, therefore ill suited for mechanical drive of a rotary drilling rig. Turbines though, in combination with an electric transmission system, are a feasible means for powering a drilling rig. Gas turbines, so called, can be fueled by either gas or liquid hydrocarbons.

The first successful application of turbo-electric power for a rotary rig occurred in 1965 when Continental-Emsco assembled a 3000-hp rig, referred to as Electrohoist III which used three 1100-hp Solar Saturn single-shaft gas turbines, each rated 22,300 rpm, full speed and connected through double reduction gears to d.c. generators. The transmission system used was the conventional Ward Leonard system which was then the only electric transmission system used to drive rotary rigs.

There is no record at this time of the A.C. SCR sys-

Fig. 9-9c. Caterpillar SCR power module. Engines (2): V16, D399. AC generator: GE Custom 8000. Tandem rating: 2650 hp without fan. Capable of single engine operation at 1325 hp. Base packaging by Caterpillar.

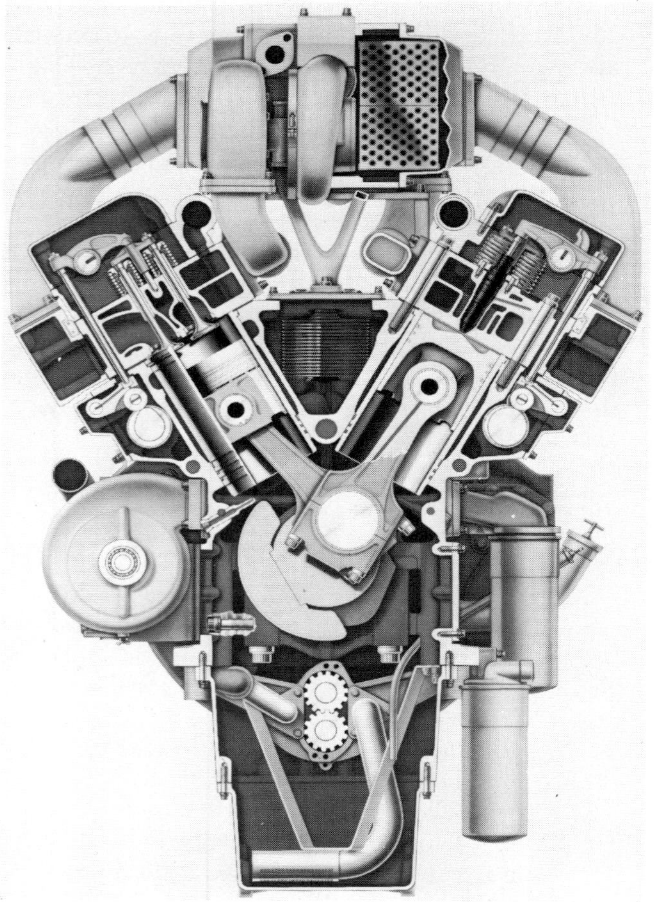

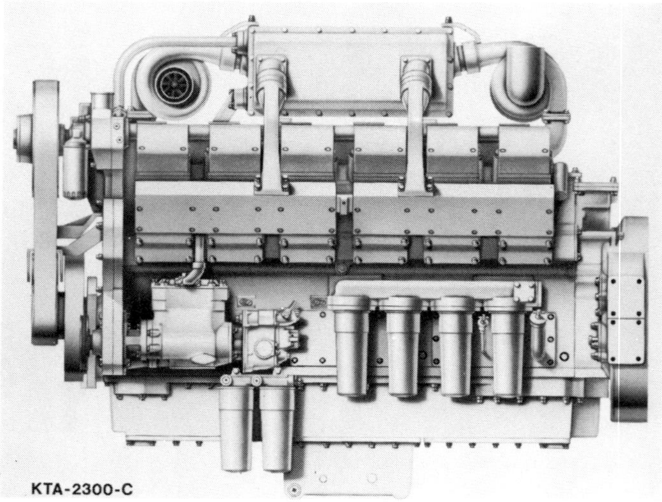

◄ **Fig. 9-10a&b.** Cummins Diesel Model KTA 2300 C.

tem of electric transmission but considering the simplicity of turbines and the low maintenance expense, use of the A.C.-SCR system would significantly reduce fuel consumption, which has been one unfavorable aspect of turbines.

The Continental-Emsco rig was sold to Dixie Drilling Co. which was taken over by Two-R Drilling Co. After the completion of two wells — 13,700 ft and 14,020 ft — the operators reported favorably on both performance and operating costs compared to reciprocating engines.

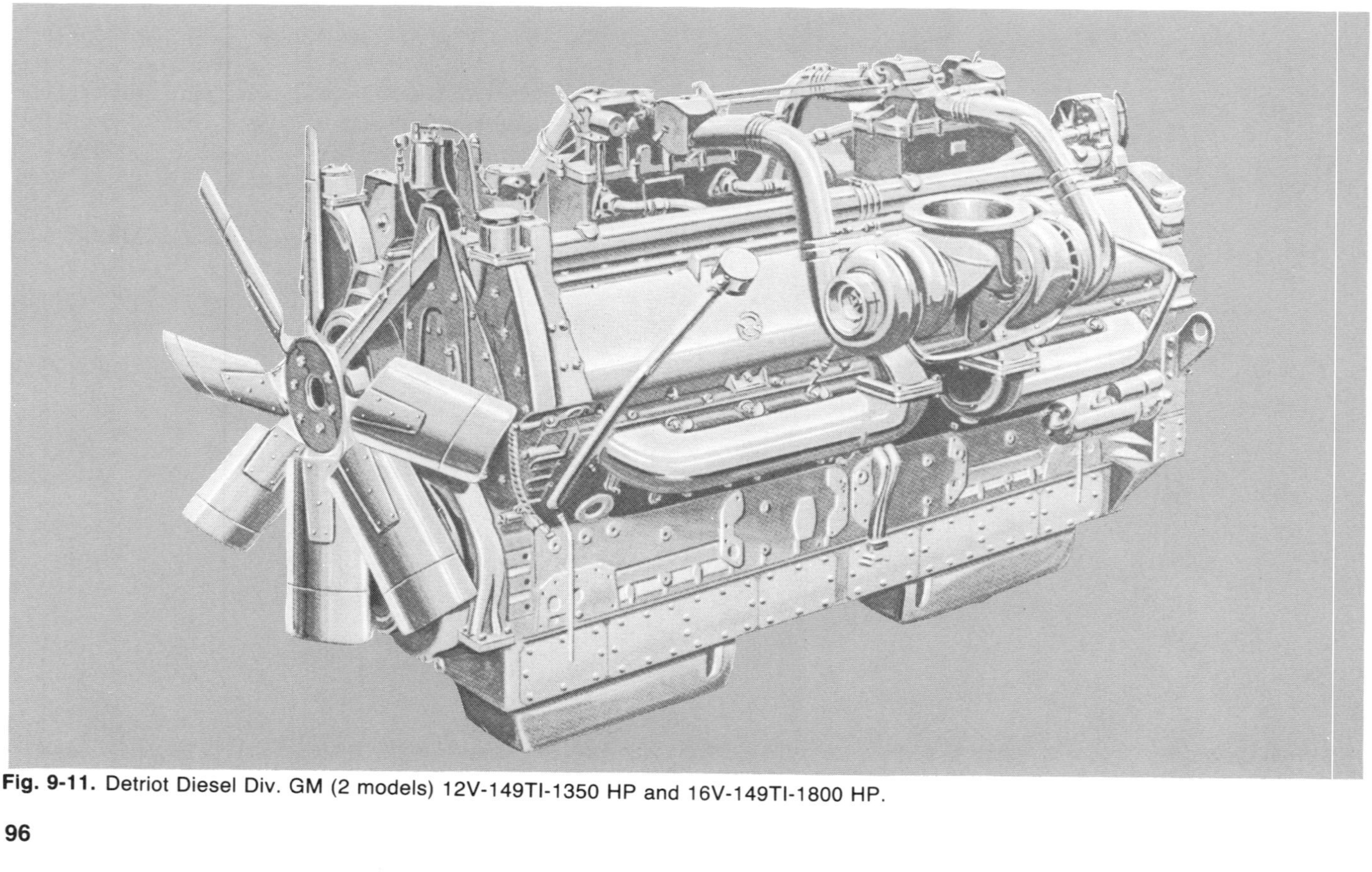

Fig. 9-11. Detriot Diesel Div. GM (2 models) 12V-149TI-1350 HP and 16V-149TI-1800 HP.

The first well was drilled using liquid fuel whereas the second well used natural gas. It was found that high ambient temperature of inlet air reduced power output significantly. Also it was learned that inlet air must be very clean and therefore requires careful filtration.

Fig. 9-12. Electromotive Div. GM Model EMD 8-645EI.

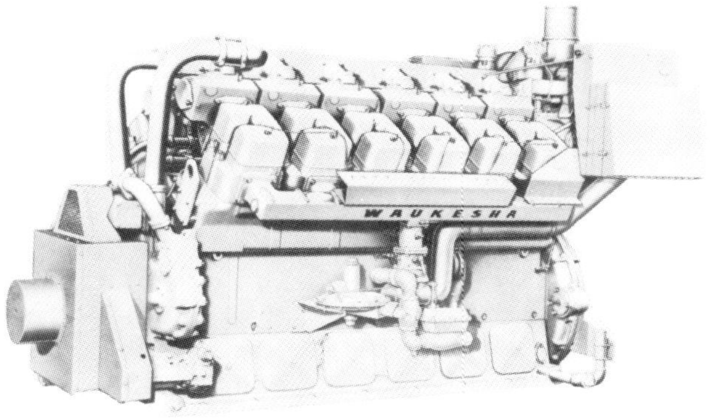

Fig. 9-13a. Waukesha V12 gas engine, model L7042GSI.

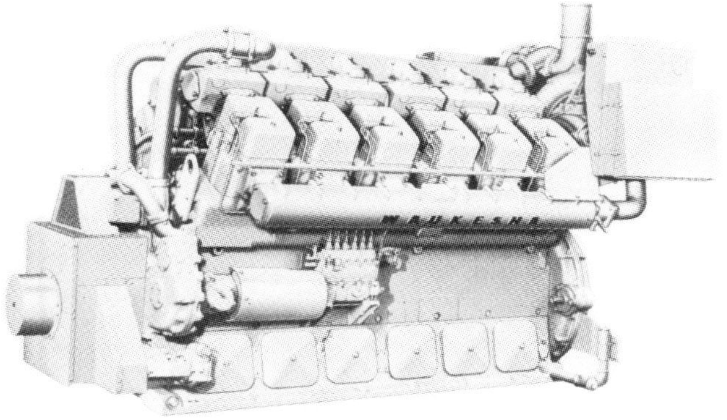

Fig. 9-13b. Waukesha V12 diesel engine, model L5792DSI.

Fig. 9-14. White superior diesel engine model PDT-8.

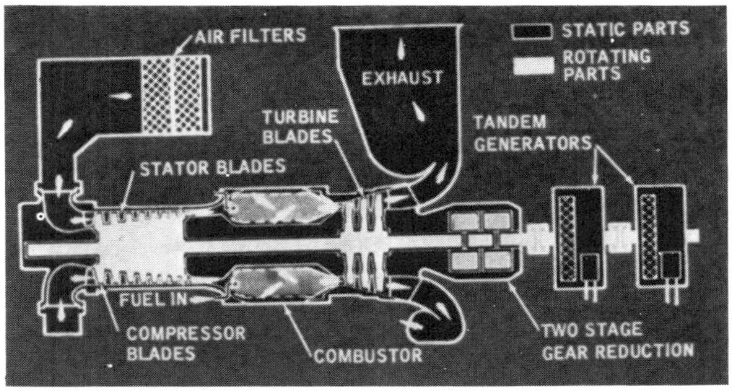

Fig. 9-15.
Hookup of Saturn single-shaft turbine to tandem-driven dual dc generators. Three of these units supply power for Two-R's Turbo Electric Rig 11.

Chapter 10
Hydraulic Circulating System

1. Chamber—air
2. Gasket—air chamber
3. Flange—discharge—and gasket
4. Head—discharge strainer chamber—and gasket
5. Strainer—discharge
6. Liner—fluid cylinder
7. Locknut—fluid piston
8. Nut—fluid piston
9. Piston—fluid—complete
10. Rod—fluid piston—HI-HARD
11. Locknut—fluid piston rod
12. Plug—fluid end drain
13. Flange—suction and gasket
14. Chamber—suction air (not standard equipment)
15. Piping—water drip to fluid piston rod
16. Baffle—mud
17. Extension—fluid piston rod
18. Nut—fluid piston rod extension
19. Guide—lower crosshead
20. Bolt and nut—lower crosshead guide
21. Capscrew—lower crosshead guide
22. Crosshead
23. Shoe—crosshead—upper
24. Shoe—crosshead—lower
25. Shims—crosshead shoe—piston end
26. Shims—crosshead shoe—crank end
27. Stud and nut—crosshead shoe
28. Pin—crosshead
29. Bearing—crosshead pin roller
30. Plate—crosshead pin retaining and gasket
31. Hose—grease line to crosshead pin
32. Fitting—grease
33. Rod—connecting
34. Bearing—crank pin roller
35. Plate—crank pin bearing retaining
36. Shims—crank pin bearing
37. Spacer—crank pin bearing
38. Crank
39. Bolt and nut—crank
40. Key—crank
41. Crankshaft
42. Bearing—main roller
43. Housing—main bearing
44. Shims—main bearing housing
45. Gear—crankshaft
46. Bolt and nut—crankshaft gear
47. Key—crankshaft gear
48. Spacer—crankshaft
49. Shaft—pinion
50. Pinion
51. Key—pinion
52. Bearing—pinion shaft roller—long housing
53. Housing—pinion shaft bearing—long
54. Shims—pinion shaft bearing housing
55. Bearing—pinion shaft roller—short housing
56. Housing—pinion shaft bearing—short
57. Retainer—pinion shaft bearing housing oil
58. Seal—pinion shaft bearing housing oil
59. Breather—frame end
60. Cover—gear case—and gasket
61. Plate—hand hole cover—and gasket
62. Plate—crank case cover and gasket
63. Plate—from bottom cover and gasket
64. Door—frame and gasket
65. Cock—frame oil level
66. Plug—frame crank housing drain
67. Plug—frame oil drain

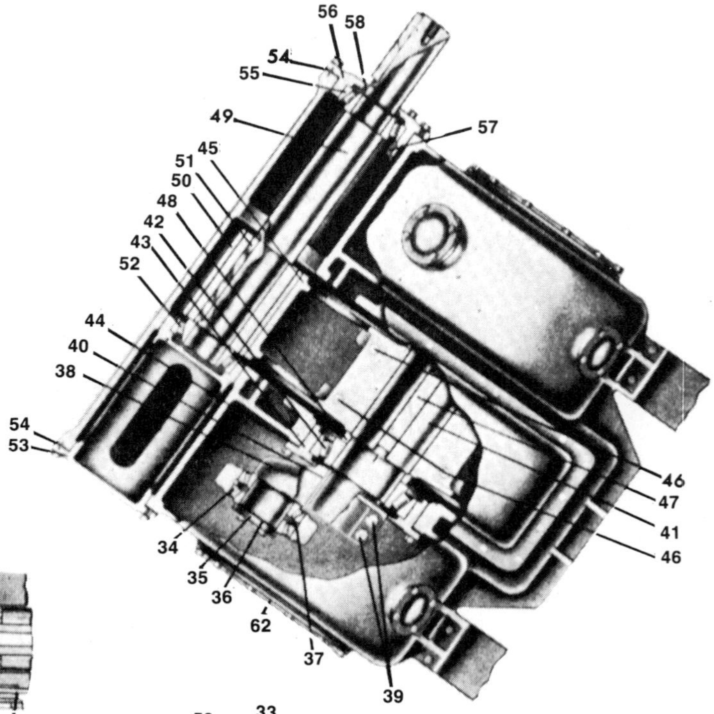

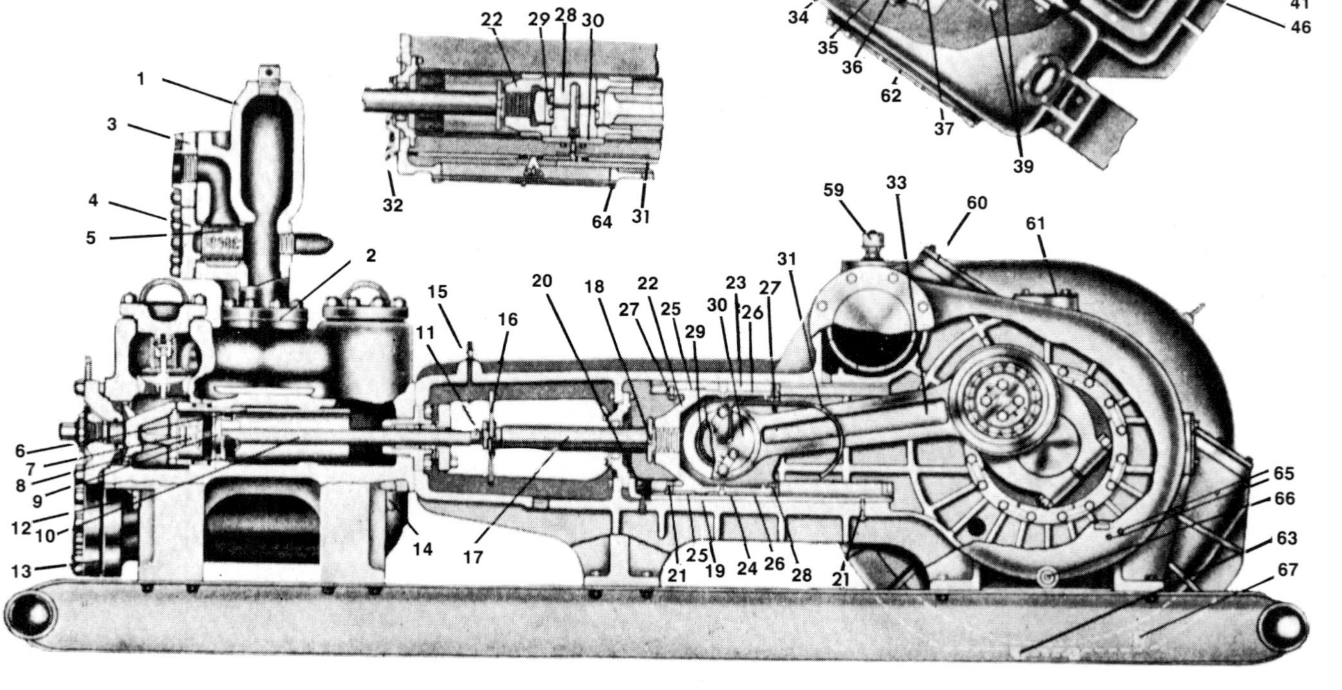

Fig. 10-1. Cross-sectional view of a duplex, double-acting slush pump.

Of equal importance to breaking up rock with the drill bit, is the removal of the resulting debris. Neither function is of any use without the other. Progress has been made and is continuing to be made in bit design to expedite rock destruction. Likewise, to scavenge the bottom of the hole and circulate out the cuttings at higher drilling rates in deeper holes, the circulating system had to be improved equally. It has been necessary to increase the horsepower capacity of the slush pump, to improve the hydraulic design of drill strings and to direct attention to bottomhole scavenging.

There are several segments of the circulating system which will be discussed in order:

1. Slush pumps, which initiate fluid flow.
2. Surface equipment, lines, hose and swivel.
3. Drill string, kelly, drillpipe and drill collars.
4. Bit nozzles as used in jet bits.
5. Return annulus, between pipe and hole.
6. Shale shaker.
7. Desander-desilter and centrifuges.
8. Degasser.
9. Mud tanks agitators and mixers.

Items listed 2 through 5 tend to restrict flow of fluid and thus they form back pressure on the pump, the amount of which varies approximately as the square of the circulating rate, for any given set of conditions. The pump does not independently produce pressure, as some persons seem to believe but rather the pump pressure produced is a reaction to the back pressure produced by flow friction.

Slush Pumps

The duplex double-acting steam pump, which was used exclusively in drilling for almost 40 years, is no longer in common use and therefore will not be discussed at this time.

The power-driven duplex double-action slush pump, which followed, made use of the technology that had been developed in the fluid end of the steam pump. There is no essential difference in the basic design of cylinders, liners, valves, pistons or rods.

An important difference, however, in a crank-driven pump, is the operation cycle of the two pistons which are 90 ° out of phase rather than 180° as in the steam pump, which produces essentially a uniform piston velocity throughout the entire stroke. Even with harmonic motion of the pistons some surges in discharge volumes would result.

Since, in crank-type pumps, harmonic motion is not achieved, these surges are further increased by two factors. First, piston movement on the outer half of the working stroke is faster than on the first half of the stroke. Second, the piston rods reduce displacement of the inner end of the cylinder. This is especially significant when small liners are in use.

This shortcoming can be alleviated to a considerable degree by use of large air chambers, or precharged pulsation dampeners, on the pump discharge manifold. Figure 10-1 is cross-section view of a typical duplex, double-acting, power-driven slush pump. Figure 10-2 is a view of fluid end of a slush pump.

The single-acting triplex slush pump has gained prominence in recent years. Triplex plunger pumps have been used extensively in industry for many years. Their basic advantages are: (1) more even discharge rates, so less pressure pulsation, (2) lower weight per unit of capacity, (3) more compact, requiring less space.

The conventional triplex plunger pump did not prove practical as a slush pump, but a modified version has proved satisfactory. The modifications that have led to success are: (1) use of slush pump type of valves, (2) substitution of liners and half pistons for plungers and packing, (3) charging of the suction pressure.

In some earlier triplex slush pumps, efforts were made to operate without a charge pump. In this instance, conventional slush pump pistons were used. Atmospheric pressure alone on the back side of the piston, on the suction stroke, was inadequate to expand the conventional piston sufficiently to obtain an effective seal between piston and liner. Expediency dictated the use of a centrifugal charge pump to assure no cavitation in the triplex slush pump operation. A piston may be developed in the future with a seal on the back side that will operate effectively at atmospheric pressure.

As of this date all five of the larger American manufacturers of rotary drilling machinery are offering a line of triplex slush pumps, ranging in input capacities from 275 hp to 1700 hp. Typical pumps of this basic design are shown in Fig. 10-3 and Fig. 10-7. The various lines of triplex slush pumps are more dissimilar than are their respective lines of duplex double-acting slush pumps.

Triplex single-acting slush pumps have higher rated operating speed ranges than duplex double-acting slush pumps of equal capacity. Maximum fluid end working pressures, for all slush pumps, are included in the manufacturer's specifications. Also the displacement, at various combinations of speed and liner size (together with corresponding safe working pressure), is available from the manufacturer. Routine maintenance of the power end of either duplex or triplex pumps is minimal except that recommended oil levels should not be overlooked.

The fluid end of a slush pump is subjected to abrasive action of fluid being handled. This may be most severe if sand and other solids are allowed to remain concentrated in the circulating mud stream. Various means of removing abrasives from mud are being used as will be described later in this chapter. Com-

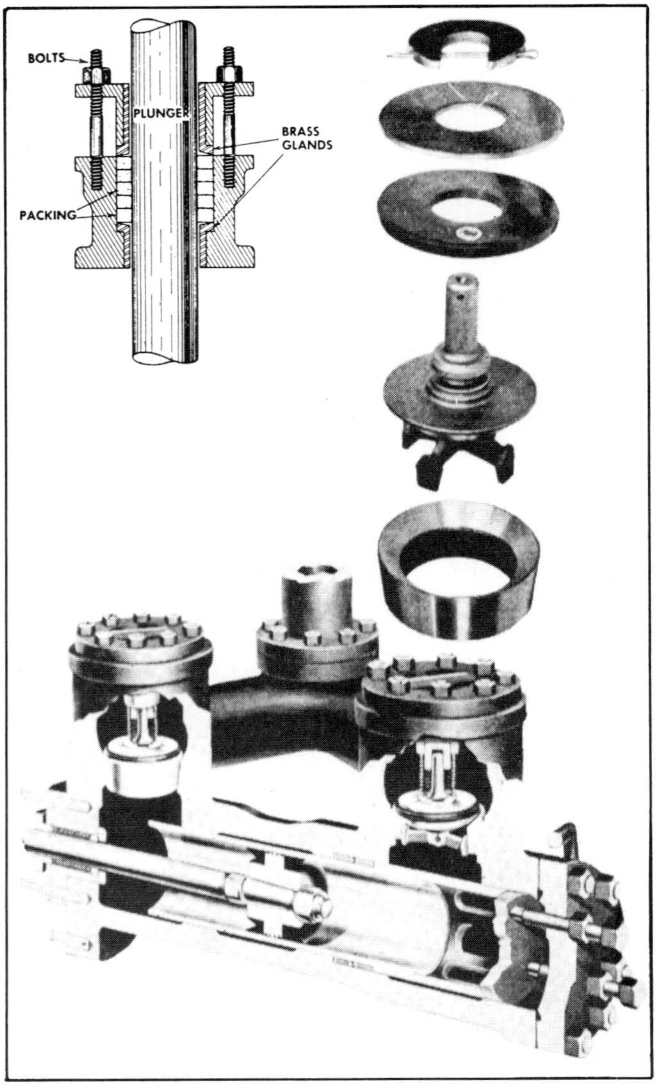

Fig. 10-2. Cutaway and exploded view of typical slush pump fluid end.

plete removal of abrasives is considered to be an impossible task.

For this reason various elements of the fluid end, as shown in Fig. 10-2, which are subjected to constant wear are referred to as expendables which must be replaced at rather frequent intervals. It is important therefore, that the fluid end be designed so that parts such as pistons, liners, gaskets, valves, seats, piston rods and piston rod packing can be replaced quickly with minimum manpower using appropriate hand tools.

High Pressure Surface Equipment

The drilling fluid being discharged from the slush pump typically flows through a vibrator hose, then through steel pipe to rig floor where it enters the standpipe and from the standpipe, usually about 40 to 45 ft high, through a flexible hose to the swivel and then through kelly and drill string, etc. The resulting pressure drop through the surface equipment usually is not a significant part of the total pressure loss but it should not be ignored.

The vibrator hose is the connecting link between slush pump air chamber and the horizontal steel mud line, usually at ground level, leading from pump area to the standpipe area. Its purpose is to alleviate vibration in the steel mud line and also to replace several pipe ells or other pipe fittings that would otherwise be required. The vibrator hose is not unlike the rotary hose in construction but is shorter.

API standard sizes are 3-in. and 3½-in. with lengths of 10, 12, 15, 20 and 30 ft. API further provides that both ends of all sizes and all lengths be

Fig. 10-3a. Gardner Denver triplex single-acting slush pump.

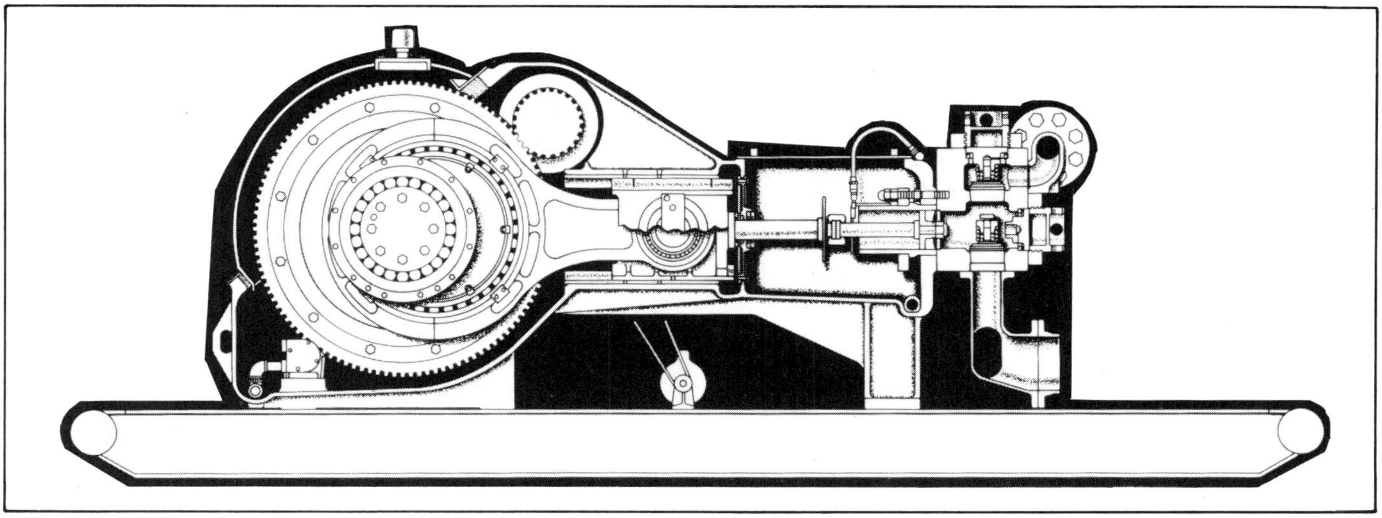

Fig. 10-3b. Cross-sectional view of G.D. triplex.

equipped with nipples having external connections threaded with 4-in. line pipe male threads. Two grades have API approval: grade C with 7500-psi test pressure, 4000-psi working pressure and grade D with 10,000-psi test pressure, 5000-psi working pressure.

The horizontal mud line from pump area to standpipe area usually is 4-½-in. OD seamless pipe with 4-in. line pipe threads or of welded construction with hammer unions at various intervals. On smaller land rigs 3½-in. OD pipe is often used. Wherever turns are required, long-sweep ells are recommended.

The standpipe, usually near a leg or other vertical member of the derrick or mast, extends a vertical distance of 40 to 45 ft above the working floor level and is equipped at its upper end with a special long-sweep gooseneck. The standpipe, like the horizontal mud line, usually is 4½-in OD seamless pipe but on smaller rigs 3½-in. OD pipe is often used.

The rotary hose connecting standpipe gooseneck to the swivel gooseneck is available in several API grades, sizes, lengths and pressure ratings. These range from 2-in. ID with lengths of 35 and 40 ft to 3½-in. ID with lengths of 55, 60, 70 and 75 ft.

The swivel gooseneck and wash pipe usually have the same inside diameter. In swivels rated 200 tons or more the passage in gooseneck and wash pipe is 3 in. In lower rated swivels the passage is approximately 2¼-in. The bore of the swivel body may vary from wash pipe bore to as much as ½ in. greater than wash pipe bore. Due to the short length of passage through swivel the pressure loss is small compared to that lost in the hose and kelly.

The friction pressure in the surface quipment usually is not a significant part of the total pressure loss in the entire circulating system. This statement pre-

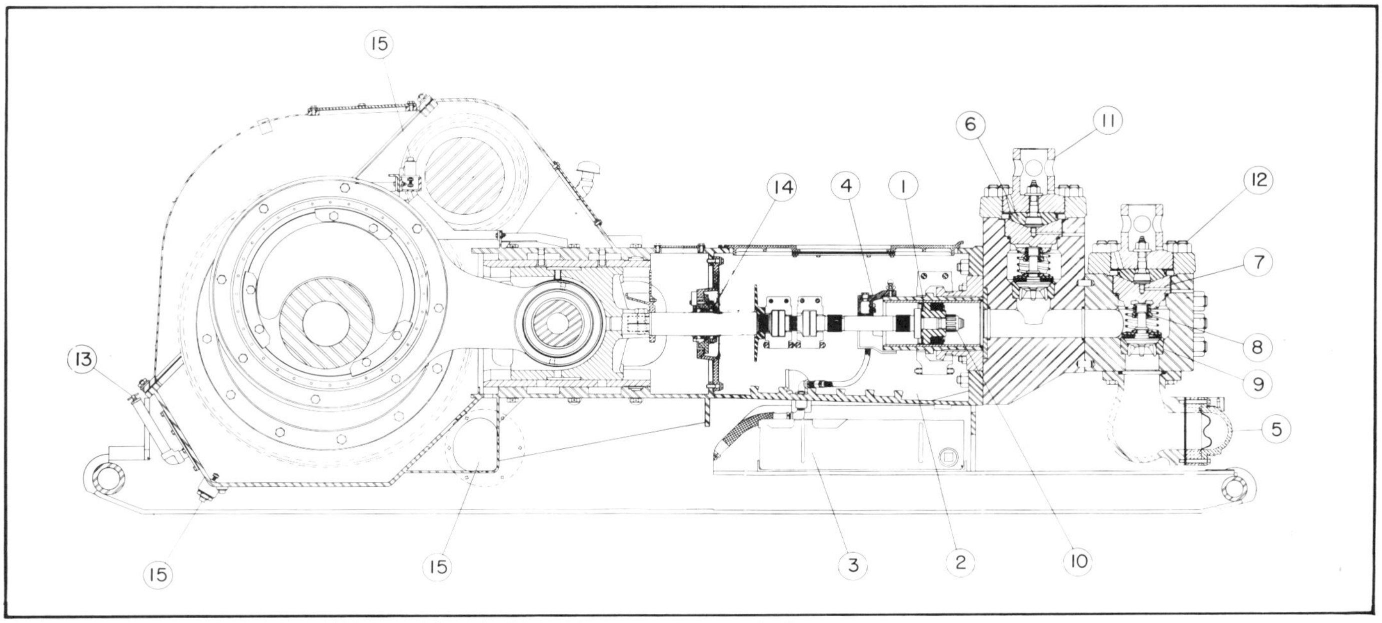

Fig. 10-4. Cross-sectional view of National Supply Div. triplex single-acting slush pump.

Fig. 10-5. Oil well Supply Div. triplex single-acting slush pump.

sumes a properly designed system.

If the circulating rate is 400 gpm or more, good practice dictates a 4-in. ID standpipe, 2½-in. ID hose and 3¼-in. bore in the kelly. At 400 gpm and 10 lb/gal mud, the pressure loss will be about 30 psi. However, if a 3-in. ID standpipe is used with 2-in. ID hose and 2¼-in. bore in the kelly, the circulating rate of 400 gpm will cause a pressure loss of approximately 130 psi, which is a significant part of the total pressure loss.

Drill String: Kelly, Drill Pipe and Drill Collars

Circulation fluid through the drill string, at great depth, usually accounts for a substantial portion of pressure lost in the circulating system. Prior to the jet bit era, it was the only significant source of friction pressure loss. The kelly, being relatively short, accounts for a very small portion of the pressure loss but even so, the larger bores generally found in hexagon kellys are beneficial.

The drill pipe, being the longest segment of the drill string, usually accounts for the major portion of friction pressure lost in the drill string. Tool joint bore is a factor but because they constitute only a small part of the entire drill string length, the loss is less than is generally believed.

The ID of the drill pipe is what counts most. A good rule of thumb, considering deep hole hydraulics, is to use the largest popular drill pipe size that can be safely fished. There was a tendency by some rig owners during the slim hole craze to switch to smaller drill pipe, hoping to stretch the depth capacity of their rigs even while drilling normal size holes. They succeeded in increasing hoisting depth capacity but hydraulics suffered, resulting in reduced penetration rates and overloading slush pumps.

The influence of drill pipe size on effective hydraulic potential is well illustrated by Table 10-2. Values of pressure losses quoted are for 10 lb mud of 3 cp viscosity and at a circulating rate of 400 gpm. They are taken from Fig. 8 of API Bulletin D-10 which appears as an appendix:

TABLE 10-2. Pressure drop through drill pipe, 400 gpm, 10 lb/gal. mud

Size, Wt. & T. J. Type	T. J. OD — T. J. Bore	Losses/1000
4″ OD 14.00 lb/ft, 1F	5¾″-3¼″	122.5 psi — 28.6 hhp
4½″ OD 16.60 lb/ft, XH	6″-3¼″	6.75 psi — 15.7 hhp
5″ OD 19.50 lb/ft, XH	6¼″-3¾″	40.0 psi — 9.3 hhp

Fig. 10-6. Continental Emsco triplex single-acting slush pump.

Fig. 10-7. Dresser Ideco triplex single-acting slush pump.

Fig. 10-8. Link Belt single deck shale shaker.

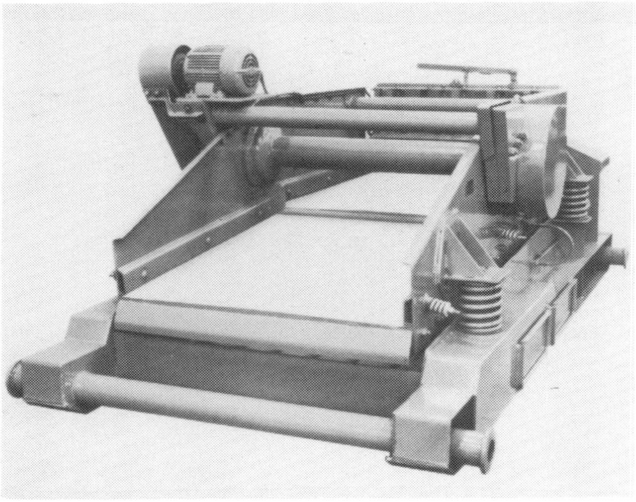

Fig. 10-9. Swaco "super screen" Shale Shaker.

Pressure losses and the hhp losses listed in Table 10-2 apply only to 10 lb/gal mud. The losses for heavier muds will be proportionately greater.

The weight of 5-in. OD drill pipe (and presumably the price) is only 39% greater than the 4-in. OD drill pipe (Table 10-2) and the OD of tool joints is only ½-in. greater. The friction pressure loss and hhp, however, for 5-in. OD drill pipe is one third that of the 4-in. OD drill pipe.

Less dramatic but nevertheless significant, the losses in 4-in. OD drill pipe are 82% greater than in 4½ in. OD drill pipe and also the losses in 4½-in. OD drill pipe are 69% greater than in 5-in. OD drill pipe.

Drill collar strings have increased in length in recent years and often are 400 to 600 ft long. At 10,000 ft drilling depth or less, the friction pressure lost in small bore collars may equal that lost in a well chosen drill pipe string. The relative influence on friction pressure loss as a function of drill collar bore is shown in Table 10-3. The values shown are taken from Fig. 9, API Bulletin D-10.

Table 10-3
Pressure Losses thru drill collars @ 400 gpm — 10 lb/gal mud

Bore	Losses/100 ft.
2″	148 PSI—34.5 HHP
2¼″	81 PSI—18.9 HHP
2½″	48 PSI—11.2 HHP
2¾″	30 PSI— 7.0 HHP
3″	21 PSI— 4.9 HHP

It is unlikely that as much as 400 gpm would be circulated through drill collars having a 2-in. bore but the relative influence of bores on hydraulic losses as shown are valid.

The primary function of drill collars is to provide a downward thrust on the bit. Larger bores decrease the weight per foot of drill collars somewhat but not enough to be given serious consideration. In small drill collars, a small bore through the pin (for strength) and a large bore through the remaining length tends to reduce the hydraulic pressure losses.

Where long drill collar strings are used, a reduction in hydraulic losses can be realized most inexpensively by increasing the bore of the collars by as little as ¼ in. as indicated by values shown in Table 10-3.

Bit Nozzles and Jet Bits

It may be an old saying but the statement is worth repeating: "The circulating mud stream does no useful work until the fluid leaves the bit." The nature of the fluid, its mass and velocity will determine the effectiveness of its purpose: to scavenge the bottom of the hole and to carry the drill cuttings to the surface.

There is much debate in recent years on whether hydraulic effectiveness can best be measured in terms of impact, mass times velocity (MV), or in terms of bit horsepower, a function of mass times velocity squared (MV^2). Currently many students of the subject favor the impact and cross flow theories.

In any event, as has been shown in field tests, *more* is not always *better*. It is possible to run into the law of diminishing returns, in which event the pump is loaded beyond actual needs, particularly if excessive pump pressures are involved. The pressure lost in turbulent flow through bit nozzles is a function of velocity and for a given fluid is proportional to the square of the nozzle velocity. As an example, the pressure drop at a nozzle velocity of 200 ft/sec using 10 pounds per gallon is about 355 psi whereas for the same mud at 400 ft/sec is about 1420 psi.

In a well designed hydraulics program the pressure lost in flow through the bit nozzles usually exceeds 50% of the total pressure lost. At great depth, where hole size usually is small, higher nozzle velocities are necessary to compensate for the lower rate of flow possible through the long string of small drill pipe.

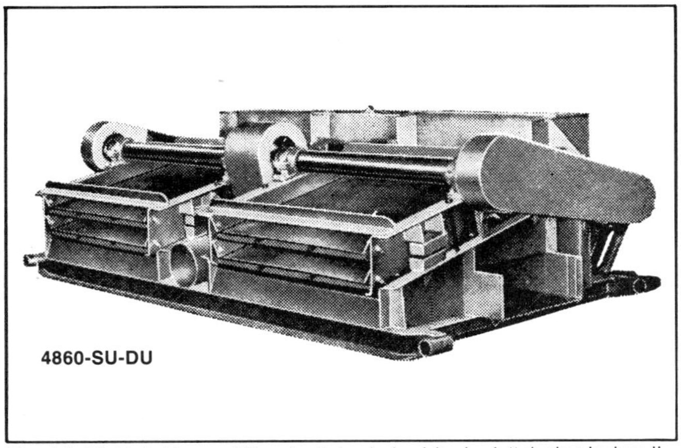

Fig. 10-10. Hutchinson "Rumbo" dual, double deck "shale shaker."

Fig. 10-11. Thompson shale separator.

Return Annulus Between Pipe And Hole

The pressure lost in annulus return flow usually is small, possibly in the order of 5% to 10% of the total and therefore need not be given much consideration from the standpoint of pressure loss.

The velocity of flow in the annulus, however, does affect the efficient removal of bit cuttings and other debris. With the improved drilling muds we now have, annular velocities over a wide range are effective in the removal of drill cuttings. Annular velocities usually fall within the range of 100 to 200 ft/min.

Shale Shakers — Mud Separators

Return mud issuing from the well carries with it many solids which should be removed before the mud again enters the pump for the next cycle of circulation. The shale shaker is a vibrating screen which usually screens out caving and most of the bit cuttings but allows silt,fine sand and liquid mud to pass through the screen and into long mud tanks where the heavier solids are allowed to settle out before entering the pump suction.

Vibrating screens have been improved and now are available in single or double deck arrangement and in a variety of sizes. In the double deck arrangement two screens are used, one above the other, both inclined about 10° from horizontal. Coarse material is removed by the upper screen thus lightening the job to be done by the lower fine-mesh screen. The screens with the supporting frames are mounted on coil springs or rubber blocks to permit the vibration induced by the vibrator, a rotating shaft eccentrically loaded and belt driven at high speed by an electric motor or similar power source. Figure 10-8, 10-10 illustrate several types of shale shakers.

Other means of separating bit cuttings from the mud have been used, such as an inclined rotating cylindrical screen, into which the well effluent enters at the higher end and flows by gravity toward the lower end. The liquid mud drops through the screen before reaching the lower end of the cylinder but cavings, cuttings and coarse solids are carried out the lower open end of the cylinder. See Fig. 10-11

Desanders-Desilters-Centrifuges

As previously indicated, the shale shaker does not remove all the abrasive material in the mud stream. The life span of mud pump expendables can be extended significantly by further treatment and removal of sand and silt from unweighted mud. Separators of one type or another have been used for these purposes.

The most common type of fluids-solids separator operates on the cyclone principle. Fluids with entrained solids enter an inverted cone, at high velocity, through a tangentially directed inlet near the top or large end of the cone. The resulting rotary motion of the fluid within the cone generates a centrifugal force which separates the fluid into conical layers. The heaviest solids and fluids move along the cone wall and drop by gravity to the underflow outlet at the bottom of the inverted cone. The lighter clarified fluids float inward and are discharged through an outlet at top center of the cone and thence into the pump suction mud tank.

The lower or small end of the cone is equipped with an adjustable orifice and other controls designed to limit discharge of solids and fluids so as to balance and control the separator action.

Cone type separators are made in nominal sizes from 2-in. to 12-in. Nominal size refers to the ID of the cone at point where fluid first enters. In most instances the cones are lined with rubber or plastic to minimize abrasive wear. Larger cones are used for removal of coarse solids and sand while only the smaller sizes effectively remove silt.

Since the capacity of an individual cone is much

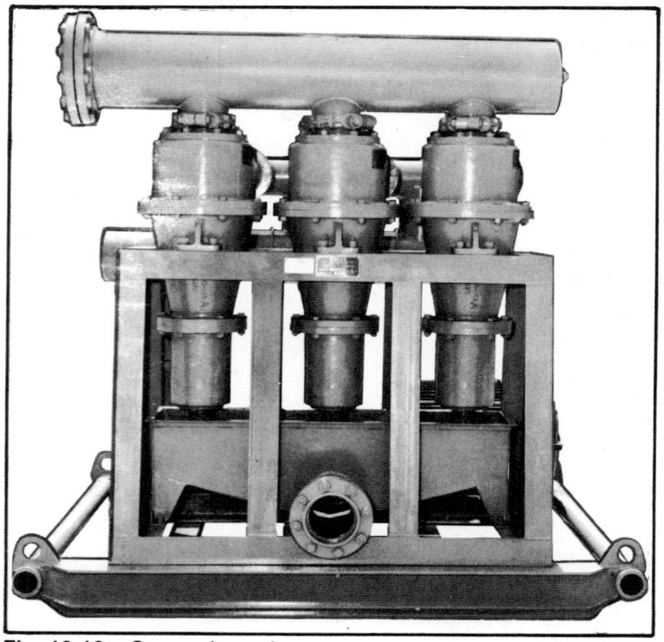

Fig. 10-12a. Swaco desander.

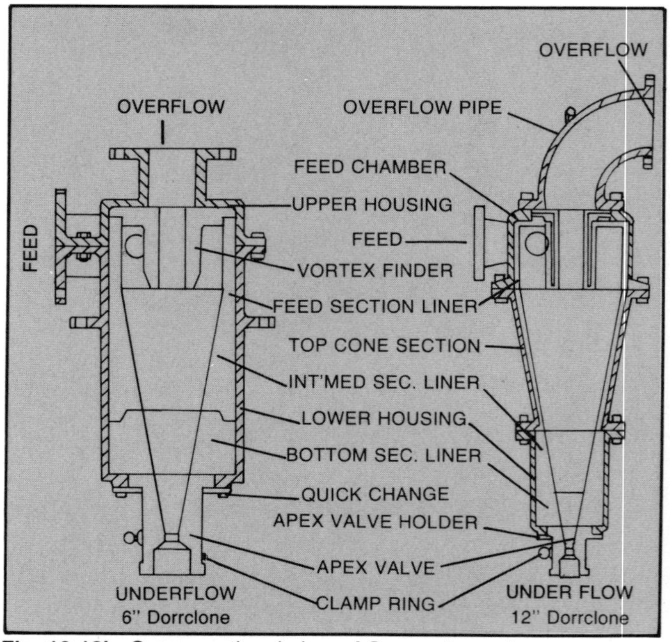

Fig. 10-12b. Cross-sectional view of Swaco desander.

less, in gpm, than the usual circulating rate while drilling, it is common practice to manifold several cones into a skid-mounted unit so as to attain the desired capacity.

If desanding unweighted mud is the only purpose desired, the unit may consist of only 2, 4 or 8 larger cones even if they do not equal in capacity the circulating rate. If it is the desire also to thoroughly desilt unweighted drilling fluid, a second unit of a greater number of small cones will be needed so capacity of the desilting unit exceeds the circulating rate. This makes certain that all the drilling fluid has gone through the desilting process at least once during each cycle of circulation.

The purpose of desanding and desilting drilling fluids is to remove abrasive substances that tend to destroy pump parts and the tooth structure of bits. Independently powered centrifugal pumps provide the drilling fluid circulation through the desanding and desilting units. See Figs. 10-12a-b and 10-13.

Another type of centrifuge is used for recovering barite from weighted drilling fluids and at the same time disposing of clays and other unwanted solids. The barite thus recovered is returned to the weighted mud system. The process is an alternate procedure to the practice of jetting pits to discard weighted mud that has become heavily ladened with clay and abrasive solids to provide pit room for fresh mud of more desirable quality.

The centrifuge commonly used for this purpose is the mechanical decanting type and is driven by a variable speed prime mover. The operating principle is illustrated in Fig. 10-14 . The mechanism consists essentially of a helical screw conveyor rotating inside a conical bowl, which also revolves in the same direction but at a slightly higher speed. The speed of rotation of the bowl creates a centrifugal force acting at right angle to the axis of rotation.

Diluted drilling mud enters the bowl through a stationary tube located in a hollow center shaft. The heavier barite particles are thrown outward by centrifugal force and are deposited on the wall of the bowl. The clays, solids and liquid, being of lower density form a concentric inner layer in the bowl.

The barite deposited on the wall of the bowl is conveyed to the small end of the bowl by the pitch and forward rotation of the helix relative to the bowl, where it is discharged through ports that are suitably placed, and returned to the mud tanks.

The clays and liquids are rejected through ports in the opposite end of the bowl. Any desired degree of

Fig. 10-13 Pioneer Manifolded cyclone unit.

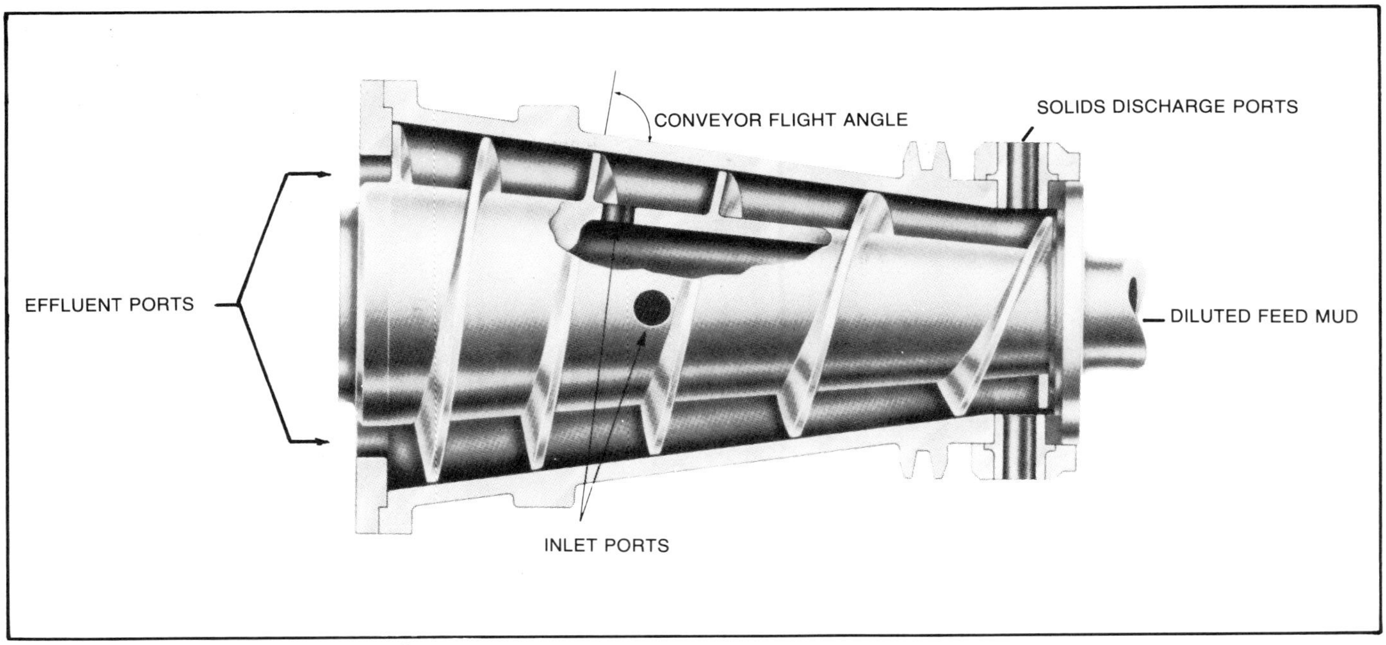

Fig. 10-14 Cutaway view of Baroid centrifuge.

separation can be effected, commensurate with the particle size distribution of the two solid phases in the system, by varying the rotating speed of the machine.

Degassers

The appearance of small gas bubbles in mud returns, commonly referred to as gas cut mud, is not an uncommon occurrence. It seldom, if ever, is the initial warning of an impending blow out. The first indication of extraneous fluids entering the borehole, while drilling ahead, is a gain in the rate of return flow and subsequently an increase in pit mud volume. This is called a kick and unless proper action is taken immediately a blow out very likely will occur.

In gas cut mud, where few pockets of free gas are present and small gas bubbles are well dispersed in the mud, it suggests that the gas is coming from high pressure formations with low permeability. These formations may be shale in which the gas present is under geostatic pressure rather than hydrastatic pressure; dense limestones or nearly impervious sandstones frequently are sources. Another possible source of gas that seems to cut mud can come from drilled cuttings while drilling normally-pressured permeable sandstones. In this instance the gas bleeds from the cuttings as they approach the surface where the pressure is reduced.

Heavily gas cut mud in the pits, at atmospheric pressure, will rarely reduce the bottomhole hydrastatic pressure by more than 100 psi even in deep wells. For this reason, well control is not a problem with gas cut mud. There are, however, two serious concerns about gas cut mud.

First, it is next to impossible to determine the mud's true density without first degassing it. More often than not, barite will be added needlessly, resulting in an effective downhole pressure which may exceed the fracture gradient, resulting in loss of mud.

Second, if the mud is gas cut 50%, for example, the capacity of the slush pump will be reduced even more than 50%, thus reducing drilling efficiency. Where gas cut mud persists, a degassing device is an essential drilling piece of equipment.

Currently there are two types of degassers available. Fig.10-15 illustrates a type used extensively for over 25 years. The device consists of a horizontal skid-mounted vacuum vessel about 3 ft in diameter and 10 ft. long which is positioned above the mud tanks, taking suction from the first tank and discharging into the second tank. The flow capacity of the degasser is designed to exceed the normal cir-

Fig. 10-15. Swaco degasser.

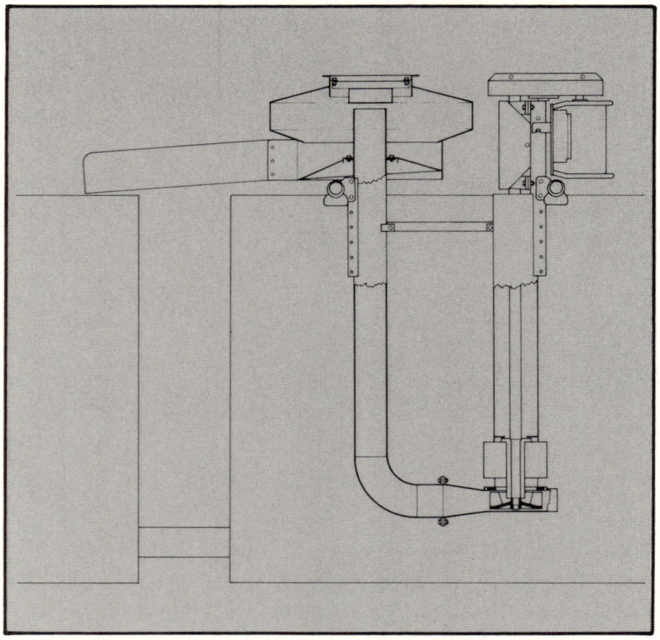

Fig. 10-16. Cross-sectional view of Drilco degasser.

culating rate, so excess mud must flow back to the first tank.

Gas cut mud is drawn into the degassing vessel by vacuum created by the discharge jet and the vacuum pump. The drilling fluid is distributed over the degasser baffle in a thin, evenly dispersed layer, and the gasses are liberated by the induced low pressure. The vacuum pump then discharges the gas through piping a safe distance from the rig. Since the discharge jet and the vacuum pump both tend to create vacuum within the vessel a balance of these forces is maintained by a float-actuated control system, which insures that the vessel neither overfills nor runs dry.

Another type of degasser of recent vintage, which operates on a different principle, is illustrated in Fig. 10-16. Its action is not dependent on the assistance of a vacuum pump. The separation of gas from the mud is effected by turbulence and impact of a thin layer of gas cut mud at atmospheric pressure. The complete unit consists of a submerged centrifugal pump and an elevated spray tank. Spiral ducting of the pump inlet causes the mud to form a whirlpool above the center of the impeller. Any gas that leaves the mud in the inlet and in the impeller moves to the whirlpool and moves upward through the pump frame.

Gas cut mud in the impeller flows up through the riser pipe to a circular orifice, through which the mud flows, forming a high velocity sheet. The mud sheet strikes the inside walls of the tank with enough turbulence and mixing to drive small gas bubbles out of the mud. The mud flows to the next pit in a trough. If advisable to pipe the gas away from the site, a gas collector and blower can be installed on the discharge trough.

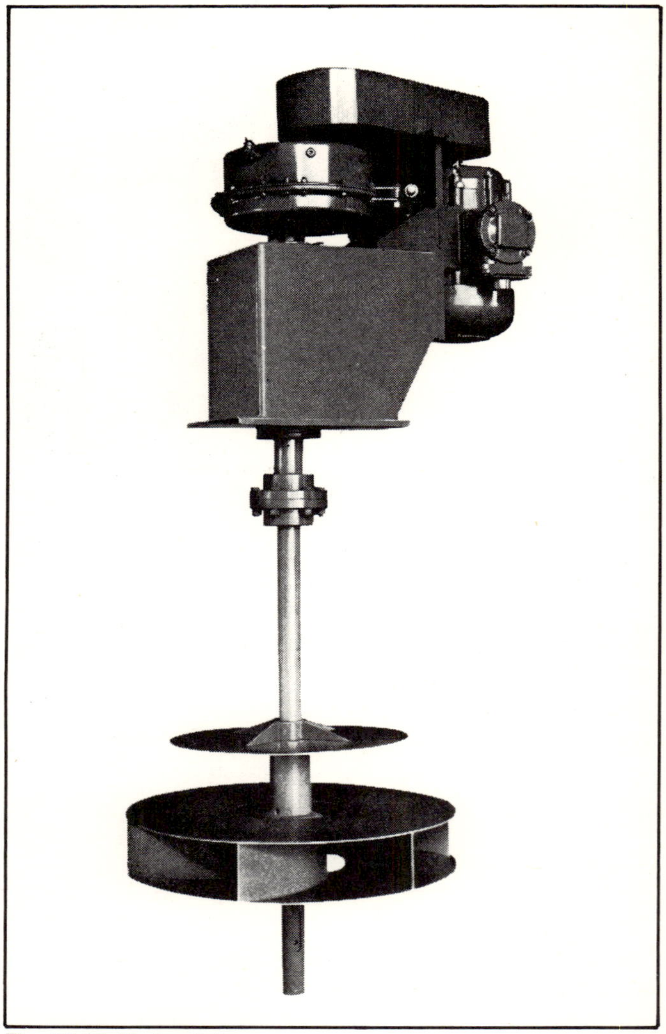

Fig. 10-17. Mud agitator.

Mud Tanks, Agitators And Mud Mixers

Steel mud pits usually are custom made to purchaser's specifications. For use with land-based rigs, highway restrictions influence designed dimensions. Widths usually approach 8 ft and the height most often is 6 ft. Lengths may vary from 20 to 30 ft but frequently are 26 ft since a tank having these outside dimensions will have a nominal capacity of 200 bbl. Three such tanks will provide a useable capacity for 450 to 500 bbl of liquid mud. If desanders, desilters or other such mud treatment is used, four tanks are often used.

The drilling fluid, as it flows from the shale shaker through settling tanks toward the pump suction, tends to establish a live stream or channel-like flow through the main body of mud, much of which remains relatively dormant. This situation, if allowed to persist, results in a build-up of solids throughout the inactive portions of the pit. In weighted mud systems this leads to waste and inefficiency.

Fig. 10-18. Jet type mixer and hopper.

Fig. 10-19. Mud mixing system unitized.

Several types of agitators are used to keep mud stirred up and active in all parts of each tank. The most common agitator used is the mud gun, several of which may be arranged alongside the several tanks and pivoted in such a manner that the high velocity mud stream issuing from the gun nozzle can be directed to almost any area of the tank.

Mud guns require high pump pressure along with a considerable mud volume in order to be effective. The overall mechanical efficiency of a mud gun is very low and for this reason they are displaced, to some extent, by mechanical agitators (Fig. 10-17). The most commonly used type of mechanical agitator employs a turbine of some sort situated near the bottom of the tank and is independently driven by a small electric or hydraulic motor, rated 10 hp to 20 hp. One or more units may be used in each tank.

When additional fresh mud needs to be added to the system, a mud mixer is employed. This device may consist of a simple jet-type mixer, such as is used by Halliburton for mixing cement. In this case high pressure mud from a spare rig pump is used to power the jet.

If the purpose is to increase mud volume, bentonite is mixed. If the purpose is to build weight, barite is mixed. If both volume and weight are needed, both additives are used.

For a simple operation of this sort, the dry mud usually is in paper bags and is manually dumped into the funnel shaped hopper. The mixed mud is discharged into the pump suction pit.

If fresh mud requirements are large, a more elaborate handling and mixing system is employed. One or two independently powered centrifugal pumps provide mud under pressure to companion jet type mixers (see Fig. 10-18).

In a system of this sort the dry mud usually is supplied from bulk storage consisting of one or more elevated steel tanks with cone-shaped bottoms so that the dry mud, with a minimum of agitation will flow by gravity to the funnel-shaped hoppers of the jet mixers.

Chapter 11 Rigfloor Accessories/Tools/ Instruments

Cat Lines

One of the more versatile accessories around a rotary drilling rig is the cat line. Traditionally the cat line is manila cordage, 1¼-in. or 1½-in. in diameter and of a length approximately equal to two and one half times the derrick height. It is reeved over a cat line sheave, suspended underneath the crownblock, with one end of the line conveniently near the cathead, on the driller side, and the other end held back out of the way near edge of the derrick floor.

The cat line is used to pick up heavy objects from one area on or near the working floor and while suspended in mid-air, these are swung or pushed to the desired spot and then lowered to rest. The cat line is used for lifting objects weighing as little as a few hundred pounds to those weighing a ton or more

One example of its use is lifting master bushings from the rotary table. The most common use is to pull in, from the walkway, single lengths of drill pipe either to be set in the mouse hole or positioned to be latched onto by drill pipe elevators.

Drill Pipe Elevators

Drill pipe elevators (Fig. 11-1) usually are of the center-latch type and can be supplied for either square shouldered or taper shouldered tool joints.

Elevators links, (Fig. 11-2) two per set, are forged steel with eyes at each end of the bars, one pair to support the elevators and the other pair to hang over the ears of the triplex hook. Elevator links are available in a variety of sizes, lengths and capacities.

Diameters of the body, between eyes, range from 1¾-in. to 3½-in. with corresponding capacities per pair ranging from 150 tons to 500 tons. The overall length of the links ranges from 36-in. for small sizes up to 12 ft for the largest current size.

Casing Elevators

Casing elevators (Fig. 11-3) are usually the side door variety and are available in either the conventional collar-type or the slip-type which grips the casing below the collar. The slip-type elevator is used only on long heavy strings of casing, usually sizes 13⅜-in. and smaller.

A special heavy duty adaptation of the side door slip-type elevator is called an elevator-spider (Fig. 11-4). One such item is used as an elevator and another like item is used in lieu of the conventional casing spider and slip assembly.

Rotary Slips

Rotary slips (Fig. 11-5) are used to support the drill string in the rotary table while making or breaking connections during a tripping operation. Slips are wedge-shaped and a set consists of three or more segments. When set in the rotary table these segments conform, on the back side, to the bore and taper of the slip bowl of the rotary and, on the inside, to the cylindrical body of the drill pipe.

The tapered slip bowl in the rotary's master bushing conforms to API specifications for diameters, length and taper. It was learned, through many years of experience, that a bowl taper of four inches per foot best provides both adequate pressure of the slip inserts against the pipe and freedom from sticking of the slips in the bowl after the load on the table is removed.

Each slip segment, whether three per unit or four per unit, embodies four principal parts: body, handle, hinge pin and set of inserts. In the one-piece slips, all so-called segments, whether three or more, are hinged together and hence tend to raise or lower into the bowl in unison so that damage to drill pipe is averted. For moderate depth wells two-piece slips are used wherein each piece embodies two segments hinged together. This type slip is somewhat lighter and more convenient to handle than the one-piece slip.

The segment body usually is made of cast steel and of honeycomb construction in order to reduce weight. The inserts, or dies as they are sometimes called, are high grade tool steel which bite into the drill pipe slightly. Segment bodies may be of two or more basic sizes, each of which will accommodate inserts to fit two or three pipe sizes. Drill pipe slips with proper inserts can be used for running casing in sizes up to 7⅝-in. OD.

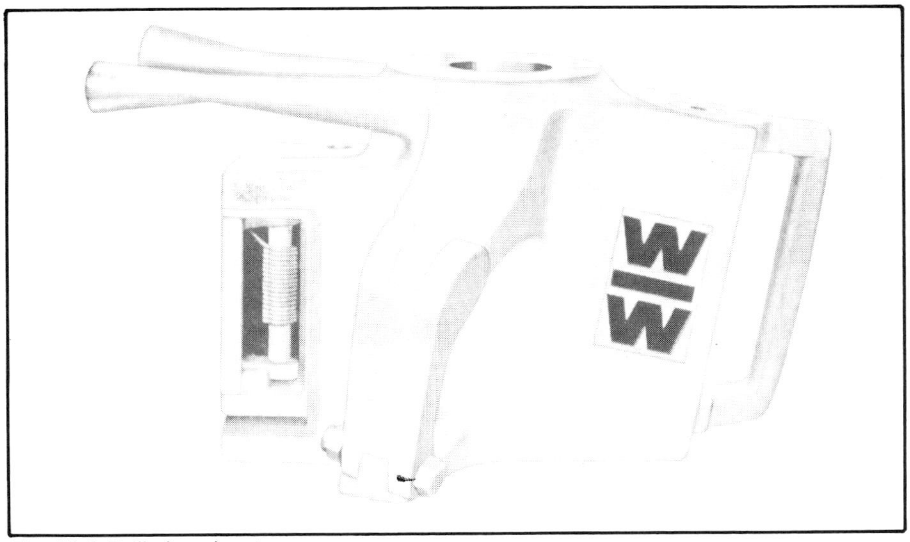

Fig. 11-1. Drill pipe elevators.

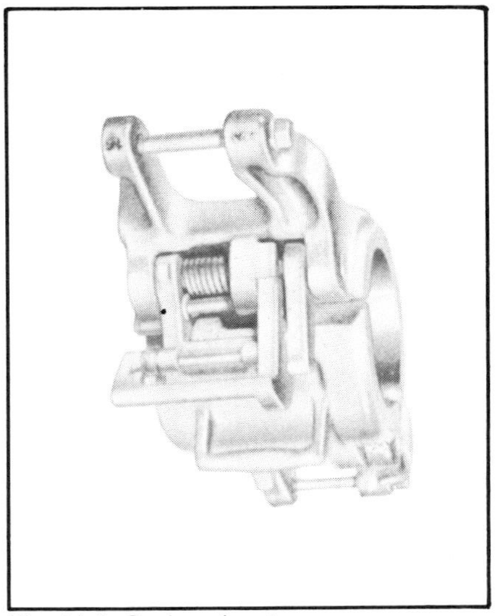

Fig. 11-3. Casing elevators.

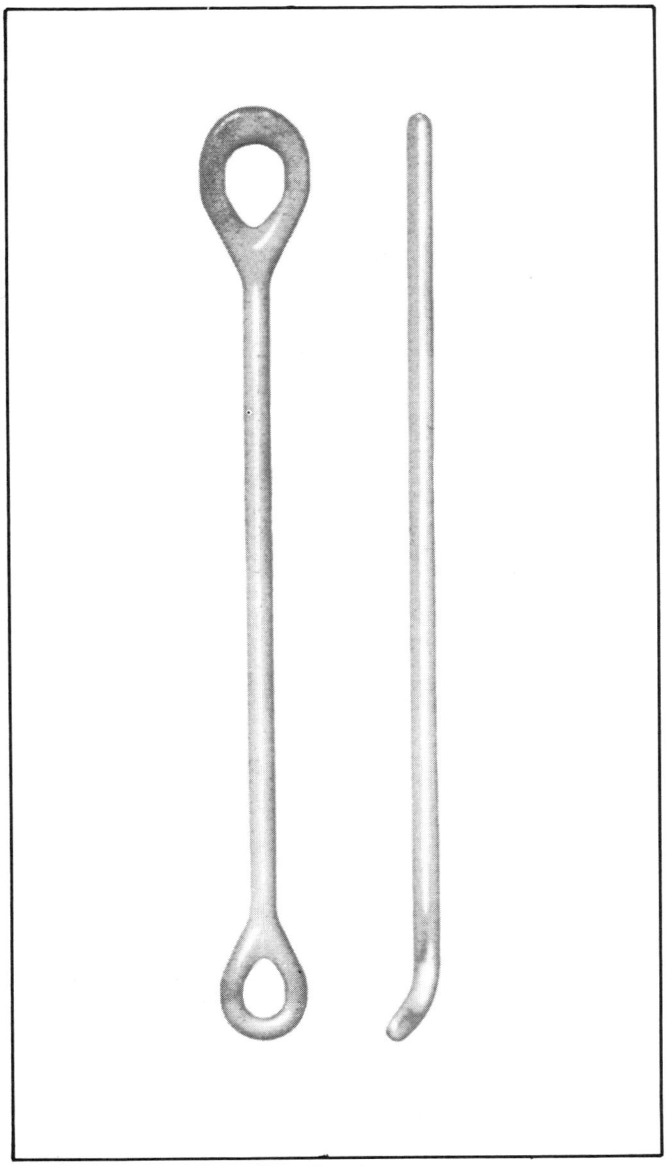

Fig. 11-2. Elevator links (two views).

Fig. 11-4. Slip-type elevator-spider.

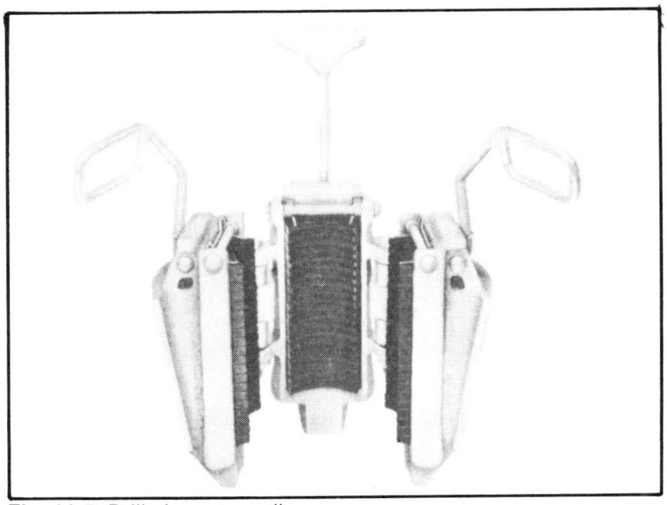

Fig. 11-5. Drill pipe rotary slips.

Power Slips

Power slips, (Fig.11-6) for use in handling drill pipe, have been widely accepted because they tend to improve working conditions and promote crew safety. The power medium utilized is compressed air (of ap-

Fig. 11-6. Power slips.

Fig. 11-7. Solid casing bushings.

proximately 100 psi) controlled by a throttle valve at the driller's position.

The several slip segments are attached to a lifting ring, free to rotate in the ring arm which extends laterally to the air cylinder alongside the rotary table. Power slips are being offered by several manufacturers and each has some special feature. All, however, sit on top of the rotary table and may or may not make use of the slip bowl of the master bushings.

Casing Spider and Slips

The original spider was of the one piece platform-type, made of forged steel with a large slip bowl to accommodate the largest anticipated casing size. Additional liners and slip assemblies were provided for smaller casing sizes to be run. This type spider and slip assembly, typical of the cable tool era, was carried over into the rotary age.

Modern casing spiders make use of the rotary table as the basic unit. Casing sizes no larger than 7⅝-in. OD can be run through the table, using the master bushing API bowl in lieu of a spider. For all casing sizes up to 13⅜-in. OD a current popular approach is to replace the table's split master bushing with a one piece casing bushing as shown in Fig. 11-7.

As will be noted, split adapter bowls are required for casing sizes smaller than 11¾-in. OD. For casing sizes larger than 13⅜-in. OD, the spider most commonly used is the hinged-type (Fig. 11-8). Shown in Fig. 11-9 is an extra long, multiple segment-type of casing slip that can be used with casing bushing (Fig.11-7) or with the hinged bushing (Fig. 11-8). An adaptation of the multiple segment-type slip can be used to support drill collars in the rotary table.

Drill Collar Safety Clamp

When handling conventional flush joint drill collars through the rotary table it always is advisable, regardless of how effective the slips are supposed to be, to tighten a safety clamp around the drill collar just above the slips before making or breaking a drill collar connection (Fig. 11-10).

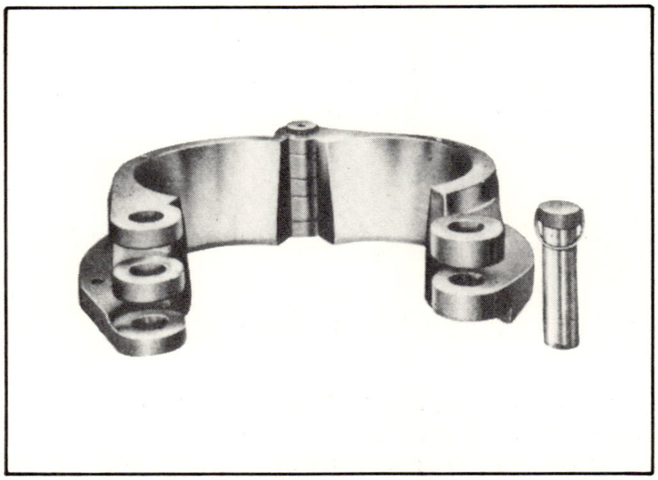

Fig. 11-8. Hinged casing spider.

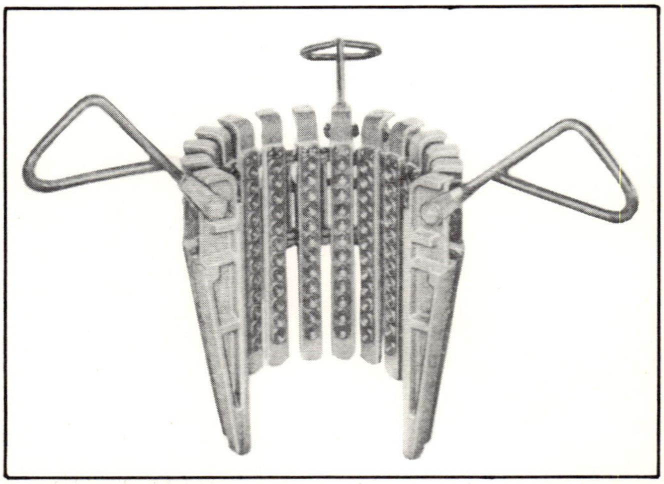

Fig. 11-9. Casing slips.

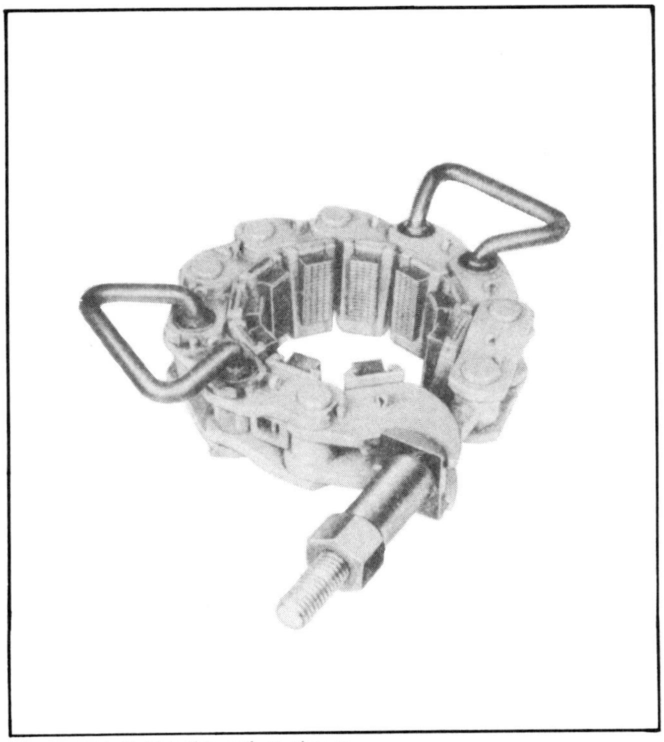

Fig. 11-10. Drill collar safety clamp.

Fig. 11-11. Bit breaker.

Bit Breaker

Rock bits are often driven tight or broken loose with a sledge hammer. A better practice is to make use of a bit breaker (Fig. 11-11). The thick upper plate, which rests in the square of the table's master bushing, has cutouts to permit passage of the bit cones to the lower deck or plate.

The cutouts provide a reasonable fit around the bit body or lugs so as to prevent rotation in the locked rotary table. The bit may then be broken out with tongs gripping the drill collar.

Tongs and Power Tongs

Tongs are to the men on the derrick floor what the pipe wrench is to the plumber — a means of making and breaking joints in pipe. Rotary tongs are manual tongs (Figs. 11-12a & 11-12b). The range in size varies from that required for tubing to sizes suitable for casing. The tongs may each weigh from 200 lb for smaller sizes up to 750 lb for larger sizes.

Each tong can cover a considerable range of pipe sizes by changing the lug jaw (Fig. 11-13). In use, tongs are suspended to permit them to be swung to the center of the floor for use and swung out of the way when not being used. When making up pipe, the general practice is to use a spinning line or chain in conjunction with the cathead on the drawworks to make up the pipe initially. To tighten, a rotary tong with a cat line is used.

When coming out of the hole with drill pipe, the practice is to use two tongs, one anchored as a back up, and a cathead jerk line connected to the other for breaking the joint. Larger sizes suitable for casing are available, but they are usually obtained on a rental basis.

The power tubing tong provides a rapid means of making joints in that they eliminate the use of spinning lines and provide a means of tightening the joints to a specified torque. The capacity of the power tongs is large enough to permit them to be used on drill pipe, and they are being employed to a considerable extent in that manner. (See Fig. 11-14)

Power tongs are reversible and have two speeds — high for spinning and making up; low speed for breaking out. One manufacturer of power tongs points out they can be used for light drilling, such as drilling out a plug.

Power for the tongs is provided either from existing air or hydraulic power sources on the rig, or trailer or skid-mounted compressors, or hydraulic power units.

Whether the tongs used are manual or power, it should be remembered that the pipe used in oil and gas wells is subjected to severe usage from tensile stresses and from corrosive and abrasive action. It is, therefore, desirable that the tongs do not crimp, distort or unduly scar the pipe.

It is suitable that the tongs have the following

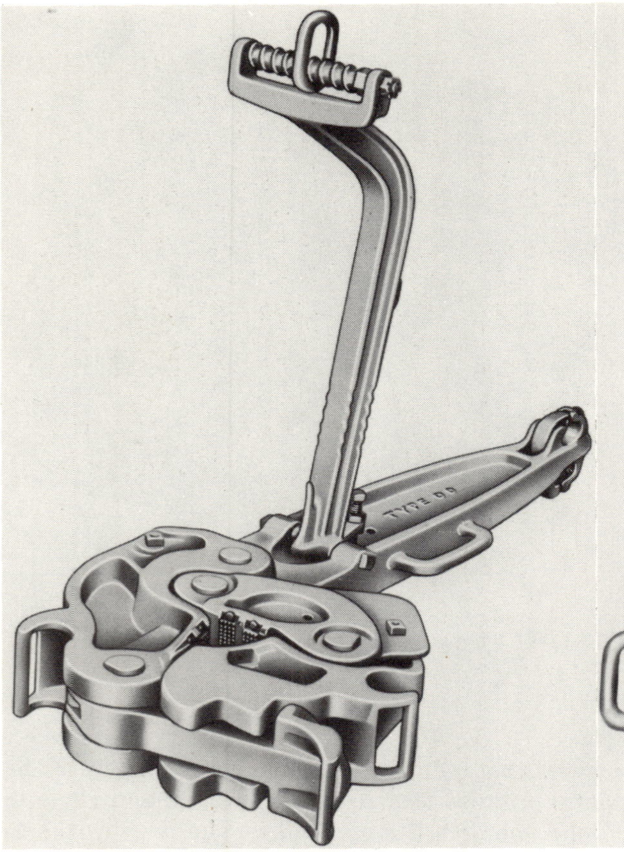

Fig. 11-12a. Manual drill pipe tongs.

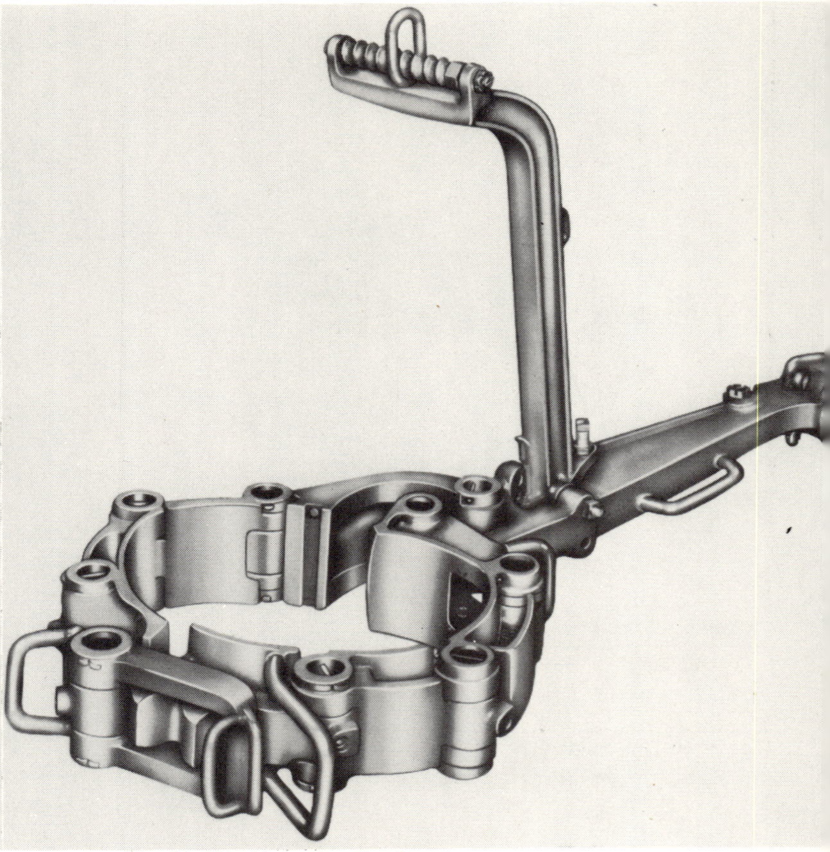

Fig. 11-12b. Manual casing tongs.

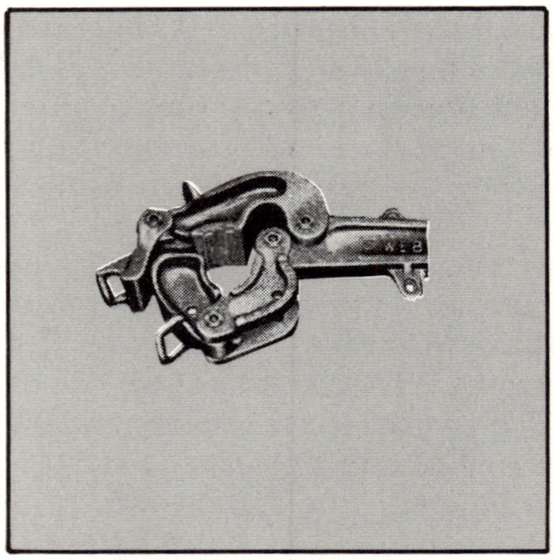

Fig. 11-13. Rotary tong lug jaws.

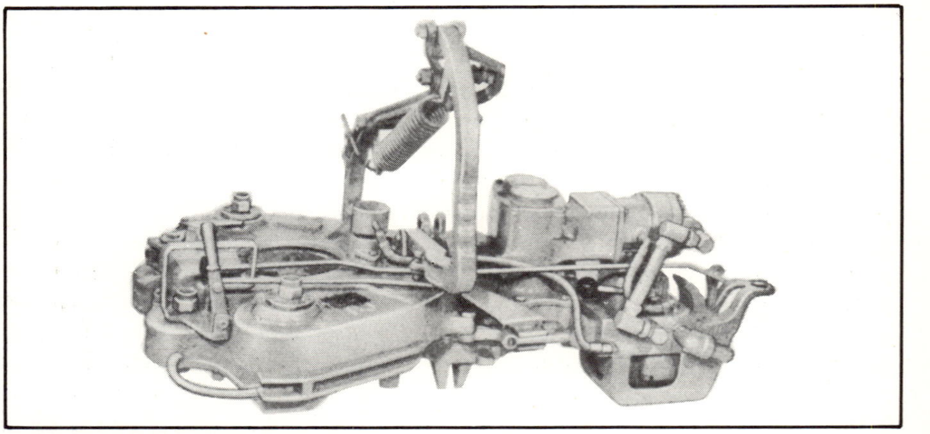

Fig. 11-14. Hydraulic drill pipe tongs.

features: (1) they should be easy to latch and unlatch, yet provide a positive lock while in use; (2) they should be suitable for a wide range of pipe sizes by changing a minimum number of parts; (3) they should be balanced properly for ease and safety in handling.

It was pointed out that tongs may represent a considerable investment. It should therefore be mentioned that major tongs manufacturers maintain repair facilities for their tongs where they rebuild or replace worn or broken parts, thereby giving the tongs an extremely long life.

Kelly Spinner and Auxiliaries

In fast drilling situations, time saved adding lengths of drill pipe to the drill string can affect significantly the overall time required to drill a well. Many years ago someone conceived the idea of a mouse hole situated on the far side of the rotary table, away from the drawworks.

A single length of drill pipe is placed in the mouse hole sometime before drilling the kelly down. When ready to add the next length of drill pipe to the drill

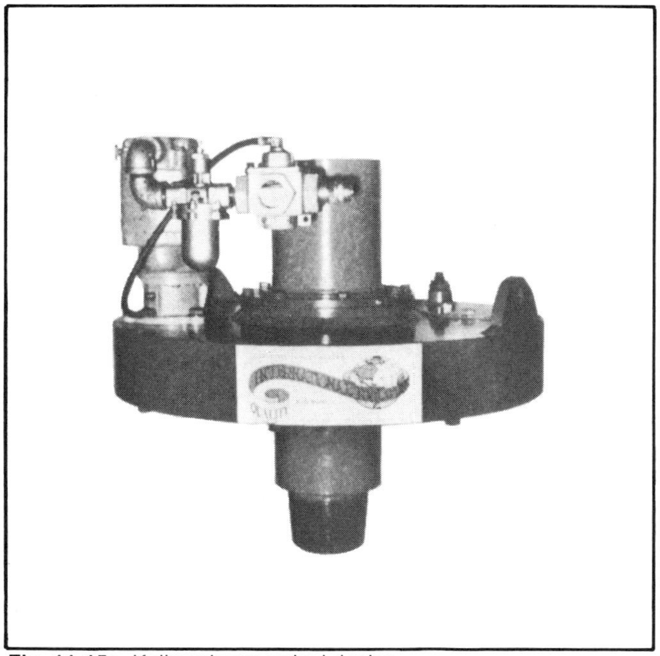

Fig. 11-15a. Kelly spinner-swivel device.

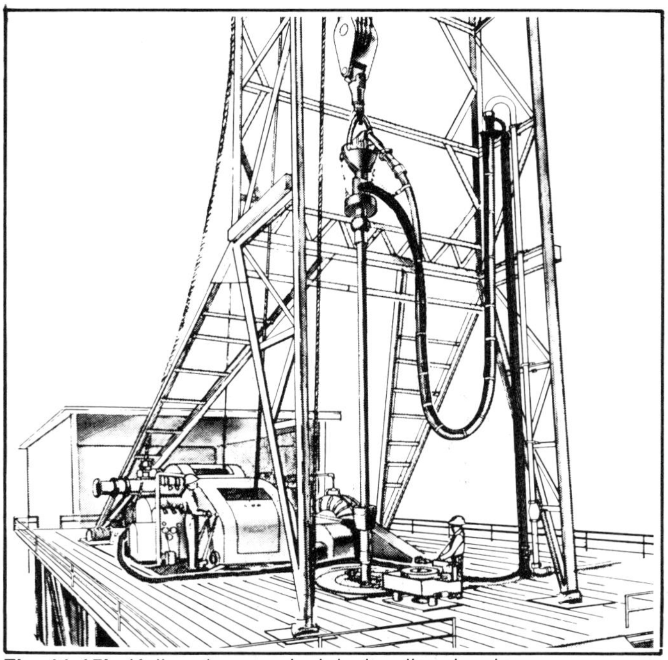

Fig. 11-15b. Kelly spinner-swivel device, line drawing.

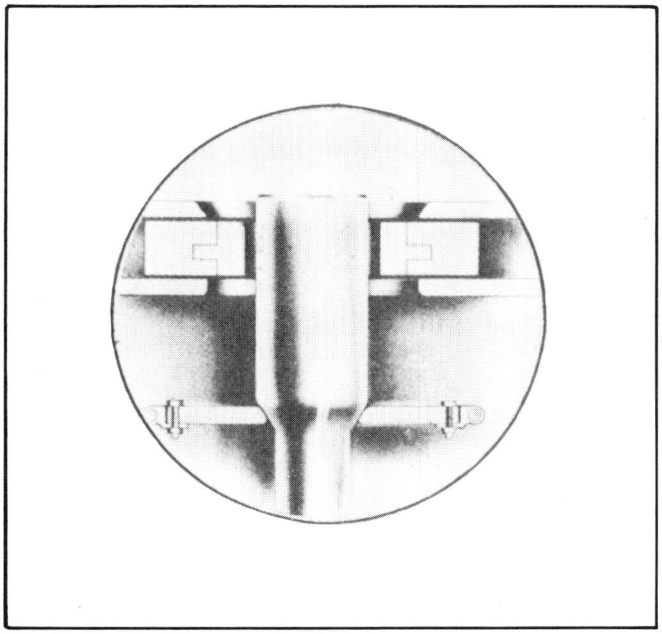

Fig. 11-16a. Backup supporting tool.

Fig. 11-16b. Kelly stabber.

string, the kelly is broken off and swung over to the mouse hole and screwed into the single pipe reposing there. The usual spinning and tonging of the connection is then employed, thus departing from the previous practice of placing the kelly in its rat hole, picking up the single, screwing it into the drill string, lowering it to near table top and then picking up and connecting the kelly.

Although mouse hole connections save considerable time, it does require quite a lot of manual effort and subjects the floormen to some safety hazards which a device known as kelly spinner can alleviate.

A kelly spinner is a power-driven unit, attached to the lower part of the swivel, which can rotate the kelly with considerable force. Usually it is driven by a vane-type air motor. The mass or weight of the swivel, hook and block provides sufficient inertia.

To take full advantage of the kelly spinner certain auxiliaries are available for installation at top of the mouse-hole: back-up pipe supporting tool and kelly stabber. The drill pipe single when set in the mouse hole is both supported by and backed up (against rotation) by air actuated mechanisms.

The kelly stabber employs an air hoist with

Fig. 11-17. Kelly and pipe straightener.

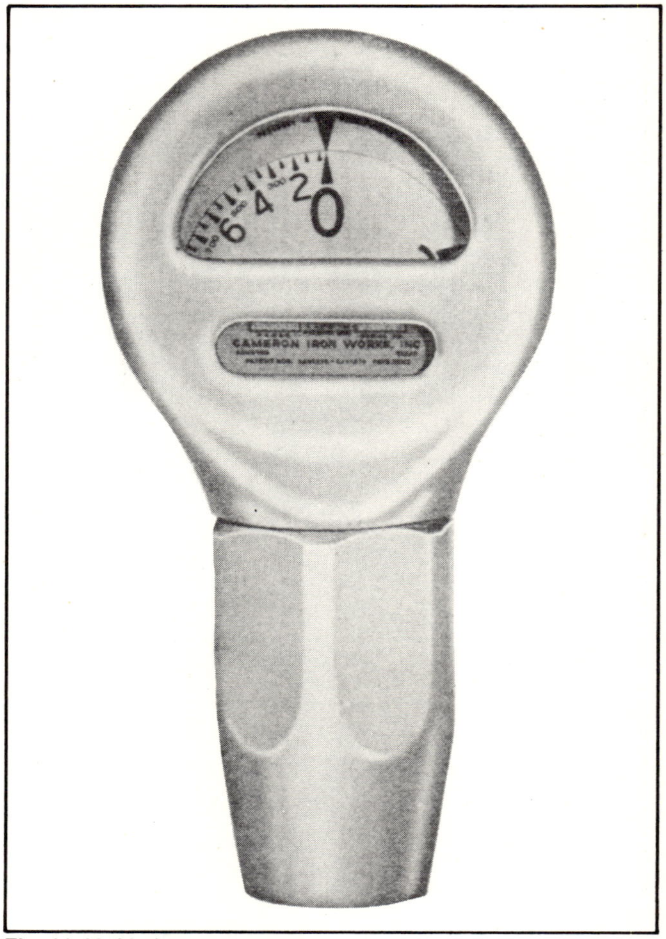

Fig. 11-18. Mud pressure gauge.

wireline and hook that can be placed around the kelly. When power is applied the kelly is drawn over to a position directly above the box of the single. When lowered to where threads are engaged, the spinner is started and connection of kelly and mouse hole single is completed. Thereafter the kelly and its single are picked up and added to the drill string.

All operations described above are controlled by air throttle valves at the driller's station. Fig.11-15 shows the kelly spinner and auxiliary equipment. Fig. 11-16 is closeup view of the back up supporting tool and kelly stabber.

Kelly and Pipe Straightener

It is important that the kelly joint be straight. High speed rotation of a bent kelly creates vibrations that are destructive to several rig elements. Kellys become bent due to various causes: moving from one location to another; being picked up from any horizontal position or being laid down. To a somewhat lesser extent, drill collars and drill pipe become bent due to buckling under too much weight or from being dropped.

The portable straightener shown in Fig. 11-17 utilizes the principle of the hydraulic jack, applying the force built up by a manually operated pump to bend the kelly, drill collar or drill pipe as it hangs suspended through the rotary or lies horizontally elsewhere. The straightener illustrated is a self-contained unit, light in weight. It has a bending force of more than 100 tons. It will handle material up to 8¾-in. diameter.

Rig Instrumentation

The art of oil well drilling is fast becoming, like other purely mechanical operations, a science of reading, correlating and interpreting the information available through the use of sensitive, relatively fool-proof instruments.

Drilling by the seat of one's pants, once an expression about as commonly used as flying by the seat of one's pants, is now about as obsolete in the art of rotary drilling as the latter is in the flying profession. Much technological advancement has been made in the instrumentation of oil well drilling and more is in the development stages.

The main control instruments are the weight indicator, mud pressure and volume instruments, rotary torque gauge and tachometer. The application of these instruments and the interpretation of the data

Fig. 11-19. Driller's console.

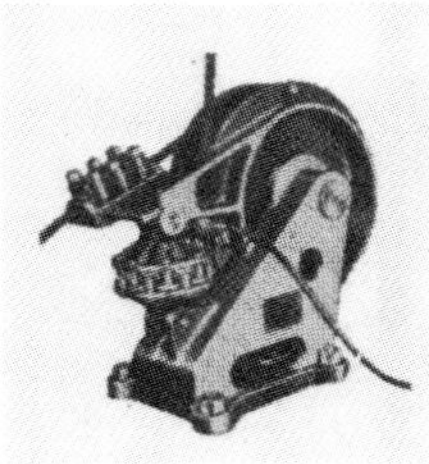

Fig. 11-20. Deadline weight sensor transducer.

obtained therefrom is an art that must be learned and effectively applied by the driller if maximum efficiency is to be obtained in rotary drilling.

Weight Indicator

The weight indicator is the primary drilling control instrument. It is used to control the weight applied to the bit while drilling, and indicates the total weight applied when running-in, hoisting, fishing and such weight measurements necessary to operate certain types of testing and specialty tools. The weight indicator is therefore a safety device as well as an instrument improving efficiency in the actual drilling operation.

The importance of controlled weight while drilling cannot be minimized, for without this control, drilling rates would be greatly impeded due to reduced bit life, as well as hole deflection and twist-offs resulting from overloading and fatigue of drill pipe and tool joints. The element of safety to equipment and men must be considered.

Overloading while fishing and/or pulling on stuck drill pipe might result in serious economic losses and jeopardy to the welfare of the crews working on the rig floor.

Hole drift and drill pipe fatigue resulting from buckling of the drill pipe within the circumferential extremities of the hole, and caving of the hole which may result from such buckling, are conditions which may be controlled by careful application of weight somewhat under established critical limits.

On the other hand, directional drilling can require intentional overloading to maintain the desired angle of deflection. Successful fishing jobs require extremely accurate bit weight and hook load indications. Automatic feed control systems operate on a constant weight principle, taking their signal from a surface weight-sensing element.

There are several different types of drilling weight indicators available to the operator. The simplest type is a mechanical unit with an integral gauge that clamps on the deadline and gives a continuous indication of hook leads. Historically, this is the kind that was once universally accepted by the drilling industry.

Because of the severe deadline whip to which instrument was subjected, plus the fact that the instrument was often so far away from from the driller that reading was difficult, unitized weight indicators gave way to remote indicators. Today, for the most part, unitized weight indicators are used as emergency, testing or standby instruments.

Another type of weight indicator has a deflection-type diaphragm on the deadline, connected by high-pressure hydraulic hose (at one time copper tubing was used) to the indicator at the driller's position. This type of instrument, a great improvement in its day, is seen today mostly on service rigs or at locations where fairly shallow holes are being drilled. It has the advantages of low cost, portability and simple installation.

Later types of weight indicators provide dials for direct reading of weight on bottom in addition to hook loads, compound pointers for amplified readings of small weight variances, and can be equipped to provide graphic recordings (see Fig. 11-19).

The most modern unit of this type takes a signal from a stroking diaphragm pressure transducer (Fig. 11-20) installed in a wireline anchor fastened to the deadline. Weight changes are transmitted hydraulically to the gauge located at the driller's position. The wireline anchor consists of a snubbing drum and lever arm. Drum is free to rotate, except as restrained by pressure transducer. Dead-line load is thereby converted to a pressure signal to actuate the weight indicator.

Rotation Speed Instruments

Generally speaking, penetration rate is directly related to weight on the bit and rotation speed. High penetration rates and short bit life will usually result from heavy weight and high rotation, while low penetration rates and long bit life will result from light weights and low speeds. Somewhere between is an optimum relationship between these variables for

Fig. 11-21. Rotary chain actuated torque gauge.

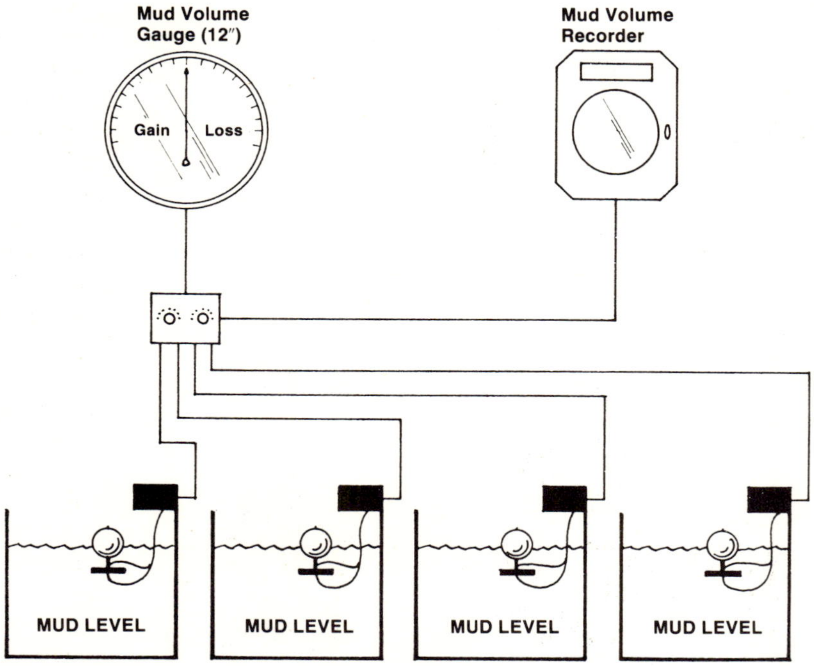

Fig. 11-23. Pit-mud level indicator and recorder.

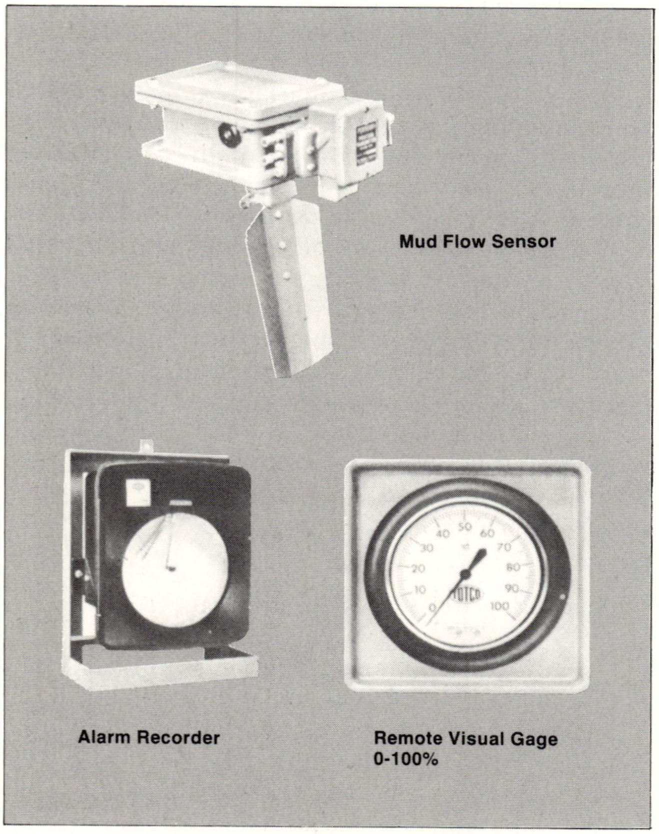

Fig. 11-22. Mud flow rate indicator and recorder.

every given condition.

Since rotation speed is an extremely important factor in this calculation it should not be dependent on guesswork by the driller. Simple tachometers are available for all rotary tables and rig designs. Many types are available to suit all rig designs. A cable-driven assembly is simple and dependable for short distances from takeoff to head. An electric type is the best solution for transmission through lengths greater than 15 or 20 ft. These may be either generator transmitter or separately energized selsyns.

The pneumatic rotary table tachometer provides for both indicating and recording the rotation speed because it is operated by air pressure which is also available to actuate a recorder pen.

Mud System Instruments

The mud system performs a number of important functions and its careful control and efficient operation is vital to obtaining effective results in rotary drilling. Besides its functions of maintaining the hole, holding back formation pressures, cooling the bit and cuttings, it supplies an additional and important function of cleaning chips from under the bit, thus improving penetration rate.

The mud pump pressure gauge is the primary mud system instrument. The gauge generally consists of a dial connected to a self-contained diaphragm unit placed in the discharge line of the pump. The gauge

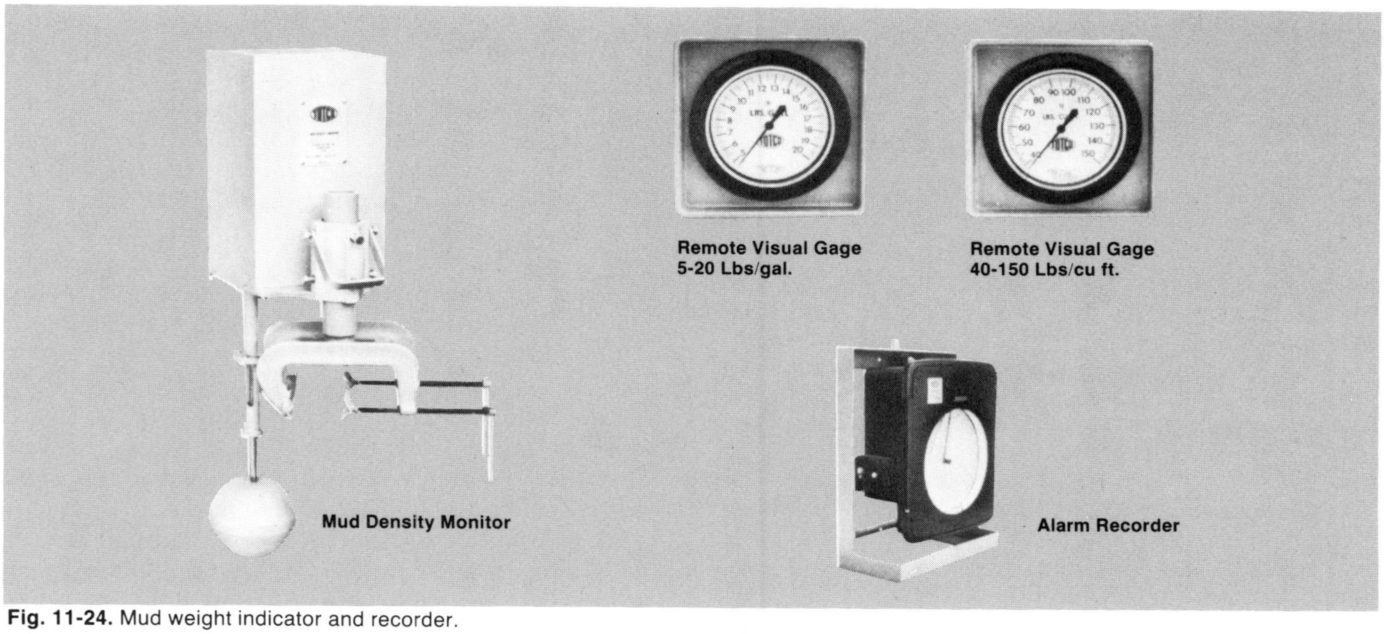

Fig. 11-24. Mud weight indicator and recorder.

Fig. 11-25. Rate of penetration recorder.

should be calibrated in large, legible figures, sensitive to small pressure changes, and damped to stabilize the pointer at the mean pressure level.

Since obtaining adequate mud volume is an important factor in supplying needs for cutting the hole, a pump speed and volume indicator can be used in conjunction with the pressure gauge to aid the driller in maintaining the pump's operation at maximum performance.

For most effective operation the pumps should be operated, without overloading, at maximum speed using maximum size liners. Any change in mud pressure should be detected immediately. A pressure rise generally will indicate a plugged bit. A pressure drop may be caused by a loss of pump speed, a cut pump valve, worn piston, scored liner, washout, lost circulation, etc.

In addition to the mud system instruments which are useful guides in actual hole making techniques, there are others that are equally important in safeguarding against possible blowouts.

Light weight, low viscosity drilling fluids are conducive to rapid drilling rates. It is natural, therefore, in the interest of economy, to drill with the lowest weight fluid that will reasonably control the downhole pore pressures likely to be encountered.

If the margin of safety is inadequate, a kick may occur. A kick is simply the influx of extraneous fluids, either gas, oil or salt water. Kicks, if undetected or uncontrolled even when detected, will surely lead to a blowout.

If a kick occurs while normal drilling is in progress there will be an immediate increase in rate of mud returns at the surface and subsequently a gain in pit mud volume. A mud flow indicator is most useful for early detection of a kick. Several instruments, both

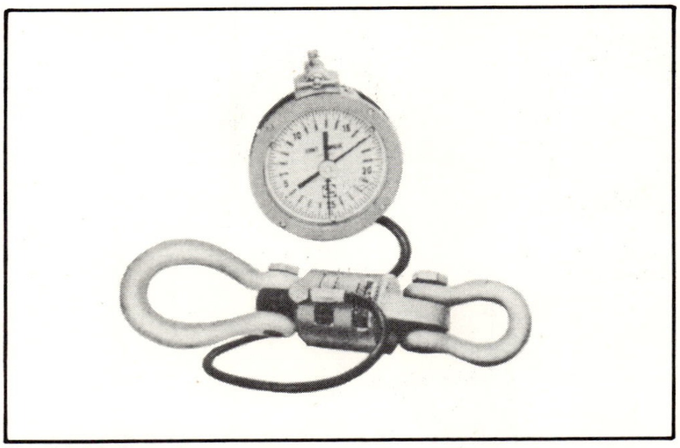

Fig. 11-26. Tong torque gauge.

indicating and recording, are now offered by various manufacturers (Fig. 11-22).

Pit level instruments, in addition to the mud flow indicator, will not only confirm that formation fluids have entered the hole, but also in what volume before the well is shut in. This information is valuable in the formulation of plans to circulate out the kick fluids while maintaining bottom-hole pressure slightly in excess of formation pressure (Fig. 11-23).

One other useful mud system instrument is the mud weight indicator and recorder. It is intended to provide the driller, on a continuing basis, with information vital to drilling operations and well control (Fig. 11-24).

Torque Gauges

The most valuable service of the torque gauge is to minimize the danger of twist-offs. Improvements in torque instruments during recent years, however, now afford the operator other valuable information in coring, reaming, milling, washing over, directional and deep hole drilling, as well as bit condition determination. (See Fig. 11-21)

Because the gauge differentiates between the work required to overcome hole friction and that done by the cutting tool, the modern torque gauge enables the driller to determine exactly what his cutting tool is doing regardless of hole friction and rotational speed.

An especially designed instrument for adaptation to chain driven rotaries consists of an idler sprocket mounted underneath and deflecting the tight side of the rotary chain. A load sensitive element is also connected to the lever arm to which the idler sprocket is attached. With any build-up of torque, increased tension is placed on the chain, forcing it down. In turn, this depresses the idler sprocket, increasing hydraulic pressure in the load-sensitive element which is transmitted to a dial at the driller's position.

A similar device is adaptable to shaft driven rotaries, consisting of a special spider with flexible spokes which drives the rotary clutch. The spokes deflect under load, controlling the output of a pneumatic gauging element. This particular equipment is not readily adapted to field installation but may be installed as original factory equipment.

The gauge is designed with a rotatable dial, permitting the driller to zero out friction on drill pipe while rotating off bottom. Net work being done by the bit can then be measured and bit condition can be detected from green to dull, locked cutters or out-of-gauge wear.

Auxiliary Instruments

These instruments do not primarily report information concerned with cutting the hole, but information related to maintaining the hole, equipment performance and safety.

Rate-of-penetration measuring and recording devices come in several types. One gives an intermittent count of time per increment of hole made; others record rate of bit penetration in feet vs unit time. These data are useful to geologists and petroleum engineers for study and evaluation (Fig. 11-25).

Make-up of drill collar and tool joints can be uniformly controlled by use of a tong torque gauge. Such instruments eliminate feel and guesswork in joint tonging. Proper make-up is especially critical on drill collars. This instrument is adaptable to casing tongs and can be of value in maintaining uniform and proper joint make-up (Fig. 11-26).

Chapter 12
Drill Rig Operation

Unusual drilling operations naturally are most newsworthy. Much publicity therefore is given to offshore operations, to extra deep wells and to wells drilled in arctic regions or in other hostile environments. The vast majority of wells drilled, however, are more routine, land-based operations of reasonable depth. This chapter is devoted to these more common drilling practices with occasional reference to some of the special situations.

Personnel On A Drilling Rig

The personnel on a drilling rig consists of a group of operators who have certain skills and functions which are uniform in the business of drilling a well. These operators are artisans of a sort who ply their trade as do carpenters, bricklayers, etc., not caring so much for whom they work as they do for the opportunity to work on a rig.

A drilling rig operates seven days a week, 24 hours a day, and a full staff must be on the job at all times so the rhythmical teamwork-type of operation may be accomplished without delays. Drilling rigs are generally staffed, for a tour (pronounced tower) or shift, by a crew of 4 or 5 men. Four-man crews are used on small shallow drilling rigs and 5-man crews on larger rigs.

Each 5-man drilling crew consists of a driller, a derrickman, a catheadman and two floormen (rotary helpers).

The Driller is in charge of his crew, and in charge of the rig's operations during his tour. He reports to a foreman who is usually called a toolpusher.

In his position as boss of the rig and the crew, the driller must be responsible for safe, efficient operations at all times. He always has a background of experience in performing the jobs of those men working under him; he knows their problems and what is required of them by having done this work himself. The relationship between the driller and his crew is such that the driller will often take his crew with him when he changes jobs.

Good judgment and alert attention to detail are probably the two prime requisites of a driller. They are necessary to safeguard the lives of his fellow workers and to protect the huge investment of his employer.

Specifically, it is the driller's function to operate the drawworks in the hoisting of the drilling string (drill pipe and tools) from the hole and in running it back into the hole, as well as maintaining the proper characteristics of the machinery while drilling with a bit on bottom.

The weight on the bit, the speed of the rotary table, the pressure and volume of the mud circulating fluid are some of these characteristics.

The Derrickman works high in the derrick or mast while making a trip; that is, when the pipe is being hoisted from the hole or run back to bottom. Stationed on the derrick platform known as the monkey board, which is 80 ft above the derrick floor, it is his function among others to seize the upper end of the 90-ft stand of pipe as it comes out of the hole and pull it over into the rack. He also has other duties.

He also tests the mud to maintain its properties in accordance with the program specified for the well. It is his responsibility to assure the proper functioning of the mud pumps; to check their operation frequently; to replace worn parts, and to provide the pump's efficient operation at all times.

The Catheadman. The term catheadman is applied to the third ranking member of the drilling crew. Some contractors term this man a motorman. He is usually more experienced in all phases of the work than the rotary helpers.

His job is to operate the manual catheads on the drawworks whenever they are used and to supervise service of the motors and engines for uninterrupted operation. It is he who adds oil when needed, supervises the oil changes, watches the radiator water and generally performs the function of a serviceman with respect to the motors.

Rotary helpers. The two rotary helpers are the tong operators during a trip. It is their job to see that the tongs are placed properly on the drilling string, to rack the pipe and to be ready for the next similar operation in rhythm with the machinery as operated by the driller. When the bit is on bottom drilling there are 101 miscellaneous duties for the rotary helper to perform.

Maintaining the rig's housekeeping is a large order in itself. They straighten up the floor; pick up numerous pieces of equipment; clean and paint the machinery; help the motorman with the engines; help the derrickman with the mud treatment, etc.

It was stated large rigs have 5-man crews and smaller rigs 4-man crews. The usual difference is that there is one less rotary helper or floorman on a 4-man crew rig.

The foreman or toolpusher is actually the manager of his branch of his company's operation. During the drilling of the well he must be available at any time day or night for consultation with the drillers. He is The Management of the drilling company insofar as his crews are concerned.

In today's fast moving contract drilling economy, the toolpusher must have at least a smattering of knowledge on labor laws, personnel relationships, chemistry (for mud control), hydraulics (for circulating fluid), internal combustion engines, etc.

He is the key man in any drilling contractor's organization. He usually has authority to spend money to keep his rig in operation, cooperating with his supervisor on any items of major importance. Since he does not drill a foot of hole himself, it could be said that his prime function is to assist the drillers and their crews to drill the hole safely, economically and promptly.

The foreman makes his headquarters in an office at the rig known as the dog house, where he must keep accurate records of daily drilling performance, purchases of materials, accident reports, and other paperwork which must go daily to the main office for coordinating with other company accounting and records.

Here he also interviews salesmen from the supply houses, consults with the production superintendent and other personnel from the producing company and is available by telephone to the outside world which may be many miles away. When he leaves the rig, he usually keeps in touch by mobile telephone.

The foreman's responsibilities cover every operation of the rig including clearing and staking out the drill site, moving in the rig, rigging up and tearing it down when the job is finished.

Drilling superintendent and assistant. Depending upon the size of a contracting organization, there may be an assistant drilling superintendent as well as a drilling superintendent. These men primarily aid the toolpushers.

The drilling superintendent usually has authority to authorize expenditure of larger sums of money than do the toolpushers. Usually there is a limit placed by top management on the amounts that any foreman of whatever rank may spend without top management's approval.

Preparation For Drilling

Prior to any activity at the well site, its location, on paper at least, will have been chosen, based on geologic considerations and the spacing pattern if one has been established. A surveyor will then stake the location as chosen. If he finds that the topography is unusually bad or that physical improvements are likely to interfere with drilling operations, he also will stake an alternate and more suitable location nearby.

In any event, some dirt work will be required, such as leveling for the substructure and grading for the mud tanks. Also the reserve mud pit must be dug. The rig's foundation usually consists of a mat made of 3-in. × 12-in. planks. An access road from a public highway into the well site also must be prepared, including bridges or culverts if required.

Moving in/ rigging up. The trend toward deeper wells has been accompanied by larger and heavier drilling rigs. It is not uncommon for a complete land-based outfit to weigh hundreds of tons, with individual items, such as drawworks or slush pumps, weighing 25 or 30 tons. Transportation equipment has kept pace, however, so it is possible to move these large units readily.

It is general practice to move in most of the rig items and to offload them at strategic points, with reference to the well site, before the crews are called out. Assembly of the rig requires the services of both manpower and heavy gin (hoisting) trucks.

Much initial work is done during daylight hours only. When the rigging up nears completion the crews break tour and start working around the clock. Moving in and rigging up involve a considerable expense.

Where conditions permit, a considerable savings in both time and money can be realized by skidding the rig from one well location to the next to be drilled. When skidding a rig it is possible to leave blocks strung up with much of the machinery and equipment left in place. Skidding rigs is practical only on relatively flat terrain where a continuous development program is in progress.

An adaptation of the skidding technique is being practiced in the deserts of Africa and the Mid-East where either rubber tire or athey-type carriers, in combination with skid beams, are used to transport an entire rig, weighing up to 200 tons, quickly and over long distances if need be. As in the case of sled-type skidding, tractors provide the pull.

One other means being used to reduce moving in and rigging up expense is the truck-mounted drilling rig. Such equipment is especially adaptable to relatively shallow, fast drilling situations.

Preliminary Drilling Operations

When the initial rigging up is completed, crews proceed with the first operation: drilling mouse hole and rat hole. The mouse hole is located close to the rotary table on the vee door side and deep enough to hold one length of drill pipe.

The rat hole is located somewhat farther away and to the driller's right of the mouse hole. It must be drilled deep enough to hold most of the kelly and far enough away from the rotary table to prevent interference wih hoisting and other rig floor operations.

The mouse hole and rat hole are usually drilled, using the conventional hydraulic rotary drilling system. In some areas, however, the surface soils are soft enough to permit washing down the hole without rotating the bit. In other instances, these auxiliary holes(also the hole for conductor pipe)are drilled with a dry hole digger before moving in the drilling rig.

The final preliminary drilling operation concerns drilling a hole for and setting conductor casing. The primary purposes of conductor casing are to provide a suitable base for wellhead equipment (to be installed later) and, with the aid of a flow nipple, to conduct mud returns to a height just below floor level so the mud can flow by gravity to the shale shaker.

The depth to which the conductor pipe must be set depends on the reliability of surface formations in the area. In some areas a single length of casing, properly cemented, is adequate, while in other areas 400-ft or more is required.

Handling mud returns while drilling mouse hole, rat hole and the hole for conductor casing presents a problem that has so many unsophisticated solutions they need not be discussed here.

Routine Drilling Operations

The hole diameter and the depth intervals for each segment of the well to be drilled are determined by the casing plan. The conductor casing must provide a suitable base, or foundation, for the support of the casing strings to be run.

The surface casing must be of sufficient length to protect all potable water-bearing formations from possible pollution. The intermediate casing, if required, is a safety measure to protect or preserve a large section of the hole drilled and to avoid possible fracturing of the upper and weaker formations in case high density drilling fluids are required for pressure control.

Production casing is run after reaching total depth and then only if production zones are indicated by electric logs, cores and/or formation tests.

The physical performance of the drilling crew is the repetitious act of drilling the kelly down; hoisting the drill string until the kelly-to-drill pipe connection is a foot or more above the rotary table; setting the drill pipe slips in the rotary table bowl; shutting down the mud pump; breaking the joint with tongs before spinning out the connection with rotary table; making up the kelly tightly in the length of drill pipe in the mouse hole; hoisting the assembly high enough to permit stabbing the pin of the new length of drill pipe into the box of the drill string; spinning in the connection and tonging it tight.

The pump then is started, the drill string hoisted sufficiently to permit removal of drill pipe slips, the drill string is carefully, slowly lowered toward bottom until the kelly drive bushing drops into place and rotation of the drill string is started.

When the indicator shows weight on the bit, normal rotational speed is resumed and after cavings, if any, are drilled up and the bit reaches bottom, normal weight on bit is applied and drilling new formation is resumed. Another length of drill pipe is placed in the mouse hole and, while drilling the kelly down, odd jobs as directed by the driller are performed by the other crewmen.

This cycle of activity is repeated time after time until the bit is dulled and it becomes necessary to pull out of the hole to change bits. Each time during the cycles of activity described above, when the bit is on bottom making hole, the driller uses the brake of drawworks to feed out drilling line at a rate that maintains the desired weight on bit.

The driller, in addition to watching the weight indicator, must also observe and control all other drilling variables such as rotary rpm, pump strokes and resulting mud circulation.

Tripping For Bit Change

One of the critical decisions a driller must make is when to pull a dulled bit. If a bit is pulled "green" the useful life of the bit is partially sacrificed. If it is allowed to remain on bottom too long, at a low rate of penetration, rig time is wasted. Worse still is the possibility of so completely wearing out the bit that cones are lost in the hole and a fishing job results.

There are no specific rules to guide a novice in making a decision of when to pull a dulled bit. It is a matter of judgment by the driller, gained from his years of experience on a drilling rig.

When the decision has been made to come out of the hole, the driller sets the drawworks brake and slows down the rotary when the bit drills off. He continues to maintain full mud circulation for about 10 min or more to assure that all drill cuttings are moved up the hole far enough so they will not settle back to bottom while tripping.

Drill string rotation and mud circulation are stopped as the kelly is being withdrawn to a point well above the rotary table. The kelly is then broken off and set in the rat hole to remain there while tripping.

As in routine drilling, pulling drill pipe requires repetitious tasks of the drilling crew, each with his

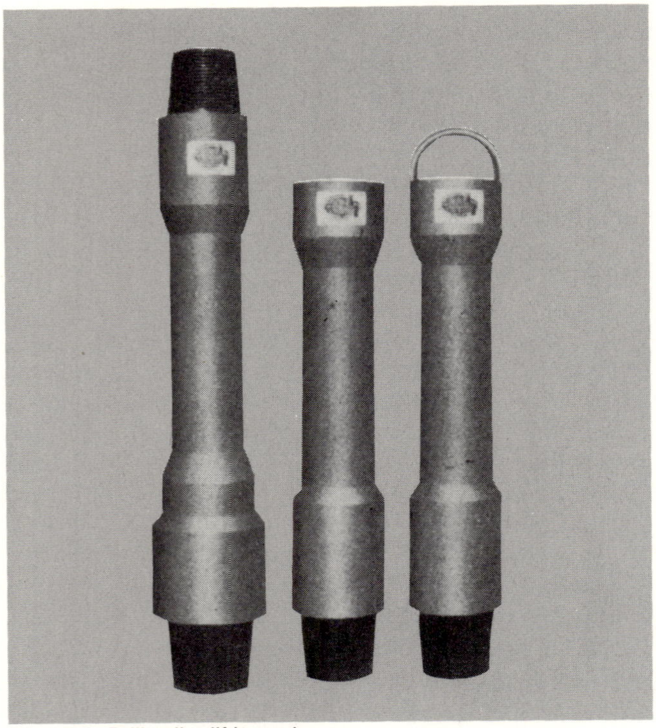

Fig. 12-1. Drill collar lifting subs.

special duty. The driller operates the hoisting machinery; floormen handle the drill pipe slips and tongs. They also set back the drill pipe stands to the racking platform, while the derrickman handles the upper end of the stand, unlatches the elevators, and pulls the drill pipe back to the proper position in the fingerboard. The catheadman operates the breakout cathead and assists the driller as directed.

This rhythmic procedure continues time after time until the drill collar string reaches the surface. Due to the greater weight of drill collars, their larger diameter and their flush OD configuration, their handling routine is altered somewhat.

Drill pipe slips no longer are self-sufficient in support of the drill collar string. A safety clamp (Fig. 11-10) above the slips is needed. After all drill pipe has been set back, lifting subs (Fig. 12-1) will be required for handling each stand of drill collars to provide space for the elevators.

When the last drill collar clears the rotary table, the bit breaker (Fig. 11-11) is placed in the rotary drive bushing square; the table is locked; with the bit lowered into the bit breaker, the bit connection is broken by the jerk line pull on tongs latched around the drill collar.

The old bit is unscrewed manually and the new bit is substituted. The new bit, in bit breaker, is tonged tight and made ready to run back to the bottom (tripping in) thus completing a round trip.

Tripping in, like tripping out, involves repetitious activities of the crew except reversed. With the fresh bit on bottom, routine drilling can be resumed.

Running and Cementing Casing

Short casing strings, such as conductor pipe and surface casing, usually are run by the regular drilling crews. The conductor pipe—being a very short string—can be run by a single crew with possibly a few extra hands.

It's not uncommon for short pieces of conductor pipe, used primarily as a flow nipple, to be sealed and anchored by dumping ready-mixed concrete in annulus between pipe and dry hole. Longer conductors are usually cemented, using the conventional two-plug method.

Additional manpower needed for running the longer surface casing strings is provided by the crews doubling over. Cementing surface casing is always done by a special third party service company. The crews get back on regular schedule after shut down to allow time for the cement to set, or harden, sufficiently to resume drilling.

Both intermediate and production casing string require extra help and longer periods of time to complete the job. For these reasons casing crews are called in, if available in the area. Casing crews are especially trained by experience to perform all tasks relative to running casing and more or less take charge when they come on the job.

They rig up special equipment pretty much to their own preference. The driller on tour continues in control of the drawworks but the other crewmen are more or less subservient to the casing crew foreman.

Intermediate and production casing strings are always cemented by a third party service company using special mounted pumps, tanks, gauges and monitoring devices.

Rig Lubrication and Maintenance

The service life of rig machinery is directly proportional to maintenance practices. Drilling equipment operating under careless repair and lubrication methods will not compete successfully.

Rig operators are continually striving for a greater percentage of on-bottom time during the course of drilling the well, as well as increased penetration rates and quicker rig up and tear down operations. Today's keen competition, higher operating expense and increased investment have made it necessary for the rotary driller to push his effectiveness to an all time high.

Just what is demanded of the equipment? Jet drilling, longer drill collar strings, greater depths — to mention a few—produce greater loads on all machinery involved.

A 650-hp rig operating in Oklahoma was considered for 10,000-ft. drilling years ago but it would be completely inadequate for this same drilling today. In 1950, 75 days was considered good time to drill to 10,000 feet, whereas only 35 days are needed at the present. Minimum power requirement is 850-hp.

Drill collar strings of 20,000 lb. were considered adequate 25 years ago; collar strings of 45,000 lb. plus are now frequently used. Hook loads are approximately 25% greater than in the past. With these conditions of heavy loading, every corrective measure must be undertaken to insure satisfactory service life from rig machinery.

Rig Maintenance Vs Safety

From the safety standpoint alone, one can see the effect of rig maintenance. No operator has been able to run a rig in poor condition for any extended length of time without having accidents that could be directly attributed to some poor maintenance practice.

Rig Maintenance Vs Repair Time

In all but a few extreme cases, the rig with a high repair down time percentage is one where a poor maintenance program is in effect. To illustrate this, rig efficiency reports from a light (425 hp) rig are shown.

On well No. 1, drilling time was a high 71.53% of total time available for making hole. On well No. 2 only 57.61% of available time was spent on bottom. Immediately the analyst examines the record to find out why the large difference exists.

On well No. 1, repairs totaled 30.75 hr. or 7.3%: on well No. 2, repairs were 145 hr. or 28.63%. Engine and miscellaneous repairs (rotary table, crown block, traveling block, etc.) were 26.47%.

On this particular well (No. 2) the master clutch was not working properly due to faulty adjustment, and during a trip, the driller was unable to disengage the clutch, and hit the crown. The second major repair was due to improper maintenance and overloading drawworks engine No. 1. This engine had to be shopped and completely overhauled. Both mechanical failures are attributed to improper rig supervision and maintenance.

ACCIDENT	END RESULT	IMPROPER MAINTENANCE PRACTICE
Pipe being pulled in V-door — catline parted.	Pipe struck employee on shoulder; lost time accident.	Faulty catline allowed to remain in service.
Employee slipped, fell into pump belts.	Hand caught in pulley, mangled left hand.	Pump guard not properly in place.
Cathead clutch engaged swing tongs toward drawworks.	Pinned floorman to drawworks, broke both legs above knees.	Cathead clutch in poor working condition.
Roughneck slipped on oily engine floor plates.	Fell off of platform; broke left ankle.	Oil allowed to accumulate on deck plates.

WELL NO. 1	Time/hr	Per Cent
1. Total hours available for drilling	421.50	100.
2. Drilling time	301.50	71.53
3. Tripping time	54.50	12.93
4. Straight hole & lub.	34.25	8.12
5. Ream and wash	.50	.12
6. Repairs:	30.75	7.30
(a) Drawworks	8.25	1.96
(b) Pumps	2.75	.64
(c) Engines	1.	.24
(d) Drill collars & drill string	4.	.95
(e) Compound unit	—	.95
(f) Miscellaneous	14.75	3.50

WELL NO. 2		
1. Total hours available for drilling	509.	100.
2. Drilling time	293.25	57.61
3. Tripping time	44.25	8.69
4. Straight hole & Lub.	23.25	4.57
5. Ream and wash	2.50	.49
6. Repairs:	145.75	28.63
(a) Drawworks	3.50	.69
(b) Pumps	2.25	.44
(c) Engines	65.25	12.82
(d) Drill collars and drill string	1.25	.25
(e) Compound unit	4.	.78
(f) Miscellaneous	69.50	13.65

Rig Maintenance Vs Profit

Referring again to the time summary of the above wells described — the effect of this type service on profit income can be determined by comparing operating cost and rig time. The average daily operating cost for this rig on a group of 5 wells was $1168.00 per day or $48.66 per operating hour.

The effect of down time caused by faulty rig maintenance can prove disastrous to the rig runner in a short time. Note the comparison of these two 5000-ft wells. It will take two wells drilled as in No. 1 to offset the loss incurred in the drilling of No. 2.

The same emphasis should be placed on rig maintenance as is placed on making hole. Hole must be drilled before profit can be made, but if the equipment is destroyed in doing so, no profit whatsoever

	Well No. 1	Well No. 2
Operating days	26	30
Total cost of well	$30,500.00	$35,200.00
Repair cost in hours down time	1,492.00	7,085.00
Profit or loss on operation	+2,840.00	−4,245.00

will be realized. Therefore hole making and proper rig maintenance go hand in hand — if profit is to be expected.

With prices for drilling usually providing only a narrow margin between profit and loss, only those rigs that can make greater amounts of hole in a shorter length of time should be used.

Balanced Lubricating and Maintenance Program

The backbone of any good maintenance and operational program on any drilling rig is training drilling crews in proper maintenance methods. This training and education must be provided by the rig supervisors. Time should be allowed on the job so the crews can be given proper instructions. A man cannot be expected to do a good job if he isn't given the opportunity to learn how. Some methods that can increase rig efficiency are:

1. Discussing proper lubrication of rig parts in meeting on rigs.
2. Studying maintenance manuals to increase knowledge of proper lubricants and maintenance practices.
3. Discussing equipment and maintenance with factory representatives.
4. Keeping continuous records of oil changes, chain repairs, pump repairs, wirelines, engine overhauls, etc., to provide a check on equipment service.
5. Inspecting equipment periodically to determine amount of wear on parts. Many times minor repairs can save a major breakdown.
6. Providing rig operators the necessary equipment, oils and repair parts, so that the rig can be kept in proper working condition.

Key to Correct Maintenance Practice

It is not the purpose of this book to set a hard and fast rule to follow in obtaining satisfactory rig service but to show the effect of poor rig maintenance practice on profits; also to help the rig supervisor evaluate his past maintenance procedure and how it can be improved.

The toolpusher should plan maintenance practice with his drillers and crew; a practice that will be effective, time-saving and complete, with a minimum amount of planned down time.

In many cases, the supervisor only talks about maintenance and does not see that it is carried out. Equipment operating 24 hours a day must have proper care; one failure on the part of a crew member to do his job can produce sizable loss.

Since modern machinery is enclosed in suitable housing, the rig operator needs only to keep the proper clean oil in service to provide proper chain and bearing lubrication. The oil should be checked daily in order that foreign material can be discovered.

Too often the supervisor is inclined to resist down time to make necessary oil changes, thus continuing the use of unfit lubrication fluids. Once foreign matter is imbedded in precision bearings, etc., the damage is already done and clean oil then is unable to insure satisfactory service.

Filters and lubricating systems should be checked daily along with oil pumps to be sure they are operating satisfactorily.

It might be well to list common bearing failures so the operator can adjust his maintenance practices to offset them:

- Misalignment
- Inadequate or unsuitable lubricating fluid
- Faulty installation; defective sealing
- Defective bearing material

Once it is realized that a faulty bearing exists, it should be replaced as soon as possible, before the shaft is out of alignment, causing additional wear on all parts. One should be thorough in planning lubrication and maintenance operations, so it may be well to include these practices:

1. Follow rig and equipment manufacturers' recommendations on types of oils, greases, filters, and certified repair parts to use. Examine lubricants before they are used, thus insuring against adding the wrong lubricant.
2. Provide time daily for drilling crews to check all lubricating systems and oils for cleanliness.
3. Examine periodically valves and gear power end of mud pumps; check mud valves and circulating lines; replace worn parts before the start of a well to eliminate additional downtime.
4. Keep tool houses, engines, all machinery as clean as possible; don't allow waste oil and grease to accumulate (thus reducing fire hazards).
5. Operate blowout preventers daily while drilling; at the completion of each well, tear down all equipment and clean thoroughly. Drilling mud is highly corrosive. Working parts should be oiled so they will be ready for the next well.
6. Grease all suspension equipment (blocks, hook, swivel) daily. This is a good practice since these parts are exposed to the weather at all times.
7. Keep spare rig parts in clean tool houses or boxes so they can be available immediately when needed.
8. Examine derrick and rig substructures at regular intervals to discover loose bolts and pins. Check for weld fractures.
9. In most cases, standby equipment, such as electric plants, air compressors and mud pumps are

used only in emergency. Therefore they are left idle for extended lengths of time. This equipment should be started and run for a short time daily to insure proper condition when needed.

10. Maintain a regulated engine maintenance program.
    (a) Keep hourly service record on engines.
    (b) Change oil and filters at designated intervals. This will vary with type of engine and geographic location.
    (c) Have mechanic check engines at the end of each well or during the drilling of deep wells. Check for bearing and shaft wear.
    (d) Examine cooling system for corrosion and rust and analyze water to be used in cooling system, using chemicals if necessary to treat water.

Rig crews usually can be judged by type of housekeeping practiced. A well maintained rig will have crews that are conscientious about the machinery and are doing their utmost to complete a successful drilling program.

Summary

A sensible rig maintenance program cannot exist without the proper training. Drilling crews must be educated by supervisors before costly down time can be eliminated. They must be taught by specific examples the importance of proper lubrication and maintenance practices. They must be able to see the harmony that can exist between a well planned practice and the hole making operations.

The importance of keeping the rig in proper working order 24 hours a day cannot be over emphasized. Down time caused by improper repair practices reduces profit.

A clean, well-kept rig where lubrication and rig repairs are carried out properly and timely will produce a profit for the contractor at prices which cannot be matched by the haphazard or careless rig owner.

Formation Evaluation —Wireline Methods

In cable tool drilling it was possible to learn something about the nature of the formations being drilled by an examination of drill cuttings recovered with each bailing operation following a drilling run of a few feet.

Presence of oil or gas was apparent since there was seldom present any significant hydrostatic pressure to prevent formation fluids from entering the hole. When hydrocarbons were encountered, gas soon would appear at the surface and subsequent bailing would determine whether oil had been encountered.

With the hydraulic rotary drilling system, formation evaluation is harder. Drill cuttings not only are contaminated with cavings but are very slowly circulated to the surface. The only immediate indication of a change in formation is a change in the drilling rate.

In one sense the driller is operating blindfolded so far as oil or gas shows are concerned. If, however, microscopic examination reveals an oil saturated cutting, it is possible by coring operations and core analysis to estimate qualitatively whether oil and gas is present in the interval cored. Drill stem tests can determine the potential possibility of commercial production. Electrical logging and sidewall sampling have practically displaced coring and drill stem testing practices in deeper operations.

Electric logging was developed by the Schlumberger brothers (in France) early in this century and was put into general use in the U.S. in the early 1930s. This and other wireline logging and sampling devices have since come into widespread use.

All such operations fall in the category of third party services. It is the responsibility of rig operators to prepare the well and to assist in whatever way possible to expedite the service.

Fishing and Freeing Stuck Drill Pipe

Fishing is a word generally used to denote the operation of recovering or attempting to recover lost material in a borehole. Its usage began in the very early days of drilling when objects were recovered from a drilled hole by tying a home-made hook or spear on the end of the drilling rope and dangling the hook up and down by a pole or beam.

Unlike its sporting counterpart, fishing is the nemesis of drilling and, to a lesser degree, producing industries. Even with improved equipment, and specialized tools and services, fishing accounts for approximately 3% of the total costs of drilling. The yearly expenditure of millions of dollars in this phase of drilling has led to a highly specialized segment of an industry within an industry. To explain and discuss all the tools and techniques as applied to the hundreds of varying types of fishing operations would require a large volume; therefore, this discussion of the fundamentals must be limited to the most common problems and the generally accepted methods of solution. It must be remembered that today's accepted methods will be tomorrow's obsolete history.

Most tools mentioned are obtainable only on a rental basis, which includes the services of an expert. Quite often, however, a driller finds himself in remote areas or foreign operations with tools but no expert advice. Even with the aid of an expert, it is important for the rig operator to have a working

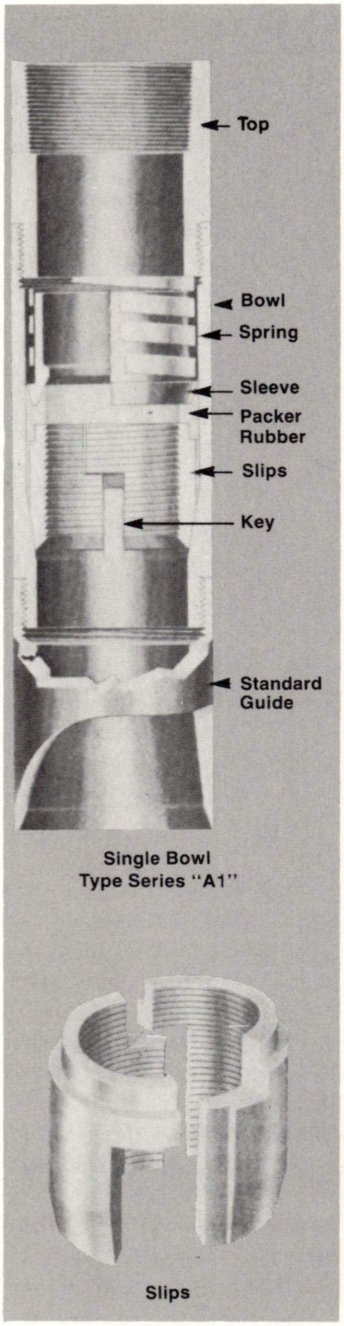

Fig. 12-2a. Single bowl overshot, Slip-type.

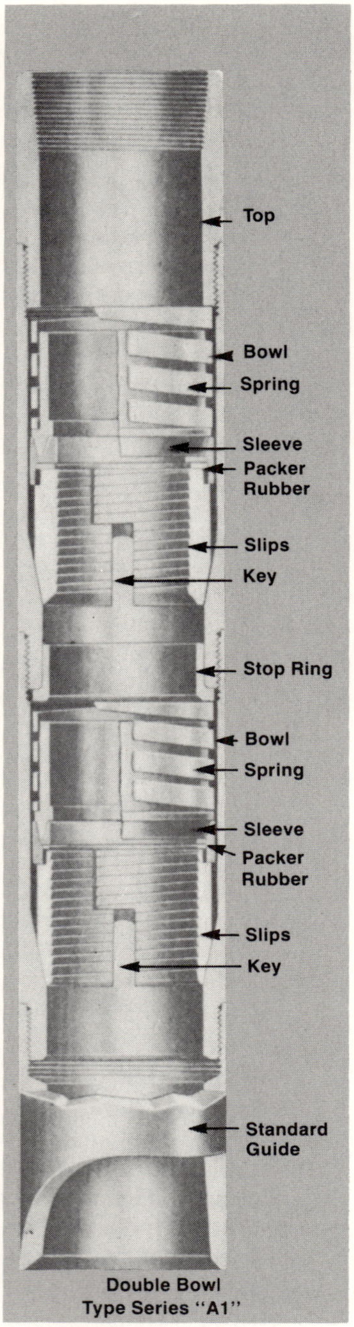

Fig. 12-2b. Double bowl overshort, Slip-type.

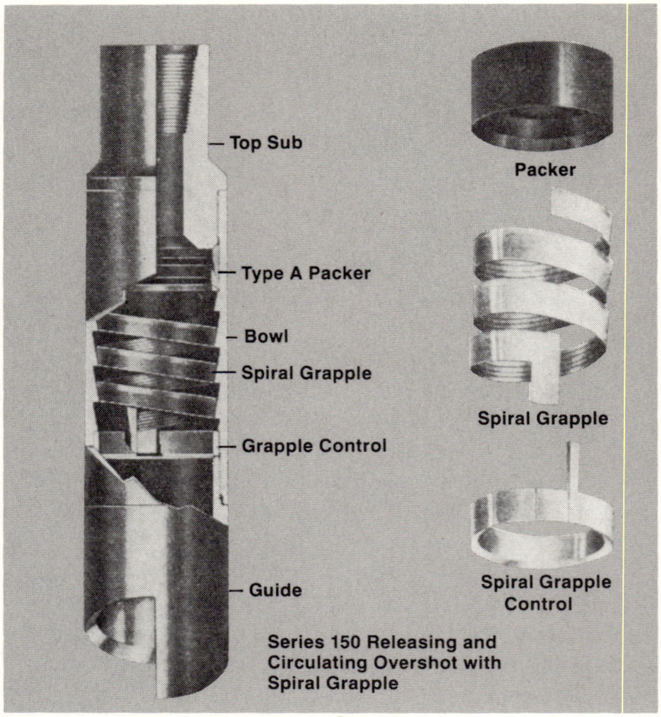

Fig. 12-2c. Single bowl overshot, Grapple-type.

knowledge of the tools that assist in coordinating the job to a faster conclusion.

For reference purposes an outline of discussions are:

1. Twist off or drill pipe failure. Requirements:
   a. Overshot
   b. Overshot and mill
   c. Overshot and crooked joint of drill pipe
   d. Overshot with knuckle joint and wall hook
   e. Taper tap and safety joint
   f. Underreamer
2. Recovery of rock bits, cones or small pieces of iron
3. Freeing stuck drill pipe. Causes:
   a. Stuck while on or near bottom
   b. Stuck lowering pipe in hole
   c. Stuck pulling pipe out of hole

1. *"Twist-Off" or Drill Pipe Failure*

The most frequent fishing job is for drill pipe as a result of drill pipe or tool joint failure. This is commonly called a "twist off" and usually occurs while drilling. The drill pipe loses its ability to stand the normal torque by excessive wear or by fatigue cracks in the metal caused from bending or tension. Pipe does not normally crystalize as commonly expressed; it either wears, corrodes or cracks.

(a) *Use of Overshot:* An overshot is the most accepted tool for catching drill pipe or a tool joint. There are several manufacturers supplying overshots but fundamentally they all operate the same. Some use slips, some grapples for the "wedge" in the bowl, but basically they are the same (Fig. 12-2).

An overshot should be standard equipment on all rigs and should be maintained in good condition for immediate use with the proper size slips and pack-off rubbers to catch any size pipe, tool joint or drill collar used in the hole.

Before dressing or assembling the overshot, it is important to examine carefully the retrieved part of the broken pipe and determine whether the top of the fish in the hole is worn, burred or split and the distance it may be from the tool joint. This inspection will assist in learning the size of the slips and pack-off rubber and whether a mill is necessary or not.

It is good practice always to use a mill control guide just below the slips to remove burrs from the top of the fish. It is important to obtain accurate measurements of the overshot and especially the distance from the

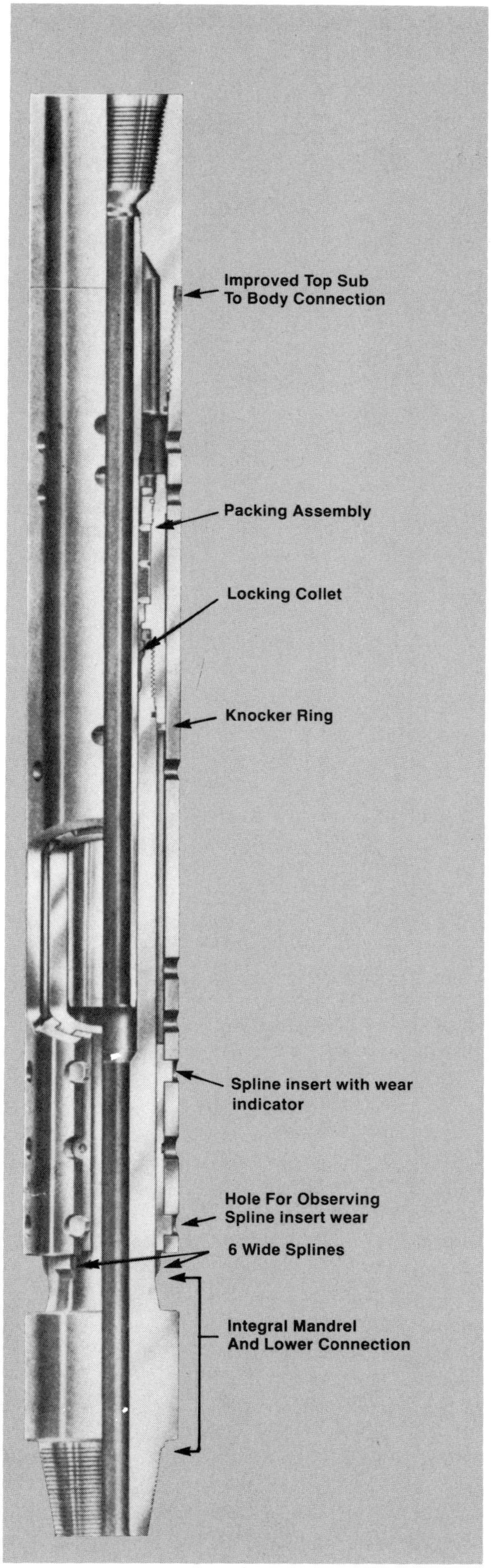

Fig. 12-3a. Bumper sub, Baash-Ross.

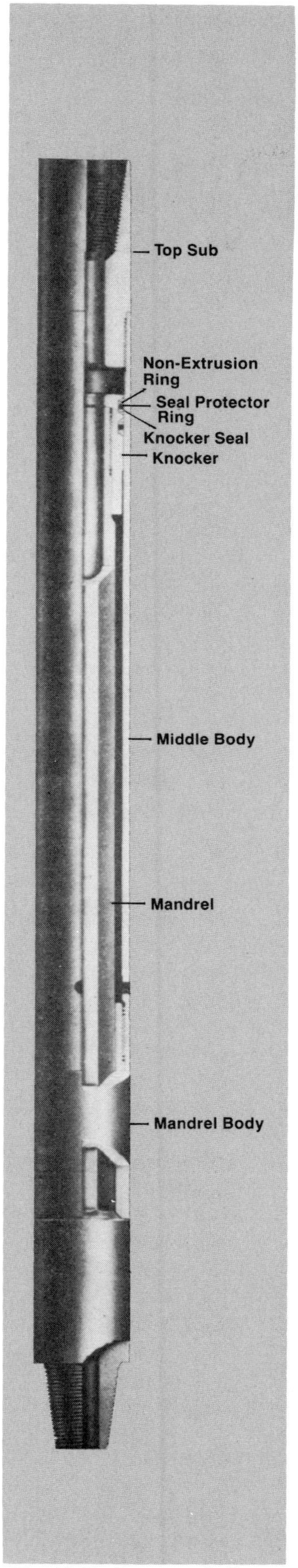

Fig. 12-3b. Bumper sub, Bowen.

Fig. 12-4a. Milling shoe, common type.

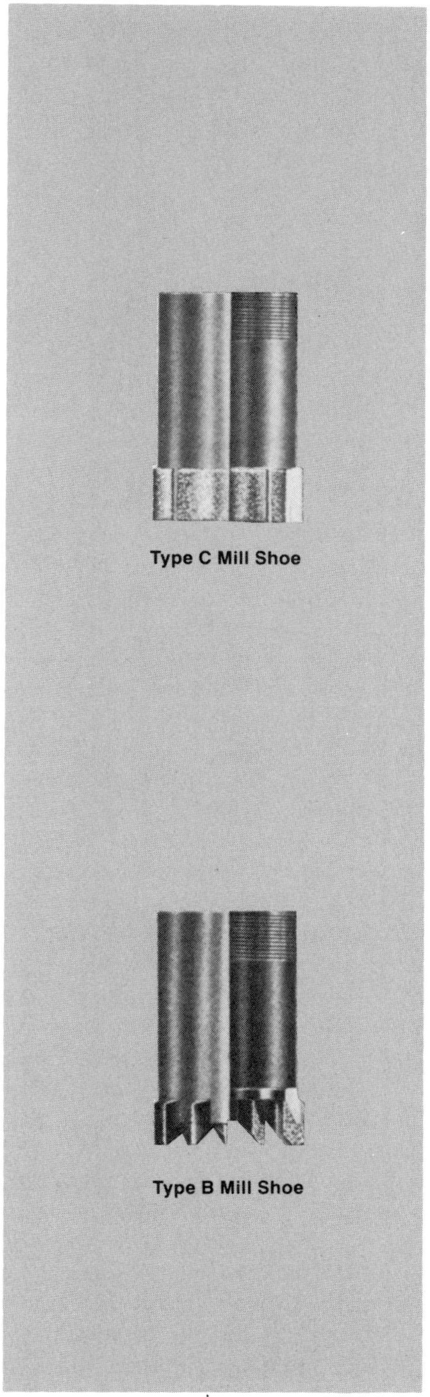

Fig. 12-4b. Milling shoe, hard-faced type.

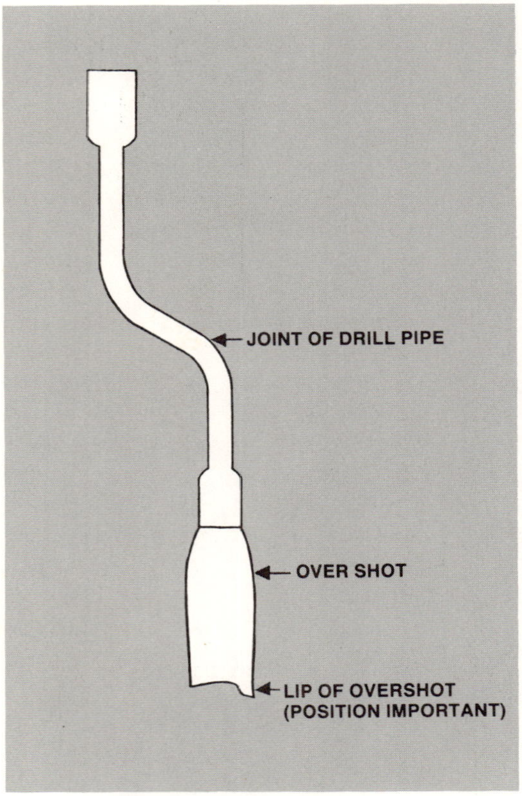

Fig. 12-5. Crooked joint of drill pipe.

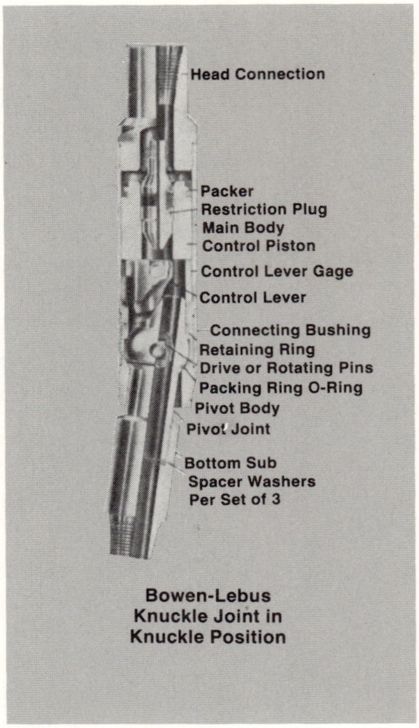

Fig. 12-6. Knuckle joint.

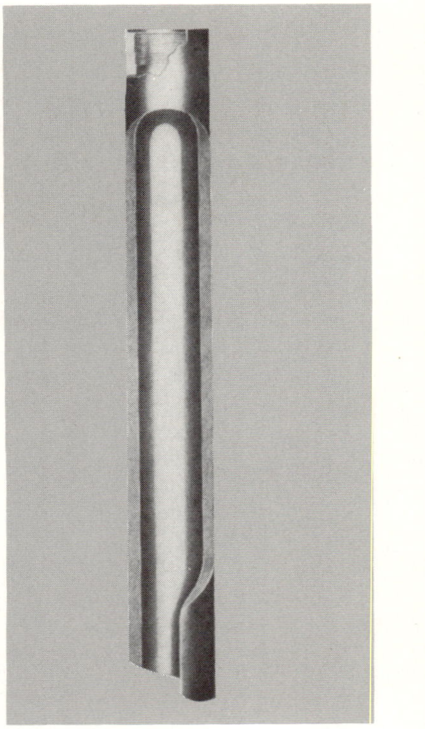

Fig. 12-7. Wall hook.

bottom of the skirt to the slips and the distance of the fish can enter the overshot after passing through the slips.

This information will be helpful while fishing. More important is finding the exact depth of the top of the fish and accurate measurements of the fishing string so the fisherman will know when he has reached the exact spot when lowering the fishing tool in the hole. Many fishing jobs have been delayed or failed because the fisherman was unknowingly probing above or below the top of the fish. If the top of the fish cannot be found and there is doubt as to its location, then an electro-log should be run to find the top.

After the overshot has been lowered to near the top of the fish, mud circulation should be obtained long enough to clean the tool. It is not good practice to circulate more than a few minutes. It is good practice to have the mud and the hole above the fish in good condition before lowering the tool to the fish.

Undue difficulties may arise if the fishing string is likely to stick at any time during the fishing operation. Ordinarily the mud is not circulated while actually fishing with an overshot. The chances are better that the pack-off rubber will hold after the fish is caught if the pump is not running while the fish enters the tool.

Actual catching of the fish may simply be to lower the overshot over the fish to the point where the top of the fish has passed through the slips and supports part of the fishing string. The fishing string is then raised, the slips or grapples wedge between the fish and the bowl, thus catching a firm hold on the fish. Circulation then should be started through the overshot and on through the entire fish to properly condition the hole before pulling out.

The catch may not be simple and may require hours of patience and careful raising and lowering, turning and feeling. This is where accurate measurements count in determining whether the overshot is passing by the fish, whether the bottom of the overshot is hitting the top of the fish, whether the fish will not enter the slips or whether the slips are not holding.

During this operation, patience, imagination and carefulness are the watch words. Irreparable damage may be done to the top of the fish by undue roughness. After the fish is caught and freed, the rotary table should be locked and the fish pulled without rotation to prevent dropping.

If the fish is stuck, all standard overshots will release their grip by having the fishing string weight applied and a turn to the right. When fishing at less than 2,000 ft, it is good practice to run a bumper sub above the overshot.

This allows a blow as well as weight to be applied in releasing the slips. A bumper sub is shown in Fig. 12-3.

b. Use of Overshot and Mill. An overshot mill (Fig. 12-4) is a cutting tool designed to run on the bottom of the overshot to dress the top of the fish to the proper diameter to allow passage into the overshot slips. Its cutting action is obtained by lowering the tool over

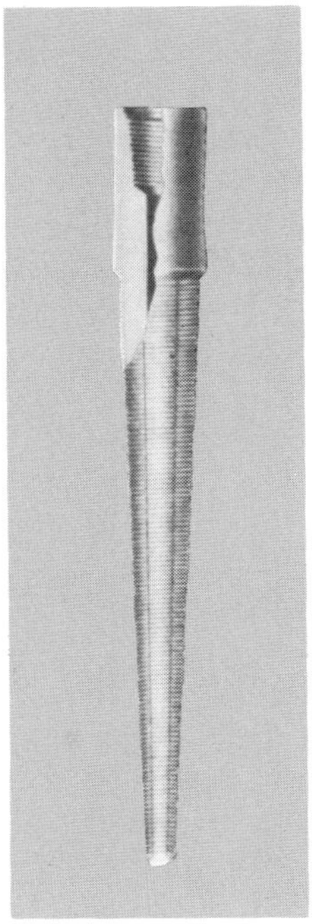

Fig. 12-8. Taper tap.

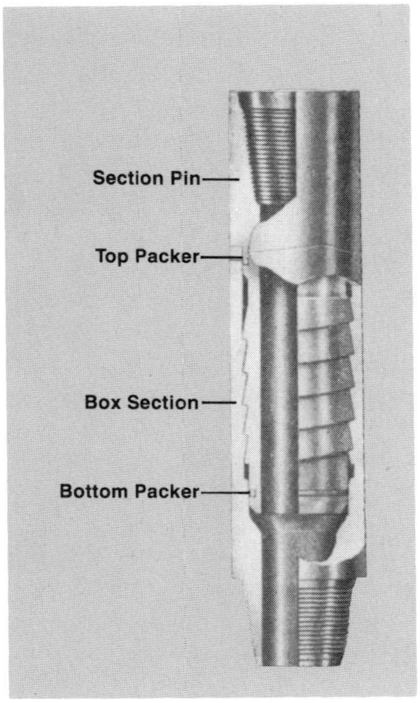

Fig. 12-9. Drill pipe safety joint.

top of the fish and then slowly rotating. Its inside diameter and cutting edges are tapered and will progressively file the outside diameter of the fish to dimension of overshot slips.

c. Use of Overshot and Crooked Joint of Drill Pipe. Quite often it is impossible to find the top of a drill pipe fish with an overshot. This is caused by an oversize hole allowing the fish to fall to one side as the overshot passes by. The first and cheapest operation is to put a double bend in the first joint of drill pipe above the overshot (Fig. 12-5).

The bent joint will force the lip of the overshot to the outside diameter of the hole. By raising and lowering the overshot with a few degrees turn each time, chances of finding the top and either going over it or changing its position to a more favorable one are good. Point of the overshot will naturally plow against the wall of the hole while running in and thus requires more care.

d. Use of Overshot With Knuckle Joint and Wall Hook. The knuckle joint and wall hook are a combination of tools (Figs. 12-6 and 12-7), used to pull a drill pipe fish out of a cavity or large diameter hole and place it in a position to be caught by an overshot run just below the knuckle joint.

This is ordinarily the tool chosen after the bent joint of pipe has failed. The knuckle or bending action is obtained by hydraulic pressure of the circulating mud. As the fishing string is slowly rotated with the mud circulating, the wall hook is forced to the outside diameter of the hole or cavity and will hook the pipe.

After the pipe is hooked, one more round of torque is applied to the fishing string, the rotary table locked, the pump stopped and the overshot lifted until the torque is released, then lowered. Weight of the wall hook and the torque should pull the fish to the middle of the hole directly below the overshot, which can be lowered over the fish.

Here again measurements are important as it may be necessary to lower the overshot 20 to 30 ft below the top of the fish in order to hook it with the wall hook. This necessitates lifting the fishing string with torque being applied to the exact spot where the overshot may then be lowered over the fish. This operation requires experience and may require hours or even days to successfully complete. A set of rotary jars is often run above an overshot to help free a stuck fish. It should be remembered that a knuckle joint will withstand jarring upward, but not downwards.

e. Use of Taper Tap and Safety Joint. Quite frequently the joint of pipe being fished will be bent from compression of the drilling string weight, making it impossible for the fish to enter the overshot above the knuckle joint-wall hook combination. In that case a good tool is a side-kicking taper tape with a safety joint on the top of the first joint of drill pipe above the tap (Figs. 12-8, 12-9 and 12-10).

The circulating hole at the bottom of the tap is plugged. A hole is cut in the side of the tap so the circulating mud pressure will force the point of the tap to the side of the hole.

The tap may then be lowered into the fish and screwed into it. After the tap has a secure hold on the fish, the safety joint can be released and unscrewed by left hand rotation. It should then be possible to go over the good joint and the tap with washover pipe and outside cutters. A cut is made at the bottom of the cavity, the troublesome part of the fish, pulled out and then the normal overshot procedure resumed.

f. Use of Underreamer. If the above procedures are a failure, the hole can be underreamed to the largest diameter possible, at least 60 ft above the top of the fish and the procedures listed tried again. The oversize hole above the fish may allow sufficient space to enable the fishing tool to catch the fish.

If all these procedures fail, whipstocking is the only answer. Occasionally an oil sand and high pressure water sand have been penetrated below the top of a fish, which cannot be caught due to a cavity. It may be desirable or even compulsory to get down beside the drill pipe to the section of the hole separating the oil sand and the water sand in order to seal the hole and prevent reservoir damage.

This has been done by cementing the top of the fish in the cavity and then drilling alongside the fish. Attempts to drill alongside without cementing have failed.

2. Recovery of Rock Bits, Cones or Small Pieces of Iron

Small pieces of iron can be recovered by running a magnet on the bottom of the drill pipe (Fig. 12-11). Circulation should be obtained to clean the cavings in the bottom of the hole to allow direct contact between

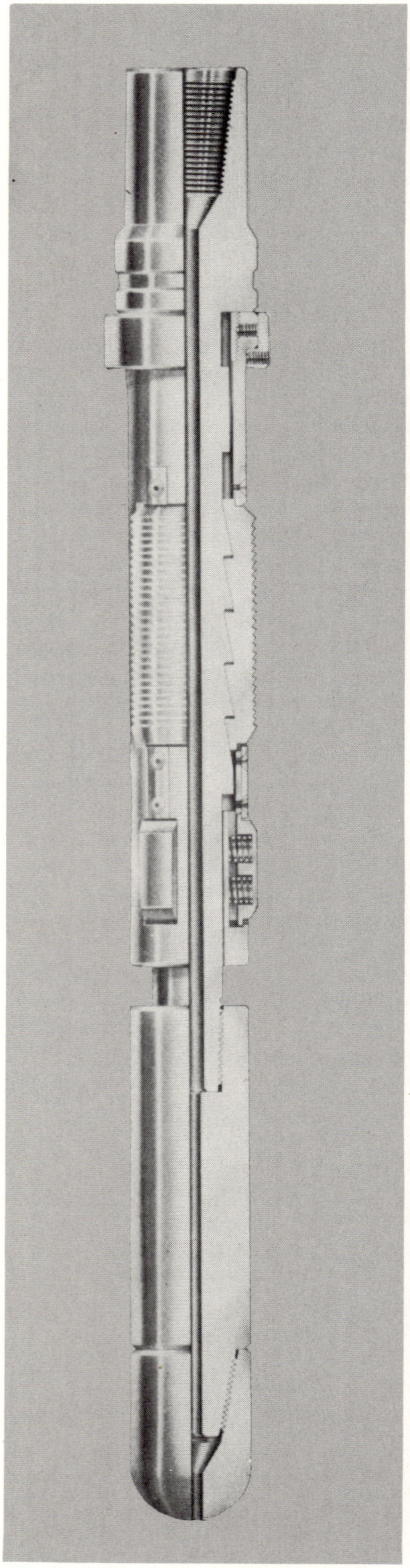

Fig. 12-10. Circulating and releasing spear.

Fig. 12-11. Fishing magnet.

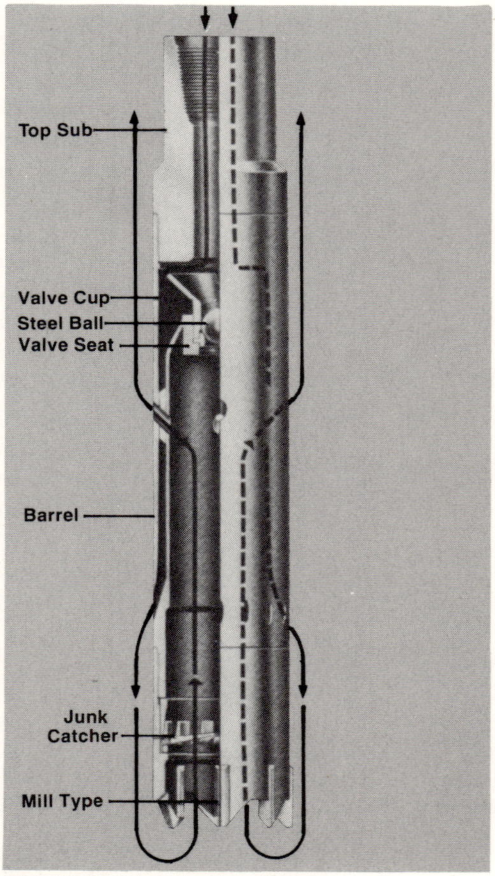

Fig. 12-12a. Junk basket, Bowen reverse circulation type.

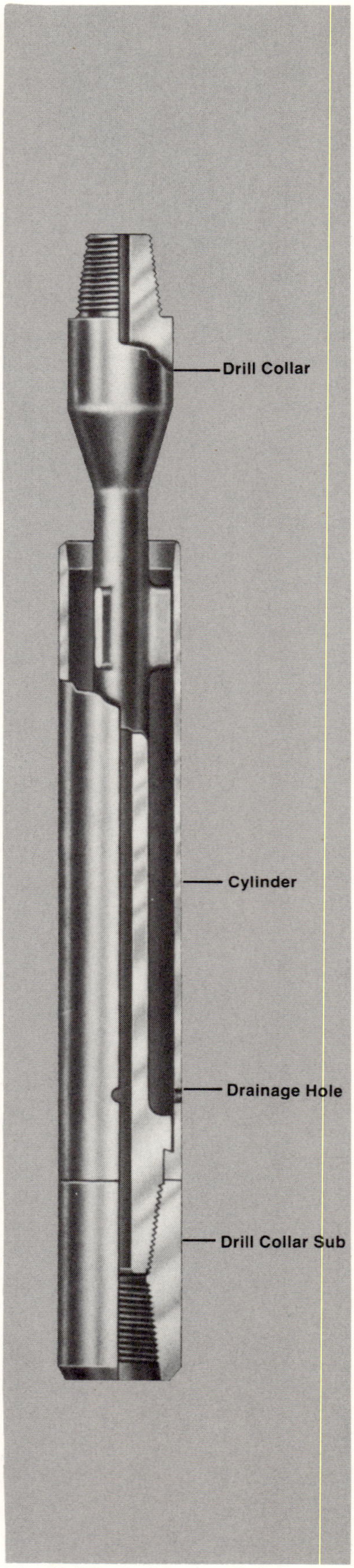

Fig. 12-12b. Junk basket, Baker string type.

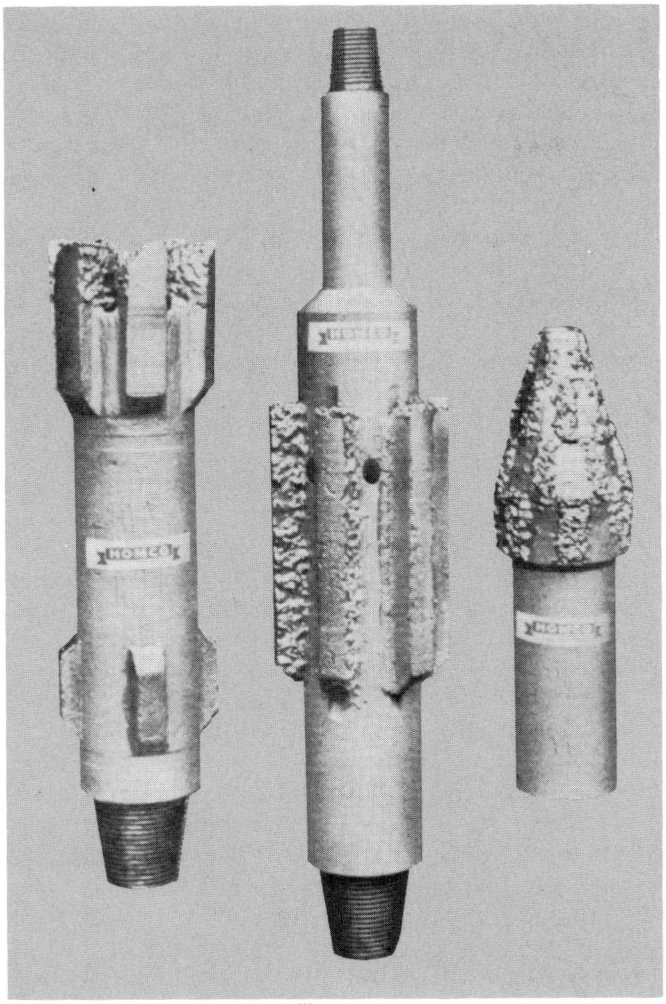

Fig. 12-13. Homco Klustrite mills.

the magnet and the iron. A basket may be used. This is simply a 2- to 3-ft core barrel (Fig. 12-12). A short core is cut thereby trapping the iron in the barrel.

There are many shapes of side baskets and junk subs designed with a pocket to catch small pieces of iron as they are pumped off bottom. They are used just above a hard formation rock bit. The secret is to break them into small pieces and alternately start and stop the pump which forces the junk to rise and fall.

This method has the advantage of allowing the driller to drill ahead once the iron is trapped and save trip time.

A rock bit, slip or similar size piece of steel must be broken up before fishing it out. This can be done with 10 quarts of nitro, with a shaped charge junk shot, with a slick joint of N-80 pipe or with a full hole mill of hard material, such as Klustrite (Fig. 12-13 and Fig. 12-14).

3. Freeing Stuck Drill Pipe – Causes and Remedies

Before the job of freeing stuck pipe can be started intelligently, a careful analysis of the reason for sticking must be made, as the approach depends upon the reason. Basically, it is usual to try to move stuck pipe in the opposite direction it was travelling when it became stuck. The most common reasons for stuck pipe are:

a. While on or near bottom. The drill string rarely becomes stuck while drilling ahead under normal conditions but it can happen. If the bit drills into a highly permeable formation causing loss of returns, then drill cuttings and cavings in the annulus will flow downward and, filtered out by the formation, they will pack around bit and drill collars, sticking the lower portion of the drill string.

One other cause of becoming stuck while the bit is on bottom is the possibility of a wash-out, or hole in the drill pipe, at some point low in the hole. If drop in circulating pressure or other surface indications go undetected, failure of the drilling fluid to scavenge the bottom of the hole will cause an accumulation of cutting on bottom that may stick the string.

It is not uncommon for the drill string to become stuck while circulating and rotating slowly, with the bit a few feet off bottom. This is a usual practice prior to running logs or for any other reason requiring the hole be cleansed of cuttings. Usually what happens is the drill pipe becomes wall stuck due to differential pressure between mud column pressure and the formation pressure in a permeable zone.

The actual sticking may occur at any permeable zone where the drill pipe presses against the wall of the hole. Movement of the pipe may have destroyed the filter cake, allowing the pipe to come into direct contact with the permeable formation, having a lower pressure than the hydrostatic pressure of the mud column. This difference in pressure, over a considerable length of drill pipe, produces a lateral pressure so great that the friction between pipe and formation prevents further movement of the pipe.

Wall sticking usually occurs when rotation or other pipe movement is interrupted, even though full drilling fluid circulation continues. The pipe does not actually become stuck until pipe movement and/or rotation is discontinued, at which moment the differential pressure grabs. When starting a trip out or making a connection, there is as yet no alternative to stopping circulation and pipe movement.

There is no single method of freeing stuck drill strings. The cause may be determined reasonably and the level at which it is stuck can be determined by the amount of stretch per unit of pull above normal weight of a free string. An effort should be made to work pipe up and down, within safe load limits.

If loss of circulation is a factor, steps should be taken to regain circulation. Spotting oil is the most common next step if the pipe cannot be worked free. This is accomplished by pumping oil down the drill pipe and around the bit and drill collars. The drill string is then usually left full of oil and displaced in small batches at about 30-min. intervals by a few strokes of the pump.

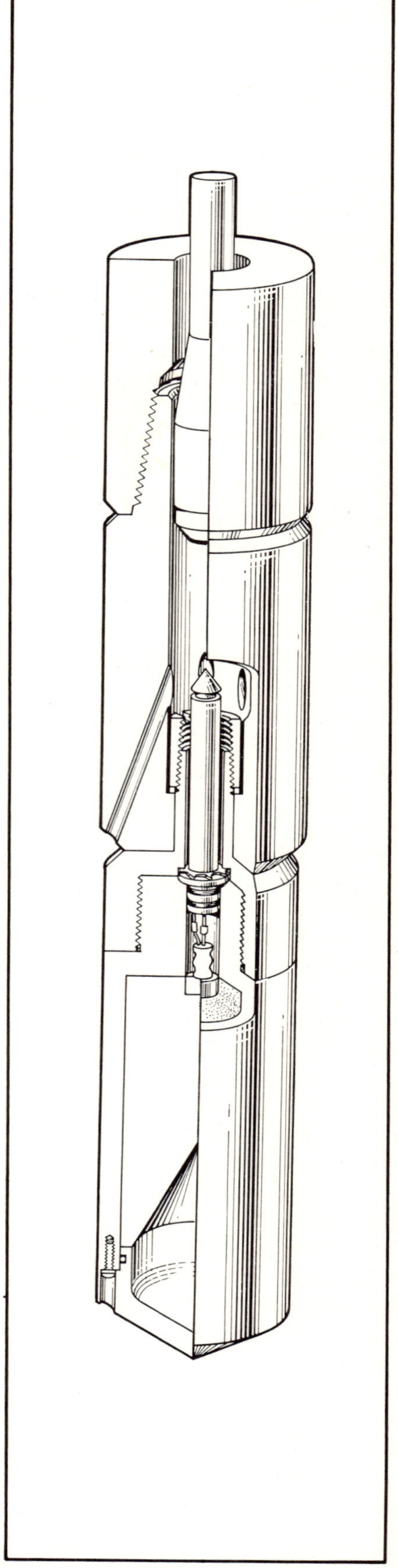

Fig. 12-14. Junk shot.

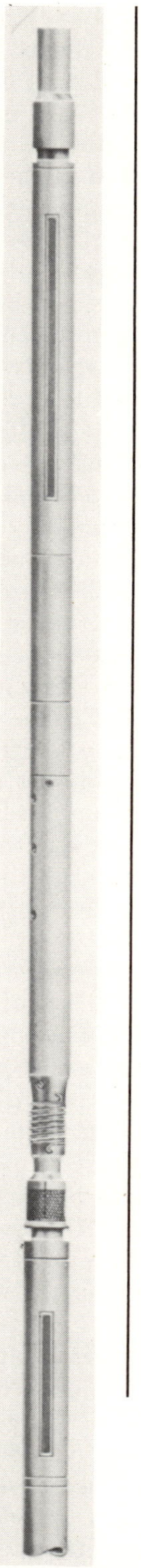

Fig. 12-15. Free point finder.

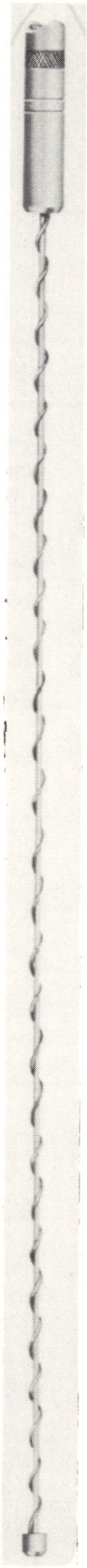

Fig. 12-16. String shot back-off.

Oil may cause the pipe to free up in a few hours or it may take a couple of days, but considering the alternative procedure it pays to have patience.

If the drill string cannot be worked free after spotting oil the next step is to run a free point indicator to determine the lowest free point from which the pipe can be recovered (Fig. 12-15). The free point indicator is run inside the drill string on an electric line by a service company. It is a long slender tool consisting of two major sections connected by a telescoping joint. Each major section is equipped either with a magnet for use in steel pipe or a friction spring assembly for use in aluminum or other non-magnetic pipe.

The telescoping connection acts as a transducer to send an electric signal to the surface when there is produced a relative movement between the two sections, each being independently affixed, some distance apart, to the wall of the pipe.

Variations in torque or tension in the string, applied at the surface, produces minor relative movement of the two sections where the pipe is free but no movement where the pipe is stuck. Several check points may be observed before the lowest free point is located.

Incorporated in the free point tool is a collar locator. Suspended below the tool is a string shot (primer cord) which can be detonated electrically from the surface. Object is to find the lowest connection in the free part of the string and then to space the string shot one joint above it.

The pipe can be backed-off by applying left hand torque and near neutral tension in the pipe at the critical point, then detonating the string shot. The free pipe is then unscrewed and recovered.

The surest procedure to recover the stuck pipe left in the hole is to run a string of wash-over pipe on the bottom of the fishing string. The size of hole, the outside diameter of the pipe, and collars in the hole, the length of the fish and the general condition of the hole will determine the size and length of the washpipe and the type of drilling shoe on the bottom.

For instance, 6-in. tool joints and 6¼-in. drill collars in a 9-in. hole would accommodate 7⅝-in. wash pipe with an 8½-in. shoe. Ordinarily, 200 ft to 500 ft of washpipe is safe. If the bit is on bottom and the fish not too long to be washed over in one try, an overshot can be run properly spaced to catch the top drill pipe joint after washing over the bit. The fish is pulled with the wash pipe.

If drill collars are in the hole, a back-off tool can be spaced properly to screw into the top collar as the washover reaches the bit. With no back-off tool, the wash pipe is pulled and a trip made with an overshot or a screw-in joint to recover the freed fish.

If the fish is on bottom and too long to wash over in one try, a back-off tool is run at the top of the wash pipe. The full washover is made and the back-off tool screwed into the top of the fish. A free point indicator

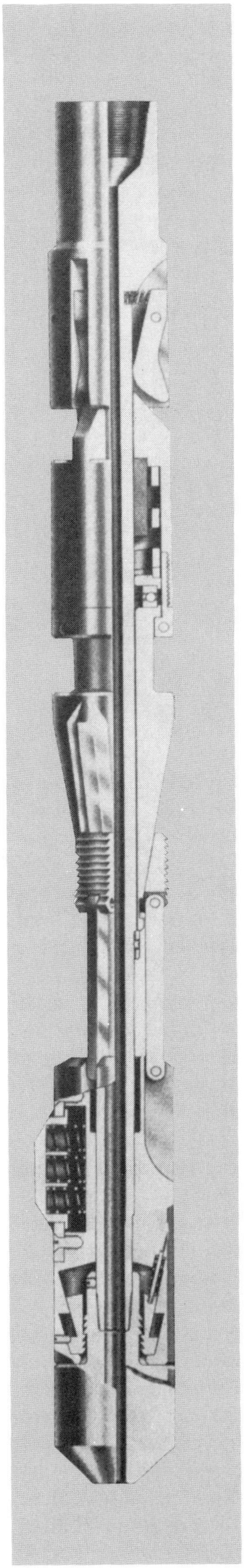

Fig. 12-17. Inside cutter, mechanical.

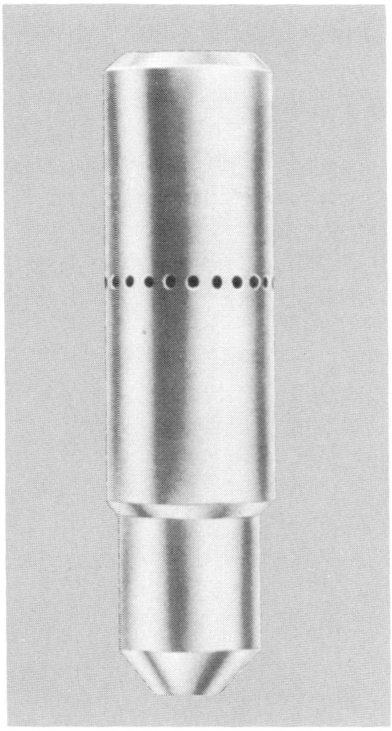

Fig. 12-18. Inside cutter, chemical.

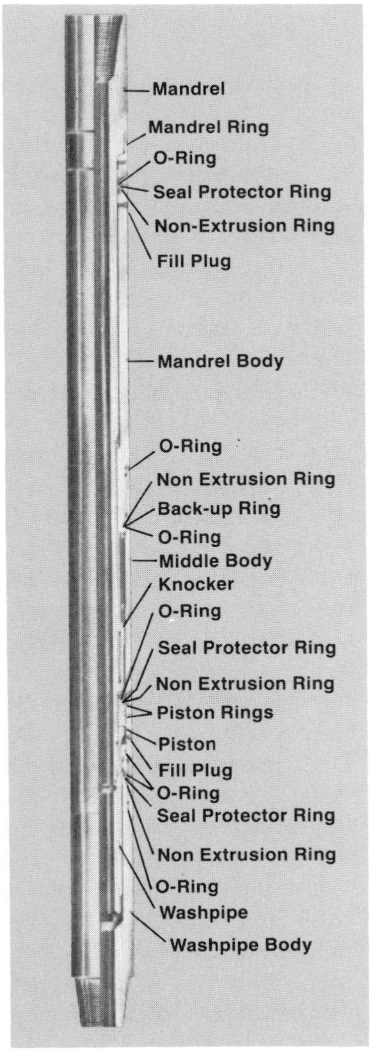

Fig. 12-19a. Fishing jars.

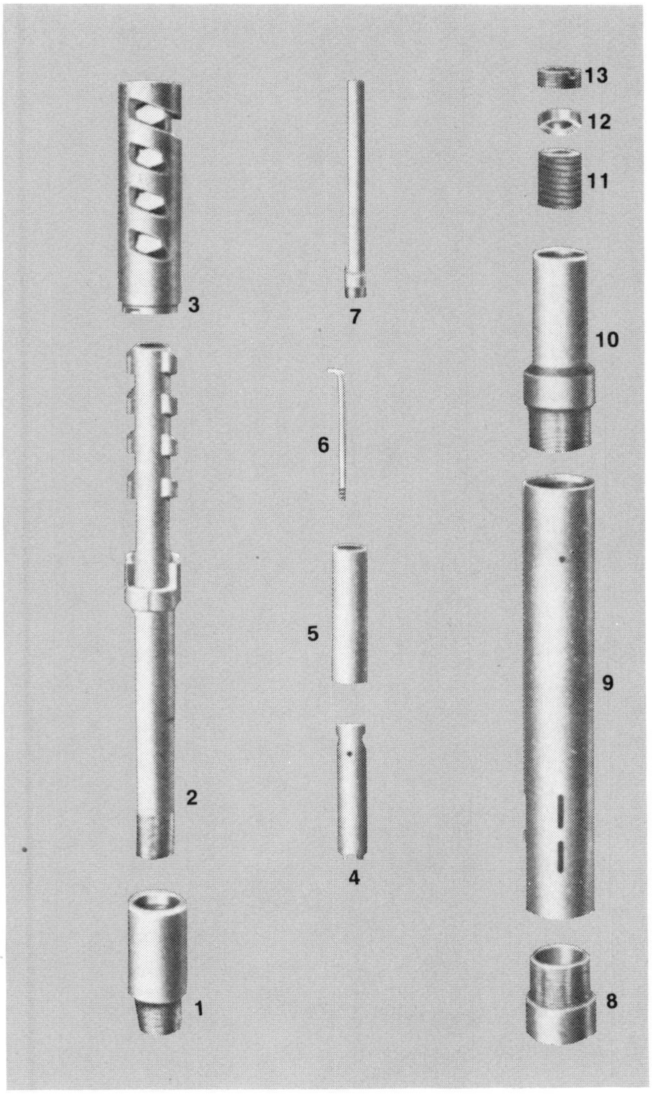

Fig. 12-19b. Oil and torque jars.

is then run and the fish backed off with the aid of a string shot. The free fish is then recovered with wash pipe and the procedure duplicated until the entire fish is recovered.

If the stuck string is off bottom and it is washed free, an overshot to catch a drill pipe tool joint is run one joint from the bottom of the wash pipe. This overshot will prevent a tool joint from falling through it. There must be one free joint of pipe at the top of the fish to allow the overshot to go over the tool joint before washing over is started, hence the necessity for leaving one free joint when backing off.

It is standard practice to run a set of rotary jars (Fig. 12-19) and a safety joint with wash pipe to assist in freeing or to get loose from the wash pipe in case it becomes stuck.

b. *Lowering pipe in the hole forcing the bit into cavings or an under gauge hole drilled by the previous worn bit.* Pipe stuck while running in the hole can usually be freed by backing off at the free point and screwing-in with a set of rotary jars and driving up-

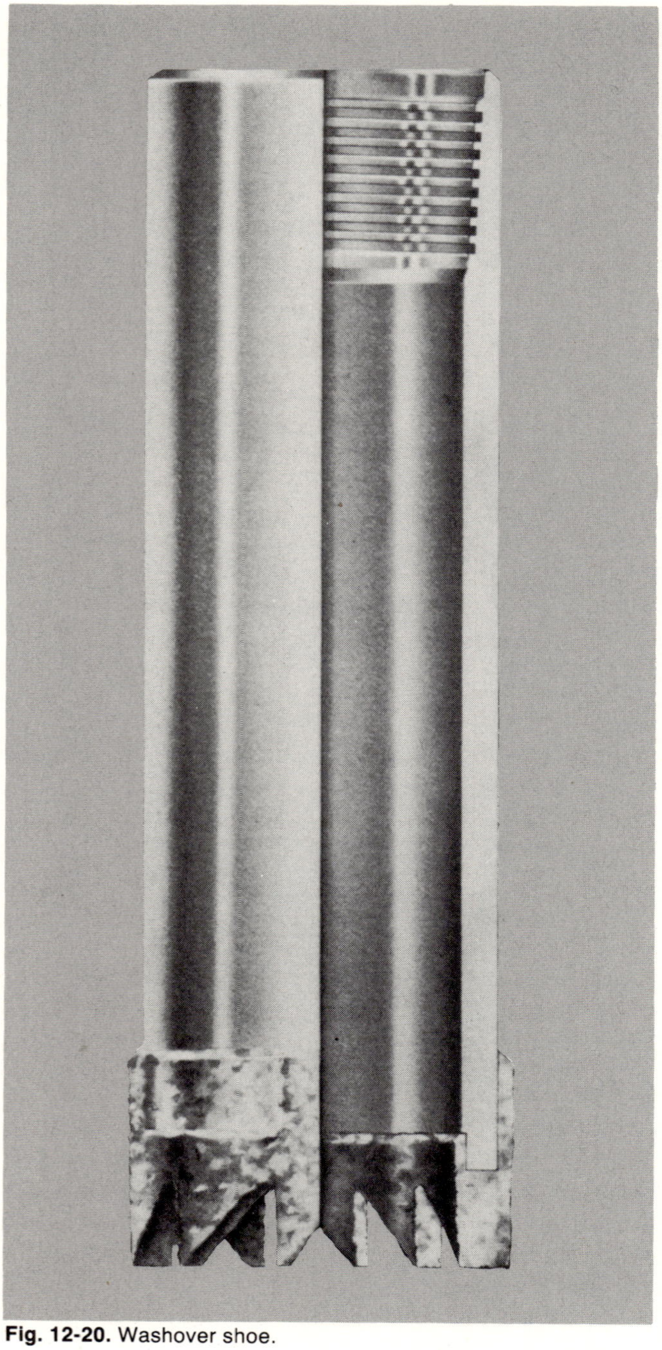

Fig. 12-20. Washover shoe.

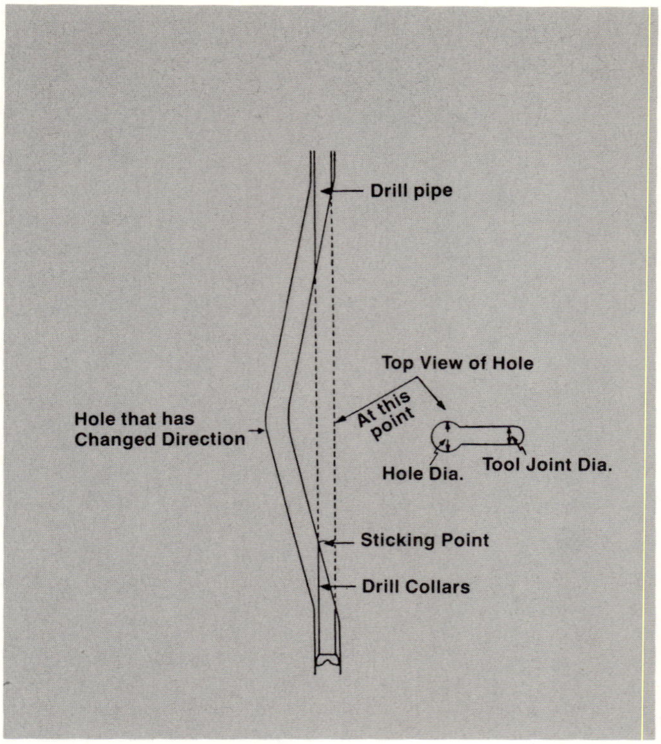

Fig. 12-21. Keyseat development.

ward. If this is unsuccessful, a washover procedure is in order.

c. *Pulling pipe out of hole.*

- An accumulation of cuttings and cavings collect on top of bit and around the drill collars until sufficient to stick.
- Pulling drill collars into a key seat.

Pipe stuck above 2000 ft while pulling out of the hole can often be jarred loose by driving down with a surface driver. If 2 or 3 hours of driving does not free it, then it should be backed off with a free point indicator and back-off shot and a set of rotary jars used to drive downward first and upward last.

An overshot should be used below the jars to insure getting loose from the fish without using the free point tools again. Drill collars should be run on top of the jars to give enough weight for the desired blow.

Pipe that has been pulled considerable distance into a tight spot and stuck is very infrequently freed by jarring either from the surface or below. The capacity of the derrick and the damage to the hoisting equipment should be considered in a jarring operation.

Drill collars frequently are stuck by pulling into a key seat (Fig. 12-21). It has been proved many times that key seats can occur in holes with deviations of less than one degree. This is due to a change in the direction of the deviation, usually 180°. Holes that have required 30 to 40 days to drill from 7000 ft to 8000 ft are likely prospects for key seats at 2000 ft to 3000 ft in many areas.

A key seat will usually give warning to an alert driller by hanging slightly for two or three trips before sticking. A change in the number of drill collars may result in sticking in a key seat the first trip out. Drill collars can usually be jarred downward and freed from a keyseat, but there remains the problem of pulling out, which may be difficult or impossible.

Safest procedure is to back off one or two joints above the stuck point with the aid of free point indicator and back-off shot. Then go in with wash-over pipe and hole-size shoe. This reams out the keyseat and eliminates it for two or three weeks of drilling and frees the stuck collars at the same time.

As indicated, a key seat is likely to develop at a

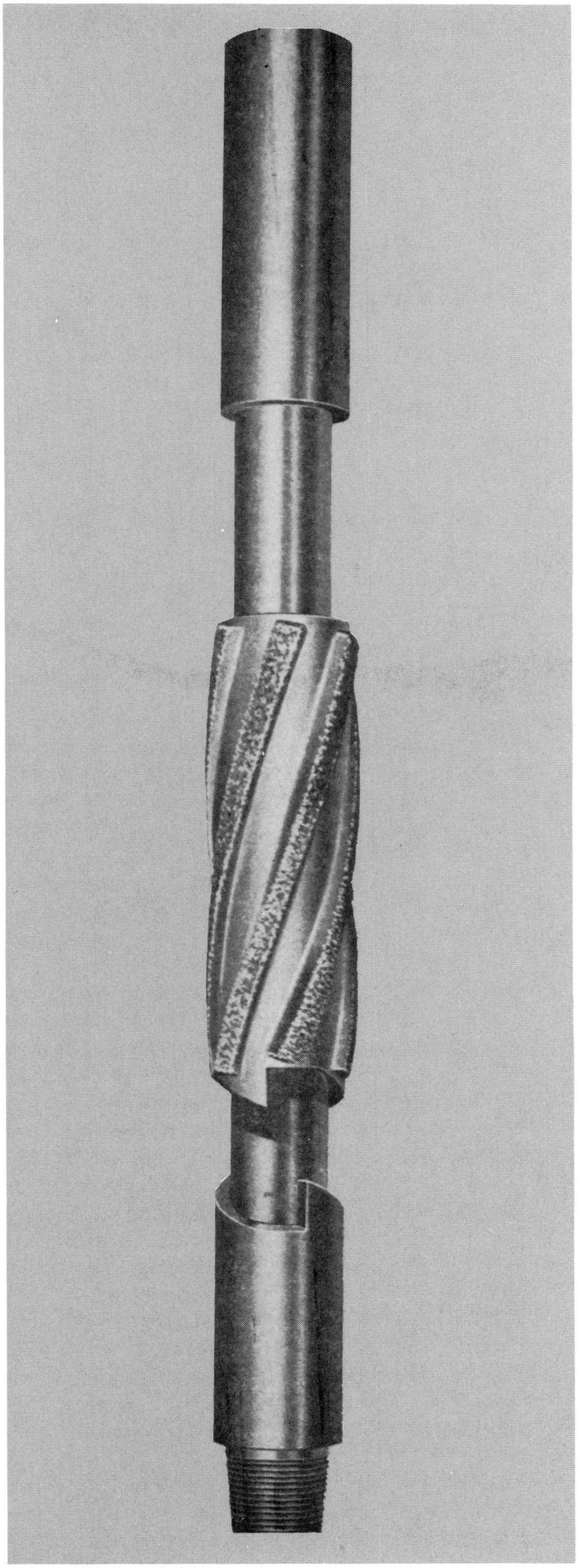
Fig. 12-22. Keyseat wiper.

relatively shallow depth when the well gets deep enough to develop a heavy drill pipe load and provided further that a dogleg exists at the assumed shallow depth.

High tension in the drill pipe causes it to bear heavily against the wall of the hole and the drill pipe and tool joints simply wear a groove in the wall that is large enough to pass tool joints but too small to pass drill collars.

Since key seats do not develop suddenly the driller usually has some warning before the situation becomes hazardous. In this instance, an ounce of prevention may be worth a pound of cure.

If development of a key seat is indicated or conditions are such as to cause one to be suspected, a key seat wiper (Fig. 12-22) should be added to the drill string between drill collars and drill pipe. It is recommended the OD of the ribs on the wiper sleeve be one quarter inch larger than the largest drill collar. This tool will detect and ream out key seat, thus reducing the probability of getting stuck.

When tripping out, the sleeve will tend to wedge in the key seat first. The driller can slack off on the drill pipe in such a manner as to jar the tool loose which then can be used to ream out the key seat by rotating with an upward thrust to assure engagement of the sleeve clutch.

The proper tools are not always available, unfortunately, and the best tools and procedures often fail, but adherence as closely as possible to the best accepted methods will result in a high percentage of success.

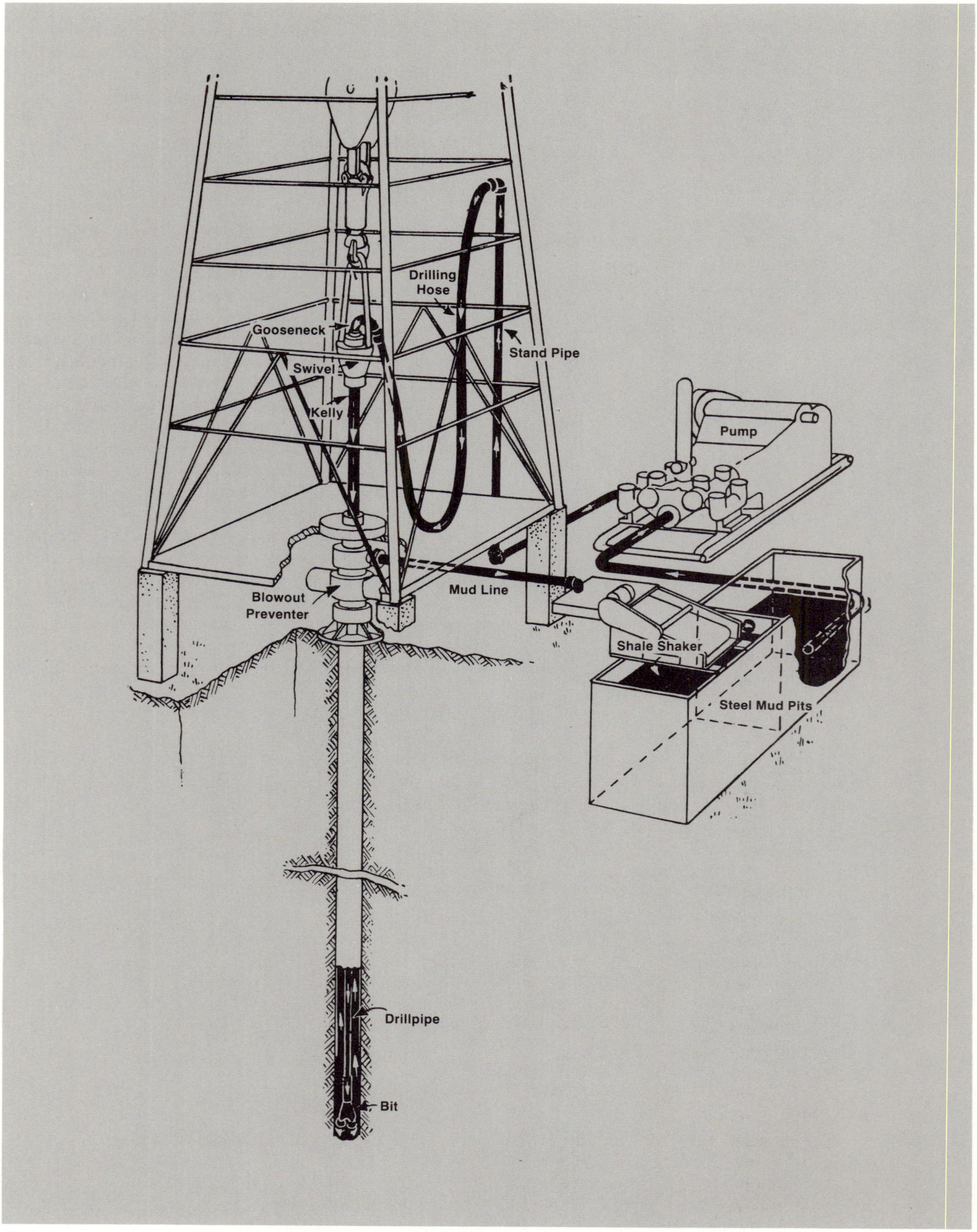

Typical mud circulating system on rotary drilling rig. Mud is withdrawn from mud pits by pump; then forced through standpipe, hose, swivel, kelly and drill pipe to bit. It then returns to the surface through the hole annulus and is cleaned of cuttings at the shale shaker and prepared for further service.

Chapter 13
Mud Programs and Mud Control

The mechanical aspects of modern drilling practice evolved over the years mainly from field experiences with trial and error methods. Some other aspects of the art are better served by scientific investigations and analysis of laboratory experiments.

Key to successful rotary drilling is not so much mechanical rotation of the drill pipe as it is washing the hole clean by fluid pumped down the drill pipe. This makes continuous operation for removal of drill cuttings possible.

First such drilling probably took place about 1882 in the Dakota Territory and was done by two water drilling contractors. They poured water down the annulus and washed cuttings back up through the drill pipe. In 1895 the rotary method of drilling was used in the Corsicana oilfield in central Texas. Most notable early well drilled by this method was probably the one drilled by Captain Lucas at Spindletop near Beaumont in 1901.

In these early wells water was pumped down the drill pipe, through the bit, and returned laden with cuttings to the surface through the annular space between the drill pipe and the walls of the hole.

Early drillers observed that some clays which became accidentally mixed with the drilling water increased the ability of the water to keep the hole free of cuttings. The drilling mud industry grew from this small beginning until it now seeks special mud-making materials in all parts of the world.

Mud Circulating System

In a well being drilled by the rotary method, drilling fluid is circulated at all times as drilling is in progress. The mud-circulation system varies some, from one area to another, but in general it follows a standard plan. For convenience it is divided into two general parts. One part consists of the mud system above ground and the other is that part of the system underground.

Above ground, or surface system, begins with mud issuing from the top of the well and flowing through a flow line to the shale shaker and by gravity to the mud pits. It is taken from the mud pits through the mud pump suction, forced out through the discharge end of the pump through mud lines to the stand pipe.

The rotary hose connects the stand pipe with the swivel. The mud passes through it and the kelly joint to go down the drill pipe. After passing through the bit, it returns to the surface outside the drill pipe, removing bit cuttings from the hole, and enters the shale shaker.

Surface System. The shale shaker or shale separator is a mechanical device for screening coarse cuttings out of the mud and removing them from the active mud system. It is one of two general types, vibrating or rotating.

Screens for each can be obtained in a choice of several mesh sizes. After leaving the shale shaker the mud returns to the pits or tanks.

These mud tanks or mud pits vary in size and method of construction. They are either constructed of steel or dug in the ground. Tanks are usually about 25 ft long, 8 ft wide and 6 ft deep. Number used is between one and four.

They are connected by large diameter pipes, usually a size between 7 in. and 10½ in. Guns are fitted, either surface or submerged, so that mud in the tanks can be stirred. Tanks often are arranged so that one or more can be taken out of the system. Where earthen tanks, or pits, are used they are usually dug by a bulldozer and are divided by earthen or wooden dividers.

In many cases the surface mud pits are designed for ease of transportation rather than for efficient mud maintenance. Their volume must be such that (1) any entrained gas can be released before the mud enters the pump suction; (2) as many cuttings as possible, which are fine enough to pass through the shale shaker screen, will settle out of mud stream and be collected on the mud pit bottom; (3) small mud losses to the formation will not deplete mud in the pits; (4) spots of mud not up to specifications will mix with enough good mud to cause no hole trouble, and (5) enough mud is available for keeping the hole full when the drill pipe is pulled. For wells deeper than 8000 ft the mud pit volume usually is between 300 bbl and 400 bbl.

Adjoining the mud tanks on the side away from the drilling rig is a large earthen pit used to hold discarded drilling mud in reserve. Any drilling mud which must be so discarded is jetted or pumped into this reserve pit. Use of reserve mud and ability to get it back into the circulating system often means a substantial savings.

The pump is usually a two-cylinder duplex power pump being driven from the same engines that drive the drawworks and rotary, or from a separate engine used only for pump motivation. Sizes vary for this pump but in recent years there has been a decided trend toward larger volume pumps requiring greater amounts of power. Some pumps on the market require from 1400 hp to 1600 hp for capacity operation. Mud discharged by the pump passes through a mud line usually made of extra-heavy pipe to the stand pipe that is placed in one corner of the derrick. Attached to it is the rotary hose, the other end of which is connected to the swivel. This hose and stand pipe arrangement makes it possible for the swivel to be raised and lowered while mud is still passing through it.

Underground system. At this point the mud enters the underground portion of the circulating system. It is pumped down the drill pipe, through the drill collars, out of the bit nozzle into the hole that is being drilled by the bit. It is then forced up the hole on the outside of the drill collars and drill pipe. It reaches the surface and moves through flow line to the shale shaker.

Time required for each circulation of this continuous operation varies according to how big and deep the hole is, amount of mud in the surface circulating system and size and speed of the pump being used. Amount of power expended in moving mud through this system at various rates and in several states of flow is receiving detailed study.

Drilling Fluid Properties

A drilling fluid consists of a base liquid, usually water, into which are dispersed solids which range in size from a true solution, through the colloidal range, to suspended particles visible to the naked eye. Solids which are colloidal in size are decisive ones in a drilling fluid. Dissolved solids, usually salts, affect mud properties chiefly by their effect on the colloids. The larger solids affect the mud properties by supplying material from which colloids are made.

All drilling fluids have practically the same properties but it is the varying magnitude of these properties which is of great interest to the drilling mud engineer. Drilling fluids have weight, viscosity, gel strength, wall building and filtration rates, and electrical resistivity.

When little effort is needed to control these properties the mud is thought of as a normal drilling fluid. As more effort and materials are needed to keep mud properties in a desired range, the mud becomes more abnormal and special. Normal muds are as difficult to find in the field as they are to define. In actual practice, mud engineering has come to mean the proper use of special drilling fluids to accomplish special objectives or to overcome abnormal hole conditions. Special objectives may be faster bit penetration or greater protection to the producing zone. Abnormal hole conditions could be extreme gas pressure or long salt sections.

Normal muds. The compounding and control of muds having normal properties is a relatively simple matter. Where no extraordinary conditions are expected in a well such a mud will become stabilized so that its properties are in the range needed to control hole conditions. The formations being drilled will consist chiefly of shale and sand and the chief problem is viscosity control. Amount and condition of the colloidal clay entrained in the mud is of paramount importance.

Two general types of chemical treatment are in wide use for this control. One of these treatments makes use of water soluble polyphosphates for reducing viscosity. The polyphosphate is used either alone or with quebracho.

If wall cake and filtration control are needed, extra amounts of colloidal clay are introduced to the system. Wyoming bentonite contains a high percentage of such clay. The other chemical treatment for ordinary water-base drilling muds makes use of caustic soda and quebracho as a viscosity reducing mixture. This treatment is effective under conditions more severe than the phosphate treatment.

Upper portions of most wells can be drilled and maintained by the use of a mud having normal properties. In many cases native muds produced by the formation solids dispersing in water are adequate. When normal chemical additions are begun, however, treatment should be chosen that will not hinder the compounding of special muds to combat abnormal hole conditions, which may be encountered as the well is deepened.

For example, if an anhydrite section is expected that will necessitate large additions of barium carbonate, the preliminary treatment should not include phosphates, large quebracho treatments or enough caustic to create a high pH. All of these suppress the reaction between barius carbonate and anhydrite.

Special muds. There are many special muds. In general, they differ from ordinary muds in that they are resistant to a particular abnormal hole condition or a combination of such abnormal conditions. Here are some of these muds with their particular advantages and disadvantages.

Lime-base muds. These are water base muds which have been treated with large amounts of caustic, quebracho and lime. The order of addition is caustic, quebracho (or other organic thinner such as ground lignite or calcium lignosulphonate) and lime. Amounts vary but a typical treatment would make use of 2 lbs caustic, of 1½ lbs organic thinner, and of 5 lbs of lime added to each barrel of mud in the circulating system.

Caustic goes into the mud stream with enough chemical thinner to keep viscosity of the mud workable. After one circulation with probably half of the caustic and quebracho having been added, the lime is started as a solid material through the hopper. The total amount of lime is added during one or two mud circulations and in three circulations all of the material will have been added.

The mud probably will have gone through a highly viscous stage but will reach a stable state of low viscosity. Gel strength will probably remain high for about 8 to 16 hrs after which the mud will exhibit low thixotropic properties. Such a mud can tolerate large amounts of contaminating salts. Its particular ability, however, is to remain fluid when its solids content has been raised high. Lime base muds weighing 17.5 lb/gal often have a funnel viscosity of around 45 seconds API.

The great weakness of this mud system is its tendency to solidify when subjected to high bottomhole temperatures. This high temperature gellation is affected very little by bottomhole pressures but is acceleratd by increased amounts of either lime or shale solids.

Lime-treated muds. These muds are a compromise attempt at overcoming the high temperature gellation problem encountered when using lime-base muds. They differ from lime-base muds only in degree. Caustic and organic thinner and small amounts of lime are used for control. This system has overcome the worst aspects of heat solidification but the small treatments must be supervised closely and these muds are not nearly so resistant to contamination by salts encountered in the well as are the true lime-base muds.

Emulsion muds. Oil in water emulsions are used as drilling fluids and exhibit several desirable qualities. Oil is being added to any of the regular or special water-base drilling fluids with good results. No special mud properties are necessary and in many cases emulsifying agents found naturally in a drilling fluid are effective enough to hold the oil in a tight suspension after it has been mechanically dispersed by the vigorous stirring it receives while circulating through the pump and down the hole. Special emulsifiers now in general use are surfactants, ground lignites and sodium carboxy methyl cellulose.

Oils generally used in emulsion type muds are crude oils and diesel fuel. Any available oil having a gravity between 25 and 50 API can be used.

Crude oils containing large amounts of low boiling point components become uneconomical because of extensive evaporation losses. Emulsion muds vary in oil content between a low of possibly 1% and a high of around 40%. Muds containing 8% to 12% of oil by volume have exhibited very stable properties easily maintained with low filtration and wall cake characteristics. Such muds allow faster penetration and reduce downhole friction.

A well-dispersed emulsion mud creates no fire hazard itself but the mixing oil held in reserve constitutes some danger and oil spilled around the rig or allowed to remain on the mud in the mud pits is somewhat dangerous. Good emulsion muds are a little more expensive than similar water-base muds without the oil.

The chief objection to this drilling fluid comes from the geological department. The oil used, particularly if it is a crude oil, sometimes masks the staining, fluorescence and odor of the formation cuttings. Non-staining and non-fluorescing oils alleviate this condition. This objection is rarely raised except on wildcat wells.

Inhibited muds or muds with inhibited filtrates have become popular recently. Large amounts of dissolved salts are added to the mud. A high pH is usually thought necessary for best inhibiting results.

A typical inhibited mud would be treated with caustic and quebracho in the usual manner except the amount of caustic would be increased to possibly 2 lb/bbl of mud. To this is added a combination of sodium chloride and calcium sulphate. Quebracho, or another organic thinner, is used in initial stages of the treatment to maintain viscosity and gel strengths of the mud in a workable range.

The amount of salts used will vary but 6 lb/bbl of calcium chloride and 2 lb/bbl of calcium sulfate produces a filtrate containing 15,000 or 20,000 ppm sodium chloride and a hardness of about 200 ppm calcium carbonate. Such muds are designed to reduce formation swelling caused by the filtrate.

Where producing sands contain large amounts of shale or clay, an ordinary filtrate swells the shale or clay, filling the space between the sand grains, and thus sealing the producing section immediately adjacent to the bore hole.

Filtrates containing appreciable amounts of dissolved salts inhibit this swelling and hold it to a minimum. Cases of increased production have been reported where an inhibited mud was used and the need of formation fracturing has been eliminated in some instances.

These muds have certain inherent disadvantages. Detailed electric logs with ordinary equipment are not obtained in them. They require special attention and, unless oil is used with them, low mud weights cannot be maintained.

Increases in viscosity are hard to make because the clay added for this does not swell in such a mud. Salt destroys the natural wall building properties of the clays present in the mud and so filtration control must be maintained by the use of non-clays such as starch or sodium carboxy methyl cellulose, which is more efficient for this purpose.

Gypsum base muds. A specialized type of inhibited mud has been used in the northern Rocky Mountain area, through the Dakotas and up into Canada. This mud contains large amounts of calcium sulfate in the form of either agricultural gypsum or plaster of paris. To a regular water base mud composed of clay and water is added 2 lb/bbl of gypsum. Thinner usually is not added and a mud results having high initial and high 10-minute gels. Filtration properties are controlled by addition of organic colloids such as starch or sodium carboxy methyl cellulose. When starch is used it must be protected from fermentation by the addition of a preservative. Paraformaldehyde and Dowcide G are the two most commonly used preservatives.

These muds are fairly stable, economical to maintain and their filtrate does not swell to a great degree the bentonite streaks found in these regions. High gel strengths, particularly the initial gel, cause the muds to hold and recirculate any gas which might become entrained. For the same reason, fine abrasive sand often encountered in the Rocky Mountain area remains in the mud where it causes excessive pump wear.

Oil-base muds. All of the muds described so far have had water as their liquid phase. Another liquid which can be used as the dispersant in a drilling fluid is oil. When oil is used the materials to be dispersed in it must be oil soluble and capable of imparting to the fluid proper characteristics for making a usable drilling fluid.

Most of the oil-base muds being used at the present time are compounded in central storage tanks and taken to the oil well in a liquid condition. Some sacked material, however, can be obtained to add to a special oil to produce a good oil-base mud.

One such mud uses for its base a furnace oil having certain specifications. To a barrel of this oil is added blown asphalt and a certain small amount of unslaked lime. The blown asphalt gives viscosity and a certain amount of thixotropy to the fluid. It also provides material for filtration control. The unslaked lime gives further viscosity control and guards against fresh water entering the system to form a water and oil emulsion. Other control chemicals are used in very special instances.

Oil-base muds contain no water and so cannot swell clays or shales with which they come in contact. Much information is in the literature concerning the higher productivity of wells drilled with oil-based muds when compared with those drilled with a water mud. The extremely good lubricating properties of oil-base muds allow closer tolerances to be used in a well without great danger of pipe sticking. In many instances the intense heat from both bullet and jet type perforators solidifies the clays in a water base mud so that the perforations become plugged.

This is not true where the perforating is done in an oil-base medium. These muds are expensive and dirty to work with. Conventional electric well logging technique cannot be employed in them. Their viscosity varies greatly with their temperature.

Inverted emulsions. A mud having some of the advantages of a true oil-base mud but somewhat cheaper is water in oil emulsion. This is the inverted emulsion. Inverted emulsions can be compounded several ways. The usual method is employment of a special emulsifying agent which tends to hold the water dispersed into the oil. Oil is usually the biggest component and the water is added to it. This order of addition is important. The presence of certain salts also helps this type of emulsion.

No great amount of experience has been obtained with these emulsions but they have about the same advantages and disadvantages that true oil-base muds have. They are somewhat less stable but they are also less expensive.

Salt water. Salt water, either partly or completely saturated, is used as a drilling fluid under several conditions. In workover operations it is used as a special type of inhibited mud causing very little damage to exposed producing formations. Its weight can be varied by the addition of extra amounts of salt up to the saturation point. Saturated salt water weighs about 10 lb/gal. Salt water has no wall-building properties, however, and is very readily lost to any slightly porous formation.

Silicate mud. Sodium silicate drilling fluids were the first to drill successfully the heaving shales in the Gulf Coast area. They are composed of sodium silicate and saturated salt water. This mixture has what is described as a pickling effect upon shales which prevents them from heaving or falling into the hole. Actually, this is another example of an inhibited filtrate. The weight of these muds probably aided in keeping shales in place also. The natural weight of this mixture is just over 12 lb/ gal. It can be weighted higher by the addition of barite.

The silicate muds, however, have some disadvantages and their use is not really as extensive as it once was. The 12 lb/gal. which they weigh is their minimum weight. If a lighter mud is needed another type must be chosen. Silicate muds are corrosive and expensive. Poor electric logs are obtained where these muds are present.

Low solids. Excessive amounts of clay and shale solids have always been detrimental to a good drilling fluid. These extra amounts of clays and shales are expensive to treat, make thick dangerous wall cakes and produce an unstable mud. It has been found that it is beneficial to keep the amounts of these materials in a mud even below what has previously been considered a normal amount.

Low solids muds have shown themselves to be advantageous in promoting faster and safer drilling. When oil is added to a water base mud which contains a very minimum of clay solids, a very good drilling fluid results. There are three ways that clay solids can be removed from a drilling fluid: (1) by water dilution, (2) by circulating through enlarged surface pits, and (3) by the use of a centrifuge. When clay solids are removed mechanically a very minimum of viscosity control chemicals are needed. Reducing the viscosity and gels becomes a small problem and inexpensive.

When viscosity and gel strengths become too low due to the removal of claylike solids, they are replaced by an organic or suspending material such as sodium carboxy methyl cellulose. These materials also make a very tight oil in water emulsion when oil is added to the system. These low solids emulsion muds are very stable, easily maintained, and lend themselves very readily to drilling programs which involve large pumps and high mud circulating volumes.

Faster drilling has always resulted when this type mud system is used. On the Gulf Coast where high mud weights are necessary and a centrifuge is available, this method of controlling mud properties has been very economical. Where the clay solids in light weight muds are removed by dilution, settling, or centrifuging, the economics are not too certain.

Disadvantages of excessive dilution are apparent. Usually the excesses must be jetted to the reserve pit and often water is difficult to obtain. Large circulating pits or circulation through the reserve pits have some disadvantages such as extra cost of tanks and excessive dilution in case of hard rains. A centrifuge on a drilling rig is another piece of equipment which must be moved, maintained and serviced.

Conclusion

As wells get deeper and the total cost of drilling them becomes greater, proper selection and close control of the drilling fluid plays an increasingly important part in the success or failure of the operation.

To obtain the greatest success most economically the mud program must be well planned and then well executed. To have a good plan and then not follow it is wasteful. To try hard with no program is just as wasteful. Both are dangerous in many cases.

Close supervision of drilling fluids by a trained mud specialist has now become a routine part of today's drilling program.

Chapter 14
Effective Penetration Rates

No other single factor contributes more to lowering the cost of drilling than improved penetration rates. It properly is the aim of drilling crews to make hole rapidly; good hole that can be saved and used for the production of oil and/or gas if such is encountered.

Much progress in drilling rates has resulted from actual field experience but controlled laboratory experiments have evaluated quantitatively the effects of many variables such as weight on bit, rotary speed, bottomhole scavenging and hydrostatic pressure of the drilling fluid.

Two other factors not subject directly to laboratory tests are bit selection and choice of hole size.

Bit Selection

The several tooth structures, in combination with variations in bit geometry, result in numerous bit types, one of which is suitable for drilling effectively almost any type of rock.

Unfortunately, the specific properties of the rocks to be drilled in the upcoming bit run cannot be positively predicted. The driller, therefore, must employ all available aids in choosing the bit to be run.

Most helpful perhaps are bit records from nearby wells. Also, the logs from such wells are useful, both as a check against accuracy of previous bit choice and for correlating lithology with that of the well being drilled. Bit type selection in this sense is covered extensively in the literature.

The criterion for each bit selection must be the overall cost per foot of hole drilled, taking into account bit cost, rig time and other operating expense connected with the run.

One consideration, too often overlooked in bit selection, is hole size. It is not possible always to exercise free choice of hole size, but insofar as possible the well plan should avoid use of small roller cone bits for drilling hard rocks at great depth.

Slim-hole drilling techniques, as originally conceived, presumed that drilling costs would be reduced somewhat in proportion to the volume of material excavated. This postulation has long since been disproved. Only in shallow wells and/or in soft rocks can drilling costs be reduced significantly by reducing hole size below 7⅞ in. Perhaps this paradox deserves some discussion.

To drill a small hole at a rate equal to that possible in a larger hole requires equivalent unit bit loading. This usually cannot be realized on account of the inherent limitation on bearing sizes in small bits.

It simply isn't possible to make a small bit with stamina comparable to that of its larger counterpart. As an illustration, let us compare a 6-in. tricone rock bit weighing 30 lb with a 12-in. bit weighing 220 lb.

Bit weights are proportional, approximately, to the cube of gauge diameter. Presumably this ratio applies to the various bit components, particularly the individual cone assemblies complete with bearings and spindle.

The base diameter of the cone on a 6-in. bit is about half that on a 12-in. bit, hence the areal space for the bearing in the smaller bit is only about one quarter that of the larger bit. The permissible or reasonable bit loading, therefore, is relatively proportional to the cross sectional area of the hole, whereas the required bit loading, for equal rates of penetration, is proportional to the *diameter* of the hole.

Based on this concept, a 12-in. bit might tolerate a load four times as great as that on a 6-in. bit. Stated another way, a permissible unit loading of 5000 lb/in. of diameter on a 12-in. bit must be compared with a unit loading of only 2500 lb/in. of diameter on a 6-in. bit. Neglecting other variables, one may surmise then that the 12-in. bit will have a drilling rate potential double that of the 6-in. bit, or proportional to bit diameter. Obviously, for bits more nearly equal in size the difference in drilling rate potential is far less pronounced but it does exist.

There are practical and economic considerations which often limit full utilization of the good drilling rate potential of larger bits. The size and cost of surface and intermediate casing strings, through which the bits must be run, often limit bit size. The greater capital cost of heavier drill collars, larger drill pipe and heavier rigs may more than offset the drilling rate advantage of larger bits.

Successful use of slim hole drilling techniques is confined largely to relatively shallow areas where light and inexpensive rigs will suffice. Slim holes may also prove more economical when drilling to intermediate depths in soft rock formations, provided small casing is acceptable for completion.

A precise economic analysis of slim hole drilling techniques is difficult. In the first place, the reported saving, in overall well cost, is seldom broken down into the two major considerations, i.e.: hole-making cost and completion cost. Since it is not necessary to drill a small hole for setting small casing, the merits of each procedure must be considered individually.

The second difficulty arises from the fact that all too often the rig used for drilling a slim hole is equally suited for drilling a regular sized hole to the same depth. Use of small bits requires the use of small drill pipe. Higher circulating pressures that result often cannot be tolerated by the rig pump, hence the slim hole approach is unfairly condemned. Slim hole drilling should be undertaken only with rigs properly tailored for the job.

Bit Utilization

Weight on Bit and Rotary Speed. Roller cone rock bits are expendable. The problem is how best to wear them out. Many variables are involved. No exact formula has yet received wide acceptance so the most logical criterion remains cost per foot of hole drilled, considering cost of bit, plus cost of rig time while drilling and tripping, plus cost of mud maintenance and all other expense accruing during the bit run.

Rate of penetration becomes a very important factor in achieving the minimum cost per foot of hole drilled. Where rig rates are high and trips are relatively short, it will likely prove advantageous to increase the penetration rate, even at the expense of shorter bit runs.

It has been demonstrated in the laboratory and verified by field tests that penetration rates generally increase directly in proportion to the vertical force applied to the bit, so long as good cleaning and other factors remain constant. It has been shown that the penetration rate increases in proportion to bit rpm raised to some power less than one, and sometimes as low as ½.

Theoretically, under the perfect cleaning theory of rotary drilling as demonstrated by Maurer the rate of penetration varies as the square of weight on bit, whether expressed in total weight or in weight per inch of bit diameter. In like manner, Maurer has demonstrated, under perfect cleaning conditions, that penetration rates are directly proportional to bit rpm, other factors remaining constant.

Perfect cleaning is impossible to achieve in actual drilling operations. About all that can be hoped for in practice is uniformly good cleaning. Figs. 14-2 and 14-3 illustrate these several formulas.

Results of laboratory tests and actual field runs have been subjects of many fine technical papers in recent years. The two by Galle and Woods are of particular interest. Unfortunately, due to extraneous considerations, such as crooked hole tendencies, it is not always possible to adhere strictly to an optimum weight/speed/drilling program.

It is not necessary to have at one's disposal a well

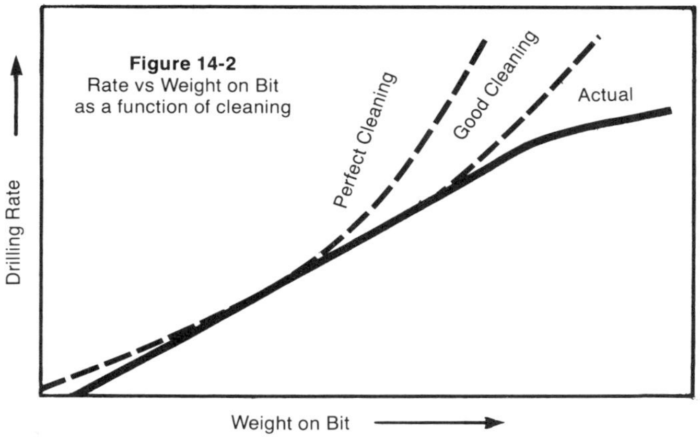

Figure 14-2
Rate vs Weight on Bit as a function of cleaning

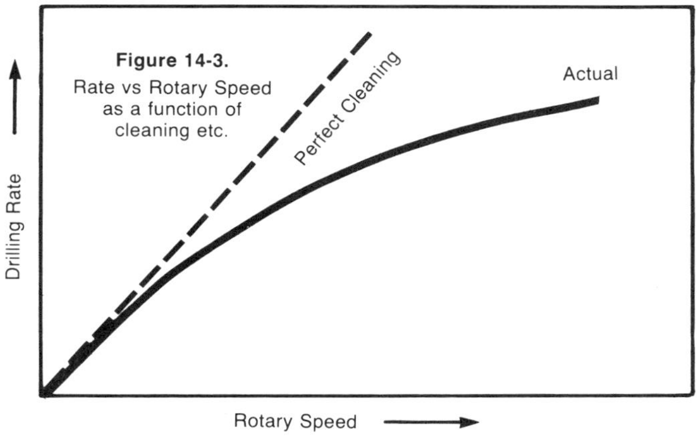

Figure 14-3.
Rate vs Rotary Speed as a function of cleaning etc.

equipped laboratory to evaluate properly the effects of weight, speed or bit hydraulics for a given drilling situation. The drill-off technique, proposed originally by Lubinski, makes most every rotary drilling rig its own laboratory.

Lubinski's original proposal dealt with the fundamentals involved. The technique employs a means for determining rate versus weight (or speed) over very short time or depth interval so that the effect of possible change in formation drillability or in bit condition is minimized, thus yielding results which may reasonably be regarded as instantaneous penetration rates.

The drill-off test procedure, as currently used in the field, involves, as a first step, the determination of a drill string slack-off or pick-up constant to be determined at the depth at which tests are to be made.

Fig. 14-4 illustrates graphically the results of one such slack-off test made to determine the constant (K) which represents depth in feet drilled (with brake set) per 1000 lb increase in suspended weight as shown by indicator. In this instance, K = .0500. Drill string constant was determined while circulating at the normal rate.

Using this constant and the indicator readings, any drilled depth interval can be calculated more accurately than it can be measured on the kelly. Expressed algebraically:

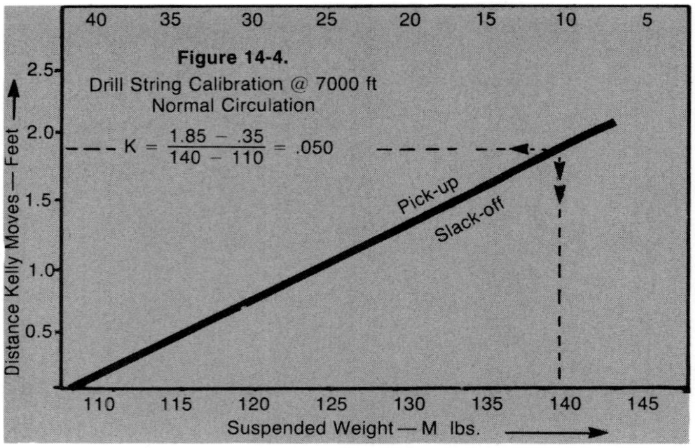

Figure 14-4. Drill String Calibration @ 7000 ft Normal Circulation

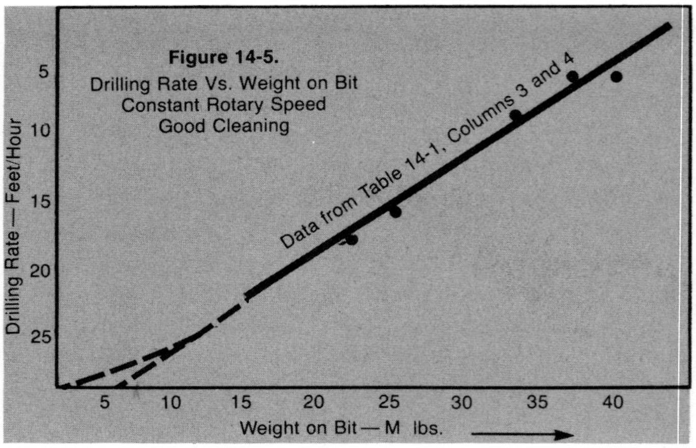

Figure 14-5. Drilling Rate Vs. Weight on Bit Constant Rotary Speed Good Cleaning

Feet drilled = K (W1 — W2)

Where W1 = initial weight on bit is 1000 lb.

W2 = final weight on bit is 1000 lb.

By recording the time in seconds required to drill off a given weight (with brake set), the drilling rate may be determined as follows:

$$\text{Rate (ft/hr)} = 3600\ \frac{K(W1\text{-}W2)}{\Delta T}$$

The weight on bit corresponding to the rate thus determined is the average weight or: (W1 + W2)/2. Test results may be tabulated as follows:

(1)		(2)		(3)		(4)
Weight on Bit		Time — Seconds		Av. Wt. on Bit		Rate — Ft./Hr.
M lbs.	M lbs. W1 − W2	Elapsed	Δ T	M lbs. (W1 + W2)/2	3600	K(W1 − W2)/Δ T
43.0		0				
40.0	3.00	21	21	41.50		25.7
35.0	5.00	58	37	37.50		24.3
30.0	5.00	100	42	32.50		21.5
25.0	5.00	157	57	27.50		15.8
20.0	5.00	240	83	22.50		10.8

Table 14-1. Note: K = .0500

Fig. 14-5 is a graph showing rates (Col. 4) versus weight on bit (Col. 3) for the test run at 7000 ft using 7⅞-in. hard formation bit and 4¼-in. drill pipe.

Similarly, rate versus speed tests may be run but when running rate versus speed or rate versus weight tests, bit hydraulics must be maintained uniform and all other variables held constant.

Although the drill-off technique has not been used extensively for determining the measure of bit hydraulics required for any given situation, it is anticipated that some work will be done in this field in the near future.

If and when a reliable yardstick can be agreed upon for measuring bit hydraulics as it relates to cleaning efficiency, it then will be possible with this technique to define some of the parameters for determining in advance the required bit hydraulic energy for effective cleaning various holes sizes over a wide range of instantaneous drilling rates.

Circulating Fluids

Hydraulics, as employed in rotary drilling, has received much attention during recent years. Possibly some aspects have been overemphasized but there remains a broad lack of understanding of fundamentals. Actually, there is still a wide disagreement among the so-called experts.

First, circulating fluids, regardless of nozzle velocity, do not destroy consolidated rock to any significant degree. The function of the drilling fluid, insofar as making hole is concerned, is simply to clean away the debris generated by the bit and remove it from the hole. Effective bottom hole scavenging depends on properties of the fluid and the rate at which it moves across the bottom of the hole.

Scavenging is improved by increases in both circulating rate and fluid density but is degraded by increase in fluid viscosity. Effective bottomhole scavenging causes hydraulic horsepower to be dissipated across the nozzles of the bit. Whether bit hydraulic horsepower is a good measure of effective cleaning is still doubtful.

Of the numerous papers to be found in the literature, the two presented by McLean in 1964 and 1965 are of particular interest because they present a new concept which seems to make a lot of sense. In his two papers, both reporting on studies of cross flow under the bit, McLean's basic conclusion is that effective cleaning is a function of QV. Simply stated this means that gpm carries just as much weight as nozzle velocity. Bit hydraulic horsepower on the other hand is proportional to QV^2, in which nozzle velocity assumes a disproportionate importance. This perhaps can be illustrated best by a hypothetical case. Suppose a 7⅞-in. hole using 4½-in. drill pipe is being drilled through hard rocks at a rate such that the hole is being effectively cleaned while circulating 400 gpm with a nozzle velocity of 200 ft/sec. Based on the QV concept the hole might be cleaned equally well, but no better, while circulating 200 gpm with a nozzle velocity of 400 ft/sec.

It would appear from the foregoing analysis, if the QV concept is valid, that it will be advantageous to use large circulating rates with moderate nozzle velocities in the upper portion of the hole and higher nozzle velocities with lower rates of circulation in the lower portion. This is no different from the bit hydraulic horsepower concept except in the degree of accent on nozzle velocity. Possibly very high velocities are less important than previously thought. McLean's work places considerable doubt on the use of bit hydraulic horsepower as being a true measure of cleaning beneath the bit.

As stated previously perfect cleaning is never realized in actual drilling practice. All efforts to achieve near-perfect cleaning soon run into the law of diminishing returns. There comes a time when the cost of providing better cleaning exceeds the benefits derived, hence the excess hydraulic energy provided is useless.

An important field for study, and one which has hardly been scratched, is to determine in advance how much hydraulic energy can be usefully employed for effectively cleaning the hole. There is little doubt that excess capital costs and increased operating expense result from providing more pump and power than can be usefully applied.

The ultimate aim of any such study would be to develop a chart or graph which could be used more or less universally in planning the hydraulics program for a proposed well.

With so many large pumps already in service, the old philosophy "the more the better" is not a very good solution to the hydraulics problem.

Effective bit hydraulics often are handicapped by ill chosen elements of the drill string. Energy dissipated through friction of fluid flow above the bit does no useful work and deprives the hole of cleaning energy. Many misconceptions still persist. Take for example drill pipe and tool joints. The restriction in the bore of tool joints has been over-emphasized.

Many people believe that 4-in. OD drill pipe with IF joints is equal, for fluid flow, to 4½-in. OD drill pipe with XH joints. The OD of the tool joints usually is the same but the pressure drop inside 4-in. OD with IF joints is 70% greater than that in 4½-in. OD with XH joints. It should be noted also that pressure loss in 5-in. OD drill pipe with XH joints is only 56% of that occurring in 4½-in. drill pipe with XH joints.

Long drill collar strings with small bores often produce 10 times as much pressure loss as does an equal length of drill pipe with which they are being run. A slight increase in bore will reduce this loss by 50% without reducing the drill collar weight to any appreciable extent.

Surface equipment — lines, hose, swivel and kelly — usually are not a critical factor since the length of the flow path through these elements is reasonably short. It must be remembered, however, that these pressure losses are additive and, in deep wells, often are the factors which limit the amount of energy available for cleaning. The graphs included in API Bulletin D-10 illustrate these features.

Figure 14-6.
Drilling Rate vs Mud Column Pressure

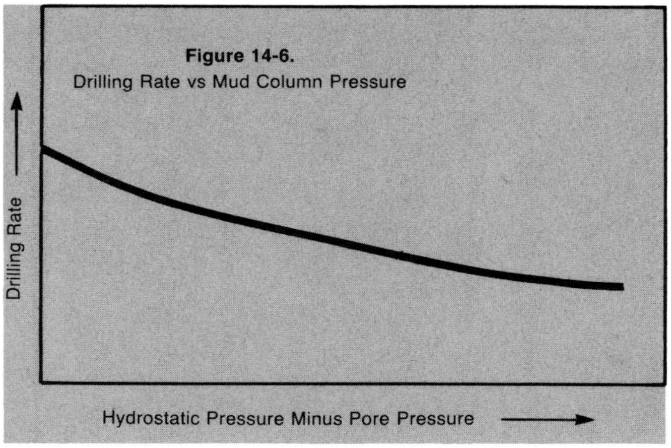

One of the more important factors influencing the rate of penetration concerns the circulating fluid and the hydrostatic pressure, in excess of formation pore pressure, which it exerts at the bottom of the hole. This phenomenon, generally known as chip hold-down, is not clearly understood but its existence is undeniable. Fig. 14-6 shows graphically a classical example of this phenomenon.

Even though bottomhole scavenging supposedly is improved by an increase in density of the drilling fluid, the hydrostatic head developed by dense fluids far more than offsets any possible benefit in its cleaning power. It appears to be impossible, by better bit hydraulics, to overcome completely the effects of hydrostatic pressure.

Two solutions, however, seem possible. First, and most effective, is reduction of density by removal of solids. This brings about the need to be ever ready to detect and control kicks if abnormal pressures are encountered. This technique, commonly referred to as drilling for kicks has proved successful in most instances.

The second or alternate solution, if low density drilling fluids cannot be tolerated, is use of diamond bits which employ a somewhat different mechanism in rock destruction.

In summary, the four major factors affecting the rate of penetration are:

1. Weight on bit
2. Bit rpm (or speed)
3. Scavenging — bit hydraulics
4. Hydrostatic pressure on bottom

This is not really anything new. Drillers have known all along that the way to make more hole is to mash down harder, turn faster, water down the mud and pump like hell.

Chapter 15
Casing Programs

Proper selection and handling of oil well casing is one of the important phases of drilling operations. Casing costs usually represent a large percentage of the total drilling and completion expenses.

In exploratory wells, decision to set casing for conclusive testing of possible producing zones is often the most important one made. Protection from caving formations, support of surface equipment and prevention of unwanted fluid production are the main functions of casing.

A casing program is a systematic plan for proper selection and handling of casing. It is the purpose of this chapter to discuss the factors which must be considered in order to have an effective program.

Casing Characteristics

Before describing methods of selection and handling casing, it is necessary to understand casing properties. There are many possible ways to describe and classify casing. In this section, the following breakdown will be used: (1) casing functions; (2) seamless steel casing; (3) other types of casing.

Casing Functions. Conductor, surface, intermediate and oil production strings are the four classifications of casing according to function. Conductor pipe is used primarily where the surface soil is weak. It makes the cellar more stable and provides a good start for subsequent drilling. Where surface is stable, this type of casing is not required. Fig. 15-1 illustrates a hole containing surface, intermediate and oil strings.

Surface pipe or casing is used with cement to protect fresh water sands from contamination and to support surface equipment such as blow-out preventers. How deep a hole for surface casing depends primarily on the depth of the fresh water sands and the expected stresses due to high pressure formations.

The minimum legal length of surface casing to be set in a given area is fixed by various state agencies.

Cement slurry is pumped through the casing and up the annulus until the slurry reaches the surface. Occasionally slurry is dumped in the annulus if it is not filled up to the top.

Intermediate, protective or water strings are set when cave-ins, wash-outs or other problems in upper formations make drilling difficult. With present improved methods of drilling and mud control, the need for the intermediate string has lessened to the point where it is seldom employed in wells as deep as 10,000 ft.

Production strings are set either above the producing formation for open-hole completions or through the zone when production will be through perforations.

In most cases all strings extend to the surface. The production string in some cases, however, extends only to the bottom of an intermediate string. In this case, the string is known as a liner. Liners are sealed to the protector string.

While use of liners can cut cost of casing, many problems such as junction leaks and less satisfactory cementing jobs (than on full length strings) limit their desirability.

Seamless Steel Casing. The most widely used type of casing is seamless steel. This pipe is manufactured from solid round bars by piercing the centers at high temperature until the desired wall thicknesses are obtained.

Seamless steel casing joints come furnished with a screwed coupling and thread protectors. Fig. 15-2 illustrates the difference between long and short couplings.

Table 15-1 describes in detail the properties and dimensions of seamless, API, non-upset casing. There are other types of seamless casing which vary from those described in the table. Information on these types can be obtained from the various manufacturers' literature.

The complete description of API casing includes outside diameter, nominal weight per foot (based on weight of 20 ft of casing including coupling), steel grade and type of coupling [long thread and coupling (LT&C) or short thread and coupling (ST&C)].

For example, 5½-in. OD, 15.5 lb per ft, J, ST&C is a complete description of a section of casing. Once this description is known, it is possible from Table 15-1 to obtain the inside diameter, wall thickness, drift diameter (maximum size tool which can be run through the casing), outside diameter of coupling which in turn is basic information necessary to determine bit size plus information about the strength characteristics of the casing. From the description given above (5½-15.5-J-ST&C) the following has been obtained from Table 15-1.

ID (inside diameter)	4.950 in.
Wall thickness	0.275 in.
Drift diameter	4.825 in.
OD coupling	6.050 in.
Collapse resistance	4,810psi

Size: OD (In.)	Weight per foot: Nominal: threads and coupling (Lb)	Weight per foot: Plain end (Lb)	Dimensions: Wall thickness (In.)	Dimensions: Inside diameter (In.)	Dimensions: Drift diameter (In.)	Coupling OD: API Non-upset, upset and buttress (In.)	Coupling Length: API Short (In.)	Coupling Length: API Long (In.)	Steel grade	Internal pressures: Mill test (PSI)	Internal pressures: Yield (min.) (PSI)	Internal pressures: Ultimate Min. (PSI)	Internal pressures: Ultimate Average (PSI)	Col-lapse: Setting depth SF* 1⅛ (Ft)	Collapse resistance (psi) (PSI)	Tension based on minimum properties: Ultimate joint strength (min.): Short coupling (1000 Lb)	Tension based on minimum properties: Ultimate joint strength (min.): Long coupling (1000 Lb)
5½	13.00	12.84	.228	5.044	4.919	6.050	6¾	...	F	1600	1810	4140	4560	2950	1810	95	...
	14.00	13.70	.244	5.012	4.887	6.050	6¾	...	H	2800	3110	5990	6650	4340	3110	139	...
	14.00	13.70	.244	5.012	4.887	6.050	6¾	...	J	3000	4270	7980	8870	5640	4270	186	...
	15.50	15.35	.275	4.950	4.825	6.050	6¾	8	J	3000	4810	9000	10000	6860	4810	211	247
	17.00	16.87	.304	4.892	4.767	6.050	6¾	8	J	3000	5320	9940	11050	8000	5320	234	275
	17.00	16.87	.304	4.892	4.767	6.050	6¾	8	N	3000	7740	10940	12160	10470	7740	273	320
	20.00	19.81	.361	4.778	4.653	6.050	6¾	8	N	3000	9190	13000	14440	13480	9190	326	382
	23.00	22.54	.415	4.670	4.545	6.050	6¾	8	N	3000	10560	14940	16600	15820	10560	375	440
7	17.00	16.70	.231	6.538	6.413	7.656	7¼	...	F	1300	1440	3300	3630	1960	1100	118	...
	17.00	16.70	.231	6.538	6.413	7.656	7¼	...	H	2100	2310	4460	4950	2440	1370	160	...
	20.00	19.54	.272	6.456	6.331	7.656	7¼	...	H	2500	2720	5250	5830	3410	1920	191	...
	20.00	19.54	.272	6.456	6.331	7.656	7¼	...	J	3000	3740	6990	7770	4440	2500	254	...
	23.00	22.63	.317	6.366	6.241	7.656	7¼	9	J	3000	4360	8150	9060	5850	3290	300	344
	23.00	22.63	.317	6.366	6.241	7.656	7¼	9	N	3000	6340	8960	9960	7640	4300	350	400
	26.00	25.66	.362	6.276	6.151	7.656	7¼	9	J	3000	4980	9310	10340	7220	4060	345	395
	26.00	25.66	.362	6.276	6.151	7.656	7¼	9	N	3000	7240	10240	11380	9460	5320	402	460
	29.00	28.72	.408	6.184	6.059	7.656	7¼	9	N	3000	8160	11540	12820	11320	6370	454	520
	32.00	31.68	.453	6.094	5.969	7.656	7¼	9	N	3000	9060	12820	14240	13160	7400	505	578
	35.00	34.58	.498	6.004	5.879	7.656	7¼	9	N	3000	9960	14080	15650	14970	8420	554	635
	38.00	37.26	.540	5.920	5.795	7.656	7¼	9	N	3000	10800	15270	16970	16140	9080	600	688
8⅝	24.00	23.57	.264	8.097	7.972	9.625	7¾	..	F	1200	1340	3060	3370	1690	1340	161	...
	24.00	23.57	.264	8.097	7.972	9.625	7¾	...	J	2700	2950	5510	6120	2540	1430	288	...
	28.00	27.02	.304	8.017	7.892	9.625	7¾	...	H	2300	2470	4760	5290	2810	2470	252	...
	32.00	31.10	.352	7.921	7.796	9.625	7¾	...	H	2600	2860	5510	6120	3750	2860	295	...
	32.00	31.10	.352	7.921	7.796	9.625	7¾	10	J	3000	3930	7340	8160	4870	3930	393	437
	36.00	35.14	.400	7.825	7.700	9.625	7¾	10	J	3000	4460	8340	9270	6080	4460	448	499
	36.00	35.14	.400	7.825	7.700	9.625	7¾	10	N	3000	6490	9180	10200	7950	6490	522	581
	40.00	89.29	.450	7.725	7.600	9.625	7¾	10	N	3000	7300	10330	11480	9580	7300	589	655
	44.00	43.39	.500	7.625	7.500	9.625	7¾	10	N	3000	8120	11480	12750	11240	8120	654	729
	49.00	48.00	.557	7.511	7.386	9.625	7¾	10	N	3000	9040	12790	14210	13100	9040	729	812
9⅝	29.30	28.04	.281	9.063	8.907	10.625	7¾	...	F	1200	1280	2920	3210	1530	860	185	...
	32.30	31.03	.312	9.001	8.845	10.625	7¾	...	H	2100	2100	4370	4860	2350	1320	279	...
13⅜	48.00	45.98	.330	12.715	12.559	14.375	8	...	F	750	1080	2470	2710	1000	560	260	...
	48.00	45.98	.330	12.715	12.559	14.375	8	...	H	1200	1730	3330	3700	1320	740	352	...
	54.50	52.74	.380	12.615	12.459	14.375	8	...	J	1900	2730	5110	5680	2030	1190	545	...
16	55.00	52.36	.3125	15.375	15.188	17.000	9	...	F	600	850	1950	2150	500	280	258	...
	65.00	62.58	.3750	15.250	15.062	17.000	9	...	H	1100	1640	3170	3520	1140	640	423	...
	75.00	72.72	.4375	15.125	14.938	17.000	9	...	J	1800	2630	4920	5470	1800	1010	662	...
	84.00	81.97	.4950	15.010	14.822	17.000	9	...	J	2000	2980	5570	6190	2630	1480	753	...

TABLE 15-1. Seamless API Casing (Non Upset)

Ultimate joint strength ....211,000 lb

Grades of steel used in API casing range (in strength) from H-40 with a minimum yield of 40,000 psi to P-110 with a minimum yield of 110,000 psi. The variation in yield and in tensile strength depends on the chemical composition, mainly carbon, and the heat treatment or tempering process used.

Table 15-1 is a condensed and comprehensive tabulation of API oilwell casing dimensions and properties. Since its original compilation, however, sizes 4½-in. OD, 5-in. OD, 6⅝-in. OD, 7⅝-in. OD, 10¾-in. OD, 11¾-in. OD, 18⅝-in. OD and 20-in. OD have been added to API specifications.

Grade F-40 has been deleted and grades K-55 and P-110 have been added to API specifications. K-55 differs from J-55 only in that its tensile strength is 95,000 psi compared to 75,000 psi for grade J-55. API specifications are revised from time to time and it is recommended that late issues of API spec 5A be consulted.

Other Types of Casing. Casing can be described as follows: (1) by function, (2) method of manufacture, (3) type of coupling or joint, (4) type of material used.

For example, types include seamless, welded and rivted casing. Casing is usually threaded but may also be beveled for welding. Some special threads other than the standard API are manufactured and used to

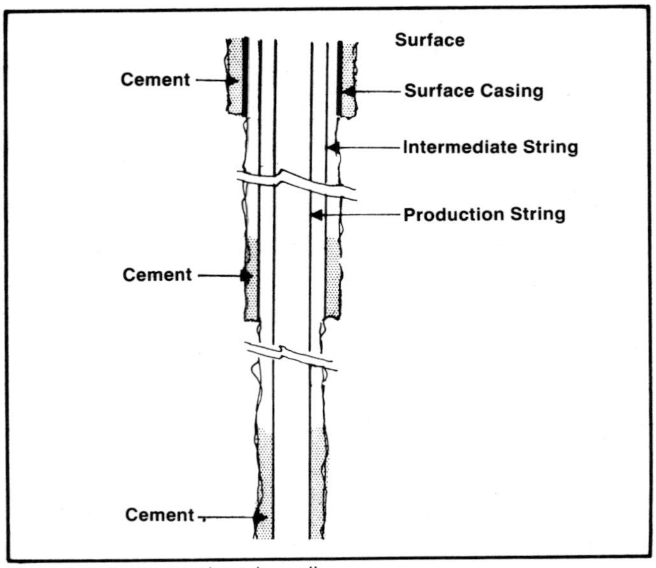

Fig. 15-1. Casing strings in well.

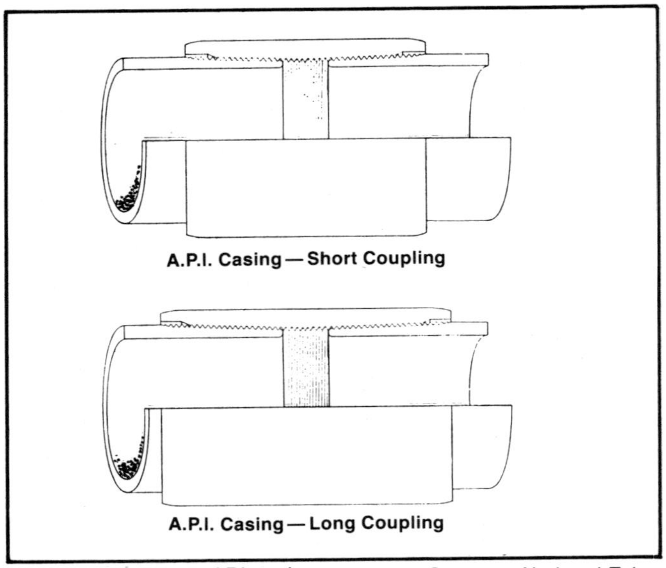

Fig. 15-2. Seamless API casing. *— Courtesy National Tube.*

improve tensile strength.

For shallow wells, less expensive casing types such as spiral welded, corrugated and riveted sheet steel are available.

Alloy steels are used where corrosion is a problem. Plastic pipe, which is easy to handle and resists corrosions, is available.

Casing Costs

Study of casing expenses is an integral part of the casing program. Cost of casing, transportation, rig time for handling and running casing and supervision are direct casing expenses. Cost of casing cement jobs could be considered as part of the casing expenditure.

Part of the expense can sometimes be recovered when a well is abandoned. This is true because it is possible to cut casing loose above the cement and bring it to the surface. If the casing is in good condition it can frequently be reused. Otherwise it has some scrap value.

Second-hand casing has been used more often in recent years due to occasional shortages. For non-critical applications even when there are no shortages, this casing has been used to take advantage of lower cost.

Casing Design

The term casing design simply means correct selection of casing. Correct, in this case, signifies (1) that the casing will not fail at any time during the life of a well; (2) that the casing size will be adequate for efficient production and (3) that the cost to satisfy the first two conditions will be a minimum.

Factors involved with proper selection of casing size are discussed later. Once the size is set, selection of casing to prevent failure at minimum cost becomes the prime problem.

A survey made by the Mid-Continent District Study Committee on Casing Programs (API) indicated that failures attributed to design practices are almost nonexistent. This means that the accepted practice method for selection of casing is satisfying the previously mentioned condition that casing must not fail. On the other hand there remains the question of minimum cost.

It is probable that in many cases overdesign or in other words over-expenditure is the result of using the accepted safe method for casing selection. This possibility was recognized by the API committee previously mentioned. Saye and Richardson reported that substantial savings were accomplished in the Elk City field, OK, by using lower than conventional safety factors. Bowers suggested that the basic assumptions in the accepted design method were unrealistic. Texter, on the other hand, indicated (in discussion of effect of tension on collapse resistance) that there are still too many factors unknown to justify radical changes in the design method.

The trend in design of casing is certainly in the direction of using less steel either by using the conventional method with lower safety factors or by considering the physical conditions which actually exist in a given case as Bowers did.

Casing Design — Causes of Failure

In order to design casing strings that will not fail, the causes of failure must be known. Texter presented a thorough study of all known reasons for casing troubles.

The principal factors considered in casing design are: (1) collapse, (2) tension and (3) bursting (internal pressure).

1. Collapse failure is the result of unbalanced external pressures. When it occurs, the casing will flatten. Collapse can occur when the fluid pressure outside of the casing is very large compared to the internal pressure. This might happen when the casing is empty and the full pressure due to the head of mud or formation water outside the pipe is exerted (neglecting counteracting buoyancy effects).

If the casing were empty at 8000 ft and a 11.6 lb/gal mud were used 0.602 psi per ft, the relative external pressure which the casing must withstand is 8000 ft x 0.602 psi/ft = 4816 psi.

Failure may also occur during a squeeze job if the high pressure exerted through the perforations affects the casing above the packer. Fig. 15-3 shows how collapse might occur during such an operation.

Besides fluid pressure failures, collapse can be caused by mechanical forces. For example, shifting formations or earthquakes can cause trouble. Dents that might result from bad handling can greatly reduce collapse resistance.

It has been pointed out that deformation of 1% can reduce collapse resistance by 25%. For this reason, casing must be handled carefully. Tension will also reduce the collapse resistance. This factor is considered quantatively in casing design.

2. Tension in casing results from (1) its own weight; (2) pulling through tight sections during reciprocation and (3) temperature changes during and after cementing.

The tension caused by its own weight is partially offset by the buoyancy in mud. For example, the weight of 12.0 lb/gal mud displaced by 1000 ft of 7-in. 23 lb/ft casing, when mud is both in and outside the pipe, is 4209 lb. This weight according to Archimedes Principle is subtracted from the weight in air to determine the effective weight in mud of the casing. In this case, it is 18,791 lb.

Casing of 18 lb/ft 5½-in. and 10,000-ft weighs 180,000 lb in air but only 141,480 lb in a 14.0 lb/gal mud. Buoyancy correction factors are available in various handbooks.

When casing, partially empty, is run into a hole the buoyancy effect is very large and the casing is floated in. In the conventional design method, buoyancy is not considered. This is one of the principal criticisms aimed at it.

When casing is cemented, the heat of hydration causes significant temporary temperature increases which cause expansion and increase compressive stresses. When the subsequent cooling occurs, if the casing is fixed on both top and bottom, shrinking will tend to occur and tension stresses will be severe. This

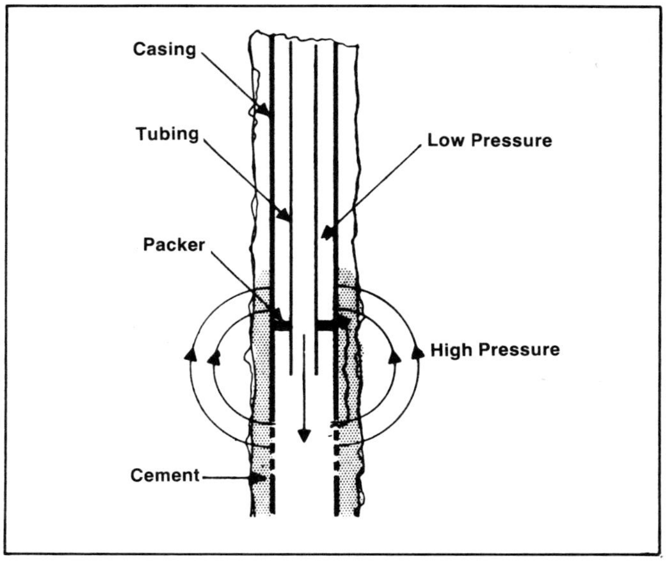

Fig. 15-3. How collapse failure might occur during a squeeze job.

in turn would affect collapse resistance, as pointed out before.

3. Bursting failure, which produces longitudinal breaks in casing, occurs as a result of high relative internal pressures. These pressures are equal to the expected bottom hole pressure less the pressure due to the weight of the fluid. After the casing is set, a pressure test with mud in the casing at the expected pressure, checks for leaks and bursting strength.

Internal pressure would be most severe at the top of the casing when gas is present inside. Some designers assume the full formation pressure (about 0.5 psi per ft) for this case while others subtract the weight of the gas. For example, if the pressure in a 6000-ft gas formation is 3000 psi, the pressure at the surface would be about 2500 psi. The actual figure depends on the average density of the gas.

Some other causes of casing trouble are:

1. Thread failures due to shocks.
2. Leakage.
3. Wear from contact with drill pipe, wire line, tubing and rods.
4. Torsion — if casing is rotated during cementing operations.
5. Improper handling.
6. Corrosion — external and internal.

Whereas the only factors which will be considered in the design method in this chapter are collapse, tension and bursting, an effective program must consider the less prominent causes of failure.

For example, if external corrosion is a problem, plans should be made: (1) to prevent it by cementing off key formations or using inhibitors: (2) live with it by using thicker casing or (3) consider cathodic protection.

Wear is another factor that receives attention of some designers.

TABLE 15-2.
Clearance Guide for Casing Strings.

O D Coupling (inches)	Clearance (inches to be added to coupling O D)
5	1.0
6	1.25
7	1.75
8	2.0
9	2.5
10	3.0
11	3.25
12	3.25
13 and up	3.50

TABLE 15-3.
Common Bit Sizes. (Inches)

Common Bit Sizes (inches)

3 3/4	7 3/8	10 5/8
3 7/8	7 5/8	11
4 1/4	7 3/4	12
4 5/8	7 7/8	12 1/4
4 3/4	8 1/2	13 3/4
5 5/8	8 5/8	15
6	8 3/4	17 1/2
6 1/8	9	22
6 1/4	9 5/8	23
6 3/4	9 7/8	

Casing Diameter Selection

The first consideration in the selection of casing is the size (OD) of the production string. No simple rules can be laid out for reaching this first decision; however, the important factors are listed and discussed below.

1. *Drilling cost.* As the hole diameter increases so will the cost of completion. This cost of large diameter holes should be balanced against expected economic advantages.

2. *Methods of production.* A well may flow naturally in its early history but may require pumping later. The hole should be large enough so that necessary equipment will fit in it.

3. *Production rates.* This is an important factor only when production rates are high enough to cause appreciable pressure drops in the production tubing.

4. *Possibilities of multi-zone completion.* Hole sizes should be large enough to handle equipment for multi-zone completion if this is a reasonable possibility.

5. *Number of intermediate strings.* If the expected drilling conditions necessitate one or more intermediate strings, the maximum size of the production string will be limited. (More detail later.)

6. *Nature of fluids produced.* This factor is important primarily because it affects the choice of production equipment. In turn, the downhole equipment and accessories limit the minimum hole size.

7. *Rig limitations.* Normally, the selection of a rig depends on the size and depth of the hole to be drilled. There are cases, however, when rig selection is limited in a given area. In these cases, the size of the hole and casing is determined by the rig capabilities.

8. *Workovers.* If experience indicates that remedial work is common, the hole size should be large enough to accommodate necessary equipment.

9. *Availability of casing.* There have been many times in recent years where casing has been in short supply. Such a shortage can and has been the determining factor in establishing production string size.

10. *Common practice.* While all factors mentioned above should be considered independently, the experience of others in a given area should also be carefully studied before reaching a decision.

11. *Exploratory vs development well.* In an exploratory well, the prime purpose of drilling is to prove up the existence of commercial zones. Frequently, uncertainty requires the setting of casing. In the interest of economy, slim hole drilling and completion may extend the exploration budget. This should be weighed against possible production problems at some later date.

When the diameter of the casing has been fixed for the production string, the selection of bit and upper casing string sizes can be determined. The method for selection of the above mentioned sizes can best be illustrated by an example.

EXAMPLE 1. What bits and casing will be used in a 10,000-ft well if the production string consists of 5½-in. casing? An intermediate string will bottom at 5500 ft and surface pipe will be run to 1500 ft.

1. *What size bit will allow 5½-in. casing to be run?*

Referring to Table 15-1 (5½-in. OD) it can be seen that outside diameter of the coupling is 6.050 in. This means that the bit size necessary to permit the casing to be run must exceed 6.050 in. A clearance beyond this diameter must be provided to account for mudcake, casing appliances such as centralizers and scratchers and unusual hole conditions (e.g. caving formations, crooked hole).

While there are no rules available which apply to every case, Hill gave a guide which correlated clearance with maximum joint size and casing length. Table 15-2 is a guide to clearance selection which takes into consideration joint size and maximum probable setting depths. Using Table 15-2 for the stated problem, a clearance of 1.25 in. is suggested for the 6.050-in. coupling (exact interpolation of data is not warranted since the table is only a guide). This means the bit size required should be at least 6.050 + 1.25 = 7.30 in.

Referring to Table 15-3 which lists common bit sizes, the 7⅞-in. bit would be selected.

2. *Through what minimum size intermediate casing string will the 7⅞-in. bit pass?*

The 7⅞-in. bit selected for the production string must fit through the intermediate casing string. The drift diameter column in Table 15-1 indicates the maximum size tool which will go through any size casing. An examination of this column shows that the 7⅞-in. bit will pass through any 8⅝-in. casing since the drift diameter varies from 7.386 in. to 7.972 in.

3. *What size bit will allow 8⅝-in.casing to be run?*

From Table 15-1 the coupling OD is 9.625 in. Table 15-2 indicates that the clearance should be 3 in. or the minimum bit size 9.625 in. + 3 in. = 12.625 in. Table 15-3 shows the next larger bit size to be 13¾-in.

4. *What size surface casing should be used to permit a 13¾-in. bit to pass through it?*

Table 15-1 shows that the smallest casing that will meet the above requirements is 16 in. (drift diameter ranges from 14.822 in. to 15.188 in.).

5. The last step in this problem is to determine the bit size that will allow the surface casing to be run.

		Reference
Coupling OD	17.000 in.	Table 15-1
Clearance	3.5 in.	Table 15-2
Minimum bit size	20.5 in.	
Actual bit size	22.0 in.	Table 15-3

Summary of:	Prod'n.	Intr.	Surf.
Casing size OD (in.)	5½	8⅝	16
Coupling OD (in.)	6.050	9.625	17.000
Clearance (in.)	1.25	3	3.5
Minimum bit size	7.30	12.625	20.5
Actual bit size	7⅞	13¾	22

The foregoing analysis of a casing program for a 10,000 ft well (EXAMPLE 1) makes liberal allowances for clearance, which may be necessary under some conditions but in most areas a more economical program and one generally followed would be as follows:

- Drill a 17½-in. hole to 1500 ft into which will be run 13⅜-54.50-J-ST&C casing, having a coupling OD of 14⅜ in. and a drift of 12.459 in.
- Drill a 12¼-in. hole to 5500 ft into which will be run 8⅝-32.00-J-ST&C casing having a coupling OD of 9⅝ in. and a nominal drift of 7.796 in.
- Drill a 7⅞-in. hole to 10,000-ft into which will be run 5½-17.00-N-LT&C casing having a coupling OD of 6.050 in. and a drift of 4.767 in.

Note: Since 7⅞-in. bits are more popular and more readily available than the next smaller sizes, it may be worthwhile to physically drift each length of 8⅝-32.00 casing with 7⅞-in. drift gauge before casing is run, then use 7⅞-in. bits.

TABLE 15-4.
Density as a Function of Pressure Head

Density lb/gal	Pressure psi/ft	Mud Weight lb/gal	Pressure psi/ft
7.0	0.363	11.0	0.571
7.5	0.389	11.2	0.581
8.0	0.415	11.4	0.592
8.3	0.433	11.6	0.602
8.4	0.436	11.8	0.612
8.6	0.446	12.0	0.623
8.8	0.457	12.2	0.633
9.0	0.467	12.4	0.644
9.2	0.477	12.6	0.654
9.4	0.488	12.8	0.664
9.6	0.498	13.0	0.675
9.8	0.509	13.5	0.701
10.0	0.519	14.0	0.727
10.2	0.529	14.5	0.753
10.4	0.540	15.0	0.778
10.6	0.550	16.0	0.830
10.8	0.560		

Casing Design — Surface Casing

Design of surface casing is frequently neglected on the assumption that cementing of surface casing to the surface provides all the strength necessary.

In the API study of casing string design factors, it was reported that many operators do consider collapse, tension and bursting when selecting surface casing. Wear and corrosion are also considered by some. One company used only J-55 grade pipe which almost eliminated the need for design.

EXAMPLE 2. 13⅜-in. surface casing is to be set to 1500 ft. Specify the casing to be ordered. Internal pressure is not expected to exceed 1000 psi. Assume conventional safety factors.

Conventional safety factors (SF) are 1.1 5 (1⅛) for collapse, 1.8 for tension and 1.1 for internal (or burst). The conventional method assumes (1) that the pipe is empty or in other words exposed to the full pressure of the external column of mud and (2) that buoyant forces of the mud on the casing are negligible. It is frequently assumed that the external column of fluid exerts a pressure of 0.5 psi/ft (9.625 lb/gal) (Table 15-4).

1. *Collapse.* At 1500 ft the column of liquid would exert 1500 ft × 0.5 psi/ft = 750 psi. The casing should be able to withstand 1.125 × 750 = 844 psi. Under Collapse Resistance column in Table 15-1 for 13⅜-in. casing, it can be seen that the first type of casing that can withstand 844 psi is the 54.50-lb/ft J grade; it has a collapse resistance of 1140 psi.

The column headed Collapse Setting Depth, SF 1⅛, was prepared from the collapse resistance data for the safety factor and mud weight conditions of this problem. Using this column the selection of the proper casing for collapse is easier. In this case where 1500 ft is the setting depth 13⅜-in., 54.50-lb/ft J grade is required. This casing could be set as deep as 2030ft, according to the table.

Conclusion: 13⅜-in., 54.50-lb/ft J grade steel will satisfy the collapse requirements.

2. *Tension.* The casing selected above comes with a short coupling that has an ultimate joint strength of 545,000 lb (Table 15-1). Using a safety factor of 1.8, the maximum weight to be placed on this casing is 545,000/1.8 = 303,000 lb and 1500 ft of 54.50-lb/ft casing weighs 81,750 lb

Conclusion: 13 ⅜-in., 54.50-lb/ft, J, ST&C is satisfactory for tension.

3. *Internal Pressure.* The selected casing has a minimum yield pressure of 2730 psi. (Table 15-1). Applying a safety factor of 1.1 means it can be safely used wherever the internal pressure does not exceed 2730/1.1 or 2480 psi. This is safer than required.

Conclusion: Specify for surface casing 1500-ft, 13⅜-in., 54.5-lb/ft, J, ST&C.

EXAMPLE 3: Here, 9⅝-in. surface casing is to be set at 2500 ft. Experience indicates that safety factors of 1.0 for collapse and 1.5 for tension are workable. Mud weighs 9 lb/gal. Internal pressure is not expected to be significant. Specify the casing which will meet the requirements. The steps to the solution can be handled by the following form.

EXAMPLE 3. Surface Casing Selection.

1. Casing size — 9⅝ in.
2. Length — 2500 ft
3. Mud weight — 9.0 lb/gal
4. Mud pressure — 0.468 psi/ft
5. Safety factors — (SF)
   Collapse — 1.0
   Tension — 1.5
   Internal — Not considered
6. Collapse stress (mud pressure × length)—1170 psi
7. Minimum casing requirements (collapse) — 1170 psi
8. Casing satisfying collapse requirements — 9⅝ in., 32.30 lb/ft, H
9. Coupling type — (short or long) — Short.
10. Ultimate joint strength — 279,000 lb
11. Maximum allowable tension — 186,000 lb (ultimate strength/ SF)
12. Tension due to weight of casing — 80,750 lb
    Casing length × weight per ft (therefore OK)
13. Interal stress expected (psi)— not considered
14. Minimum yield strength for selected casing (psi)— not considered
15. Maximum allowable internal pressure (psi)— not considered
16. Remarks — No unusual wear or corrosion conditions are expected
17. Description of casing selected—2500 ft; 9⅝ in., 32.3 lb 60ft; H, ST&C.

EXAMPLE 4. Second-hand 10¾-in., 51.0 lb/ft; N; ST&C casing available. Inspection shows that it has maintained its original thickness and the threads are in good condition. To be sure that it is used safely, the safety factor for collapse is to be 1.5, that for tension 2.0 and internal pressure 1.3. Internal pressure is not expected to exceed 2500 psi. Mud weight is 10.0 lb/gal. How deep can this casing be set?

In this problem the properties of the casing for collapse resistance (3440 psi), tension (804,000 lb) and yield strength (5860 psi) are available.

1. Check for allowable internal pressure.

$$\frac{5860}{1.3} = 4510 \text{ psi}$$

Since this is greater than 2500 psi, the casing can be used. If it were not, the casing would not meet requirements and further calculations would not be necessary.

2. Check for allowable length of casing which can be supported in tension.

   Allowable weight — $\frac{804{,}000}{2.0} = 402{,}000$

   Allowable length — $\frac{402{,}000 \text{ lb}}{51.0 \text{ lb/ft}} = 7822 \text{ ft}$

3. Check for maximum setting depth which is safe from collapse.

   Maximum allowable collapse pressure —

   $$\frac{3440}{1.5} = 2293 \text{ psi}$$

   Mud pressure — 0.519 psi per ft (Table 15-4)

   Maximum length — $\frac{2293 \text{ psi}}{0.519 \text{ psi/ft}} = 4418 \text{ ft}$

   Maximum safe setting depth for second-hand casing is 4418 ft

Casing Design — Combination Strings

Many intermediate and most production strings (for wells deeper than 3500 ft) consist of two or more sections that differ from each other in grade, weight and type of coupling. This is called a combination string. The advantage of this arrangement over single weight and grade strings is economy. Savings ranging from 2% to 20% are possible. In other words, safety factors for single section weight and grade casings are larger than necessary and result in over-designed casing.

Basic assumptions of physical conditions that will be used for design in this chapter are conservative but they are widely used. The design procedure used here is not the most rapid but it is believed that the principles of the method are more clearly illustrated by its use.

In his review of casing string design principles Hills recommends the more rapid graphical method developed by Main. Another procedure, developed by Kastor, eliminates the trial and error method used here.

Both the Kastor and the Main procedures should be studied when the fundamental ideas on casing design are understood, especially if string design is more than an occasional task.

The assumptions used in this method are:

1. Casing is empty, therefore the full head of the external liquid (mud or water) will be figured as the collapse stress.
2. The only tension stress considered will be that due to the weight of the casing in air. No consideration of buoyancy will be made.
3. Standard design conditions will mean a 1.125 safety factor for collapse, 1.8 for tension, 1.1 for internal pressure and an external fluid head pressure of 0.5 psi/ft.

Another limitation imposed only in this chapter is that no other casing will be considered other than that listed in Table 15-1. In practice other available types would be investigated.

EXAMPLE 5. Design a 9000-ft single weight and grade 5½-in. casing string. Use conventional safety factors and 0.5 psi per ft external pressure. Assume internal pressure will not exceed 4000 psi.

1. *Collapse*. From Table 15-1, collapse setting depth column, 17 lb/ft N will satisfy the worst expected collapse conditions.

2. *Tension*. 17-N-short coupling has an ultimate joint strength of 273,000 lb. With a 1.8 safety factor the maximum load which can be placed on it is 273,000/1.8 = 152,000 lb. How many feet of 17 lb/ft casing will 152,000 lb support? 152,000/17 lb/ft = 8940 ft.

In other words, the entire 9000 ft cannot be supported with the short coupling to meet the 1.8 safety factor. The designer at this stage has a number of choices:

a. He can reason that using the short coupling for the entire 9000 ft is not trifling appreciably with the safety factor, especially in view of the fact that buoyancy which would lessen the load has been neglected. The weight of 9000 ft of 17 lb/ft casing is 153,000 lb. Safety factor using the short coupling is 273,000/153,000 = 1.78.

b. He can specify that the top 60 ft must have long coupling joints.

c. He can specify that the entire string have long couplings. The long coupling joint can support 10,460 ft of its own weight (calculated in same way as for the short coupling).

3. *Internal pressure*. The minimum yield pressure is 7740 psi. The maximum allowable is 7740/1.1 = 7040 psi which is far greater than the expected 4000 psi maximum.

In the combination string, the lowest point is subjected to the greatest collapse stress. On this basis, as illustrated in Example 5, the bottom section is selected. At some point above the bottom, a lighter weight casing would be adequate to protect the string against collapse.

It is the purpose of the calculations to be described in Example 6 to locate this changeover point. At some point above the first changeover, an even lighter casing would be adequate for collapse protection. The same types of calculations are used to determine this also. At each changeover point the final check on the adequacy of a section is to determine whether the joint strength is large enough to support all the casing below it.

Whereas near the bottom of the string, tension stress is not great, at the top it controls the casing to be selected and collapse is negligible. Example 6 will be a combination string design with standard design conditions. Example 7 will be general.

EXAMPLE 6. Design a 9000 ft, 5½-in. combination casing string (assuming standard design conditions) for minimum weight. Assume internal pressure can be neglected.

Section A. 9000 ft to ?

1. Setting depth column in Table 15-1 shows that 17-lb N can be set safely at 10,470 ft whereas 17-lb J can only be set to 8000 ft. Therefore, specify for Section A: 5½-in. 17-lb/ft N casing. Make a sketch similar to Fig. 15-4 and fill in as data become available.

2. The next lighter casing listed in Table 15-1 for 5½-in. casing is 15.50 lb/ft J. It could be set at 6860 ft. (Table 15-1) if there were no tensile load, because the weight of the 17.0 lb/ft pipe places a tensile stress on the 15.5 lb/ft J casing, its collapse resistance is lowered. This means that its safe setting depth is above 6860 ft Figs. 15-5 and 15-6 show how this type of loading decreases collapse resistance for two common string sizes. Fig. 15-5 which is applicable to 5½-in. casing, will be used to illustrate this effect.

3. Assume setting depth is 6860 ft.

4. Tensile load would be (9000-6860) 17 = 36,000

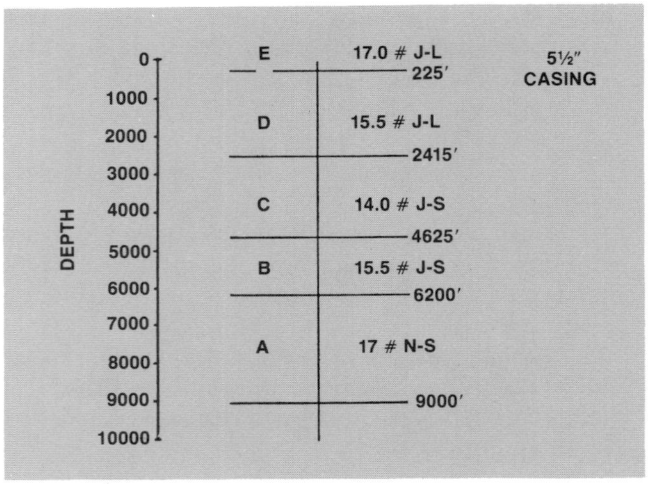

Fig. 15-4. Combination string plan.

lb (to nearest thousand).

5. A tensile load of 36,000 lb.applied to 15.5-lb J pipe results (according to Fig. 15-5) in an effective collapse resistance of 3600 psi or safe setting depth of 6400 ft for standard design conditions.

6. New assumed setting depth is 6400 ft.

7. Tensile load (9000-6400) 17 =44,000 lb.

8. From Fig. 15-5 with 44,000-lb tensile load, the effective setting depth is 6230 ft.

9. New assumed depth is 6230 ft.

10. Tensile load is (9000-6230) 17 = 47,000 lb (to nearest thousand).

11. From Fig. 15-6: 6200 ft. When calculated depth is within 50 ft of assumed depth, consider calculated value as changeover point. Tensile load is (9000-6200) 17 = 47,600 lb.

12. Final check is to determine the joint type. From Table 15-1, short coupling strength for 17-lb casing 273,000 lb. Allowable load on top joint is 273,000 ÷ 1.8 = 152,000 lb. Actual load is 47,000 lb., therefore short coupling is satisfactory. This completes design for Section A.

Section B – 6200 ft to ?

1. Section B will be made up of 15.5 lb J casing.

2. What is the length and what type coupling will be used?

3. Next lightest casing is 14.0 lb J. Its safe setting depth is 5640 ft if no tensile load is applied.

4. Assume 5640 ft is setting depth.

5. Tensile load on bottom of 14.0 lb.casing will be 47,600 lb of 17 lb casing + (6200 — 5640) 15.5 = 47,600 + 8700 = 56,300 lb.

6. From Fig. 15-5 for 14.0 lb J casing, 4900 ft is safe setting depth for 56,000-lb load.

7. Assume 4900 ft as safe setting depth, then tensile load is 47,600 + (6200 — 4900) 15.5 or 68,000 lb.

8. New setting depth (Fig. 15-5) — 4750 ft.

9. Tensile load — 47,600 + (6200 — 4750) 15.5 = 70,000 lb.

10. New setting depth (Fig. 15-5) — 4650 ft.

11. Tensile load — 47,600 + (6200 — 4650) 15.5 = 72,000 lb.

12. New setting depth: 4625 ft. This is satisfactory.

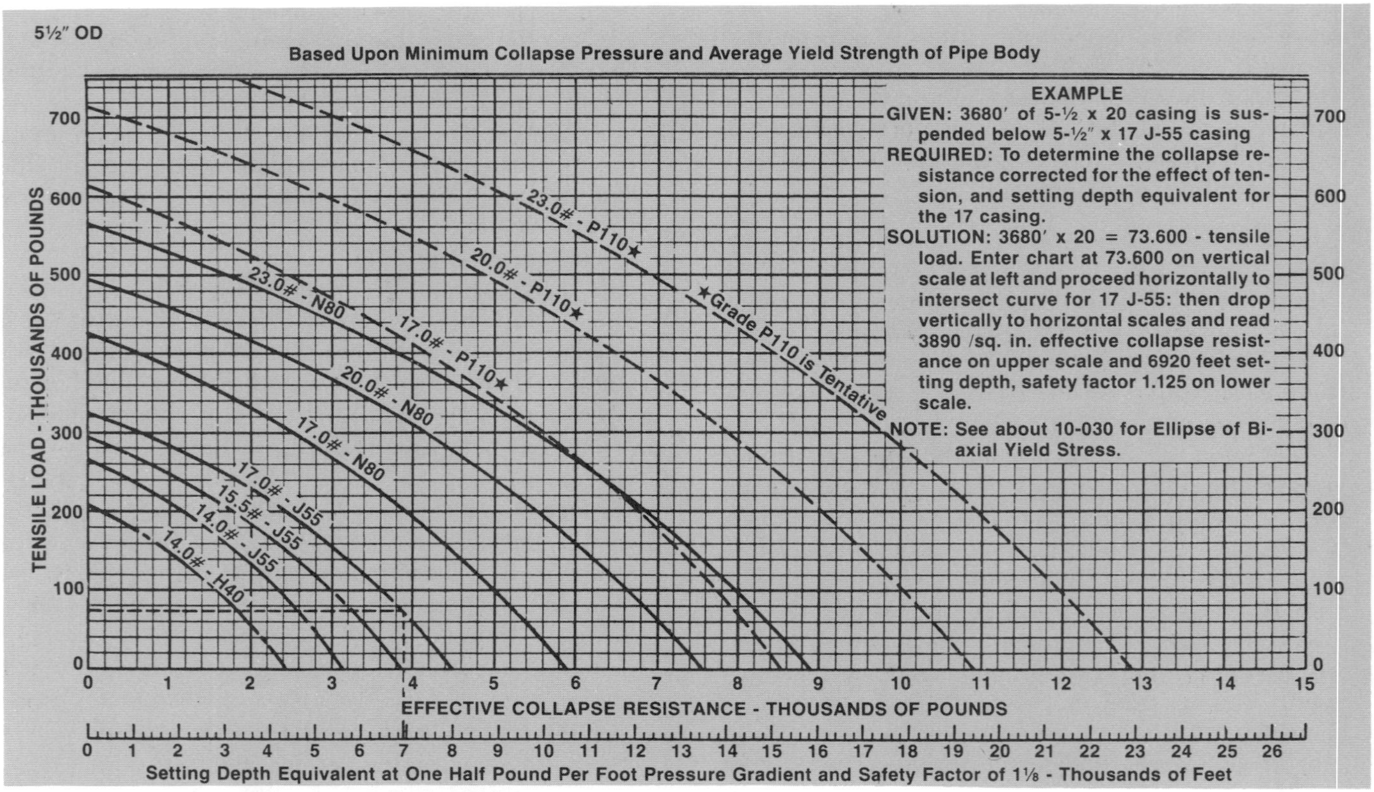

Fig. 15-5. Effect of Tensile Loading on Casing Collapse.

—*Courtesy Spang-Chalfant.*

13. Ultimate joint strength of 15.5 lb J short coupling is 211,000 lb.

14. Maximum allowable tensile load on top 15.5 lb joint is 211,000÷1.8 = 117,200 lb.

15. Actual tensile load is 47,600 + (6200 — 4625) 15.5 = 72,000 lb. Therefore short coupling is satisfactory.

Section C – 4625 ft to ?

1. Section C will be made up of 14.0 lb J casing.

2. Next lightest casing is 13.0 F but this casing only has a 95,000 lb ultimate joint strength (Table 15-1) or maximum allowable strength of 95,000÷1.8 = 52,800 lb, which is not large enough to support even the 15.5 J and 17.0 N casing used in this problem.

3. How high can the 14.0 J casing be extended? Above 4625 ft it is safe from collapse (the previous calculations showed this), therefore check its tensile limitations.

4. Allowable tensile load is 186,000÷1.8 = 103,000 lb.

5. 103,000 – 72,000 (weight of casing below 14 lb/ft) = 31,000 lb of 14 lb casing allowable.

6. 31,000 lb/14 lb/ft = 2210 ft of 14-lb casing.

7. Top of 14 lb. J-S casing is 4625 – 2210 = 2415 ft. Some companies limit the number of sections in a string to four. Others place no limit.

Section D – 2415-ft to ?

1. The next heavier casing is 15.5-lb J. The long coupling has an ultimate strength of 247,000 lb or an allowable to 137,000 lb.

2. The weight of the casing below it is 103,000 lb meaning 34,000 lb. of 15.5-lb. casing can be utilized. This is 34,000/15.5 = 2190 ft or the top of the 15.5-lb J-L is 2415-2190 = 225 ft.

Section E – 17-lb J-L

1. Allowable tensile stress — 275,000÷1.8 = 152,000 lb.

3. Allowable length of 17 J-L casing is (152,000 — 37,000)/17 = 820 ft. which is more than necessary.

This completes the design. A summary is given below:

Depth	Type
0— 225 ft	17.0 J-L
225—2415 ft	15.5 J-L
2415—4625 ft	14.0 J-S
4625—6200 ft	15.5 J-S
6200—9000 ft	17.0 N-S

EXAMPLE 7. A 7-in. casing string is to be set to 7390 ft: 10.2 lb/gal mud will be used. It is not expected that bursting pressure will exceed 3000 psi. Experience in the area indicates that safety factors of 1.1 for collapse and internal pressure and 1.5 for tension are satisfactory. Design a combination string.

Section A – 7390 ft to ?

1. Maximum collapse pressure — Fluid pressure, (10.2 lb/gal) = 0.529 psi/ft (Table 15-4). Maximum collapse pressure 7390 × 0.529 = 3900 psi.

2. Section A casing must withstand 1.1 × 3900 – 4290 psi. Table 15-1 indicates that for 7-in. casing, 23 N has a collapse resistance of 4300 psi which is

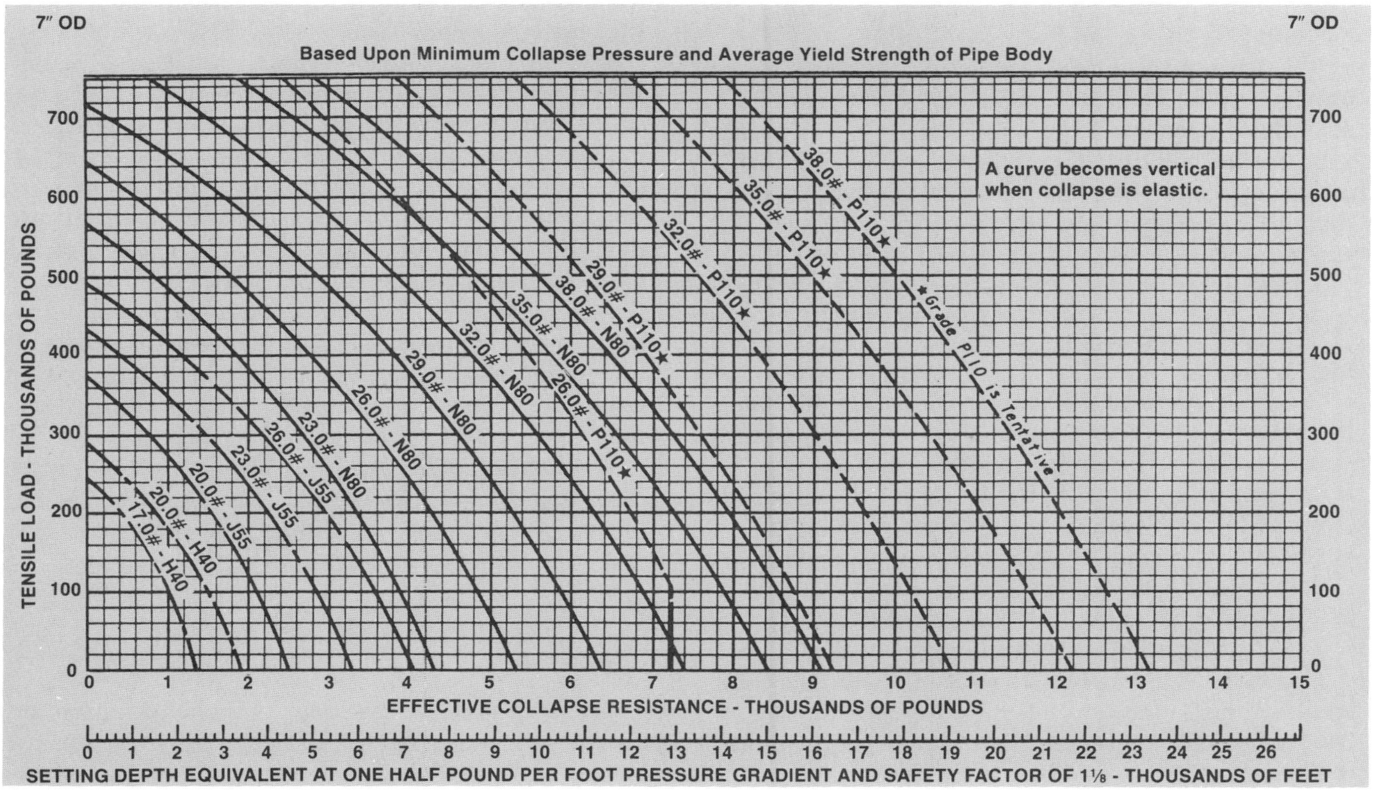

Fig. 15-6. Effect of Tensile Loading on Casing Collapse. —*Courtesy Spang-Chalfant.*

satisfactory for collapse. At what point above 7390 ft can the next lighter casing (20 J) be used?

3. Since 20 J casing has a collapse resistance of 2500 psi, it can be safely stressed to 2500/1.1 = 2270 psi. With this collapse stress the allowable setting depth is 2270 psi/0.529 psi/ft = 4300 ft providing there is no tensile load on the bottom.

4. Tensile load on bottom of 20 J casing is (7390-4300) 23 = 71,000 lb.

5. For this load, Fig. 15-6 shows collapse resistance of 20 J is 2200 psi. Allowable setting depth, considering safety factor and mud pressure, is $\frac{2200}{1.1 \times 0.529} = 3790$ ft.

Tensile load if 3790 is setting depth is (7390-3790) 23 – 82,800 lb.

6. For this load, collapse resistance of 20 J is 2175 psi (Fig. 15-6). Allowable setting depth is $\frac{2175}{1.1 \times 0.529} = 3740$ ft This is within 50 ft of assumed value (3790 ft) therefore, the changeover point from 23 N to 20 J casing will be set at 3740 ft.

7. The tensile load at the top of the 23 N section (7390-3740) 23 = 84,000 lb. Ultimate strength of the short 23 N coupling is 350,000 lb. Allowable load is 350,000/1.5 = 233,000 lb, which is greater than expected 84,000 lb.

8. Section A will extend from 7390 ft to 3740 ft and will consist of 7-in. 23 N — ST&C.

9. Internal pressure allowable is 6340 psi/1.1 = 5700 psi. This is greater than 3000 psi expected maximum therefore above specification is satisfactory for collapse, tension and internal pressure.

Section B – 3740 ft to ?

1. The weight and grade of Section B is 20 J (from previous work). At what depth above 3740 ft will the next change occur?

2. The next lighter casing is 17 H. Note that the ultimate joint strength is 160,000 lb, which means that it can support only 160,000 ÷ 1.5 = 107,000 lb when the safety factor is considered.

3. How deep can the 17 H casing be set? Allowable setting depth $\frac{1370}{1.1 \times 0.529} = 2360$ ft. (without any tensile load).

4. Tensile load at 2360 ft is 84,000 lb + (3740-2360) 20 = 110,800 lb. This exceeds the allowable tensile stress so the 17 H casing cannot be used.

5. How far can the 20 J pipe be extended? Ultimate strength is 254,000 lb. Allowable load is 254,000/1.5 = 169,000 lb. Weight of 20 J pipe that can be used is 169,000-84,000 (weight of 23 N) = 85,000 lb. Length of J pipe will be 85,000 lb/20 lb/ft = 4250 ft.

6. This length will carry pipe up to surface.

7. Internal pressure resistance is 3740 psi (min yield strength). Allowable pressure is 3740/1.1 = 3400 psi which is greater than expected maximum 3000 psi.

8. Section B therefore will be 20 J-S and will extend to the surface.

Overall specifications are:

0 — 3740 ft = 7-in. — 20 J-S
3740 ft — 7390 ft = 7-in. — 23 N-S

Casing Accessories

In addition to the casing itself, other items are necessary to make the handling of casing efficient. For example, in running casing in a hole the elevators and hooks must be strong enough to support the casing. Tongs and slips must be designed so that the danger of scoring and denting will be kept to a minimum.

Some other special appliances are described below:

1. Casing shoes. These are placed at the bottom of the casing to guide casing down and prevent it from sticking.

2. Float shoe. This is a casing shoe with a ball valve in it. The valve prevents backflow from the annulus into the casing, a situation which might occur when cement slurry is pumped through the casing. The slurry which is in the annulus after the cementing operation is followed by mud in the casing. Cement has a greater density than mud. The larger annulus pressure resulting from this is responsible for the backflow tendency.

3. Float collar. It is a collar with ball valve spaced one joint or more above the shoe or the float shoe.

4. Casing spider. Is a heavy metal block with a tapered hole in it to hold slips which in turn support the casing, suspended in it.

5. Casing jacks. When casing is stuck and the engine is not powerful enough to move it, jacks (screws or hydraulic) are used for unfreezing. In practice, many operators require that the top joint of casing be of a weight/ft equal to the heaviest in the combination string. This is to assure that tools that start down hole will not hang up in casing having smaller inside diameter than the top joint.

6. Centralizers. These are placed on the casing to keep the casing centered. This prevents sticking and increases the cement job efficiency by lessening the tendency of the cement slurry to channel through the mud.

7. Scratchers. This appliance consists of a ring with flexible steel wires attached to it. When the casing moves (reciprocated or rotated), the scratchers stir the mud and tend to prevent slurry channeling.

Handling and Running of Casing

A casing program necessarily includes a plan for inspection and proper handling. The best design means nothing if the casing is poorly handled or improperly assembled and run.

For example, placing a short coupling joint near to top of a string which calls for a long coupling can result in the loss of the casing and the hole. Failure to make up a joint properly could cause a leak which would be expensive to repair. These examples plus many other costly possibilities are reasons why careful attention must be given to all casing operations and why the casing program must include instructions for inspection and handling.

The API Recommended Practice; Std 5A, 5B, 5C1 contain detailed information about handling and inspection of casing. Some specific ideas proposed are:

1. Inspection of casing will be done at the mill. It should also be checked at the local shipping point and at the well. Check for dents, thread damage or other imperfections.

2. Measure the casing with a good steel tape and check the grade, weight and coupling before making it up.

3. Record the description of each joint in a permanent record book.

4. Have a drift diameter gauge to check any questionable joints.

5. In making up and running pipe, lubricate threads, lift and lower carefully, use a uniform makeup torque. API suggests three turns beyond hand tight for 7⅝-in. or larger pipe. This is a matter of judgment.

6. Use proper casing accessories such as tongs and slips to avoid damage.

7. Use drill pipe protectors when drilling ahead. In an abrasive circulating medium, wear might be severe even with the rubber protectors.

8. Improper welding on casing can cause severe stress. N-80 and P-110 casing are especially subjected to strength losses.

9. Casing landing practice. After the casing is made up, cement slurry is pumped through it to the annulus where it sets. The landing procedure is concerned with what should be done with the casing while waiting for the cement setting.

In the first choice, the tension load is increased which decreases the safety factors for both tension and collapse but will tend to prevent buckling caused by longitudinal or axial compression.

The second case (slack off) will decrease the tension load which will make the casing less susceptible to collapse and tension failures, but more likely to buckle.

The third choice is a compromise. As a result of a study of landing practices for the southwestern United States, it was recommended that in general, casing should be landed as set. The API committee further suggests that a detailed stress analysis should be made for cases where unusual conditions (e.g. very deep wells, heavy muds) are expected.

Chapter 16 Cementing

Oil well cementing is basically the process of displacing a cement slurry down the casing and up the annular space behind the pipe where it is allowed to set, thus bonding the casing to the formation. This technique is known and practiced wherever oil wells are drilled. It is a skilled operation which requires special equipment and trained personnel. The life of any successful producing well may depend on a good cementing job.

In drilling for oil it is necessary to pass through various types of formations. These may be permeable zones that contain oil, water or gas. These fluid-bearing formations must be segregated from one another to prevent migration of gas or fluids up or down hole. This is accomplished by pumping cement slurry into the annular space between the well bore and the casing. When allowed to set, the cement will form a barrier against fluid movement behind the pipe.

Oil well cementing serves other purposes including:

1. Bonds the pipe to the formation.
2. Protects the oil-producing zones.
3. Protects other formation strata such as fresh water, coal, potash, etc.
4. Helps prevent blow-outs from high pressure gas zones behind casing.
5. Protects casing from corrosive formation water or electrolytic action.
6. Seals against contamination of fresh water which people may use.
7. Shuts off lost circulation zones or other troublesome formations in order to drill deeper.

Generally speaking, cementing procedures may be classified as primary and secondary. Primary cementing involves the problems of lost circulation while drilling, and the cementing of casing after the producing formations have been located. Primary cementing is referred to as the original cementing operation which is performed immediately after the pipe has been run into the hole.

Any repairs or work-over jobs requiring the use of cement after completing the primary cementing job may be classified as secondary or remedial cementing. This would include all types of squeeze cementing jobs or other remedial measures used to stop the flow of water into the well, to control gas-oil ratios, or to overcome certain other conditions which may exist in the producing well.

Primary Cementing

There are three principal types of casing strings used in rotary drilling.

Surface casing is set and cemented to protect fresh water zones, to guard against possible cave-in of the surface formations, and to provide for initial drilling control.

Intermediate or protection strings are set inside the surface pipe to protect the hole. One or more such strings of pipe may be set.

Oil strings are set for completing the well as a producer or to test the well for production.

Another type of casing job which should be included under primary cementing is that called a liner job. A liner is a pipe, similar to casing, placed through

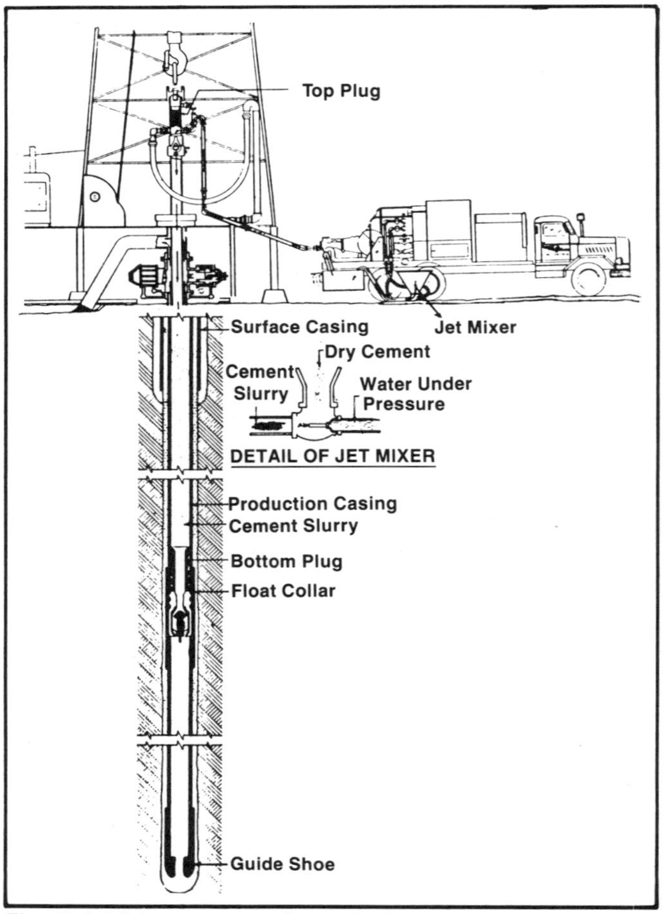

Fig. 16-1. Mixing cement and pumping into casing.

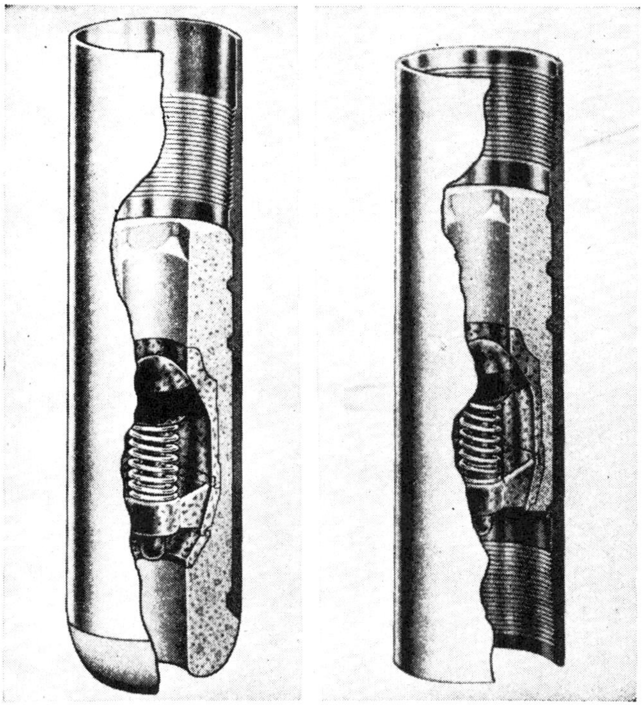

Fig. 16-2. Guide shoe, left, and float collar.

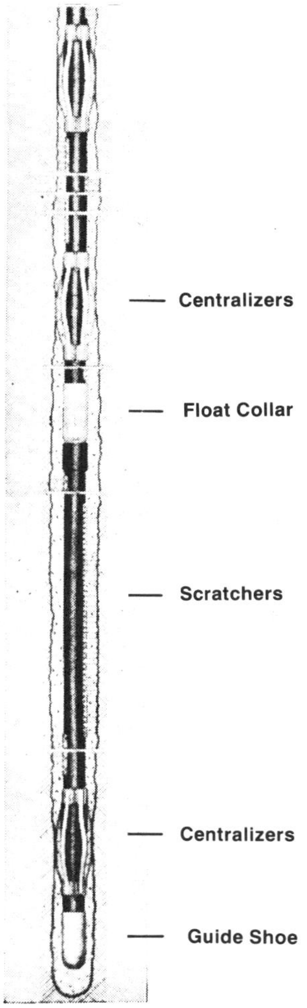

Fig. 16-3. Cementing devices

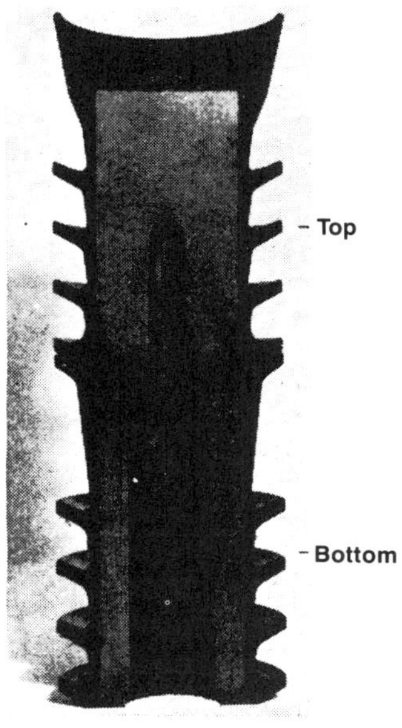

Fig. 16-4. Rubber cementing plugs.

the uncased portion of a well but it does not extend to the surface. This is a special operation common to many areas and requires a special liner setting tool run on the end of a drill pipe or tubing to allow placement of the pipe in the well.

The cementing operations on any of the previously mentioned jobs are very similar with the possible exception of the liner job. Casing, as shown in Fig. 16-1, is run to desired depth. Attached to the casing are necessary devices such as floating equipment, centralizers and wall cleaners which have been assembled at the surface.

The floating equipment normally consists of a guide shoe and float collar (Fig. 16-2) which are assembled at the bottom of the casing. The guide shoe is a rounded nose device that is placed on the bottom of the casing to guide pipe as it is being lowered into the hole. The float collar is placed some feet above the bottom of the pipe to act as a back pressure valve and to reduce the weight of the casing on the derrick as it is run.

Casing centralizers (Fig. 16-3) are used to position the pipe in the hole so that the cement will uniformly fill the annular space. Wall scratchers or cleaners (Fig. 16-3) may be of various types or designs but all are used to improve bonding properties of cement by removing drilling mud filter cake from bore hole.

When using scratchers, the pipe is either rotated or reciprocated while circulating drilling mud throughout the system. Oftentimes operators desire a water wash or a chemical wash ahead of the cement job to remove thickened mud, to reduce channeling and to insure a better bond of cement to formation.

In mixing and pumping cement slurries into the well (Fig. 16-1), dry cement is normally hauled to the well site in bulk form. During the cementing operation a bulk cement truck discharges cement into the hopper shown beside the cementing truck.

Water is circulated under pressure through a jet in the base of the hopper which carries the cement in slurry form to the tank or sump tub beside the truck. Control over the slurry weight is maintained by use of a by-pass valve on the pump truck. The cement slurry is picked up from the sump tub with another pump and displaced up to the wellhead and down the casing.

At the beginning of the cementing job casing and hole are filled with drilling mud. To prevent contamination on the interface between the mud and cement, a rubber plug (Fig. 16-4) is pumped ahead of the cement slurry. This plug is released from the plug container located at the top of the casing at the proper time and actually wipes the casing clean as it moves down the pipe.

When this plug reaches the float collar the differential pressure ruptures the rubber diaphragm at the top of the plug and allows the cement slurry to proceed down through the plug and floating equipment and up the annular space behind the pipe.

At the completion of the mixing operation a top rubber plug is released from the plug container. Its function is to prevent any cement contamination with the fluid that is used to displace the cement column down the casing. The top plug is solid and the design is such that when it reaches the first or bottom plug at the float collar it causes a complete shut-off of fluids being pumped into the casing.

Landing the top plug will prevent any further displacement of the cement slurry and provides good cement around the bottom of the casing where a good bond is required. After allowing the cement to set, the float collar and cement remaining in the lower joints of pipe are usually drilled out.

The interval between the cementing job and the resuming of normal operations is commonly referred to as WOC time or waiting on cement. This time will depend on type of job, depth, bottomhole temperature and strength the cement must develop to hold the casing in position.

Often a temperature survey is made within a few hours after the cementing job to ascertain the height to which the cement has reached behind the casing. This is possible because cement, as it sets, gives off a considerable amount of heat. By measuring the temperature of the fluid in the casing, it is possible to locate the top of the cement behind the pipe.

By comparing the actual fill up against the total volume of cement used, the actual percentage of fill up can be predicted for adjoining wells.

Secondary Cementing

Perhaps the most common operation under secondary cementing is that which is called a squeeze job. The term squeeze cementing is the physical process of applying hydraulic pressure to force or squeeze a cement slurry into permeable space of an exposed formation or through perforations of the casing or liner. This technique generally requires high pressures and is usually performed through a packer arrangement which is set above the point where the cement is to enter the perforations or formation.

The packers used in squeeze cementing may be used to remedy such problems as (1) high gas-oil ratios; (2) excessive water; (3) repair of casing leaks or perforations improperly placed; (4) isolation of producing zone prior to perforating for production; (5) remedial or secondary cementing to correct a defective condition such as channeling or insufficient cement on primary cement job, etc.; (6) sealing off low pressure formation that engulfs oil, gas or drilling fluids; (7) abandonment of depleted producing zones to prevent fluid migration and contamination of other zones or wells.

Another operation under secondary cementing is that referred to as a plug back job. This simply is the placement of a cement plug at the desired point in the well and allowing it to set up. A plug back operation is normally used in oil wells to shut off bottomhole water or to reduce the hole depth in order to produce from an upper formation. A plug back job may also be employed to whipstock or change the direction of drilling.

Cementing Materials

The basic materials that go into the manufacture of cements are the same whether it is used for construction work or for oil well cementing. Cement is produced from a properly proportioned blend of limestone and clay which is composed primarily of compounds of silica, lime, alumina and iron oxide. These raw products are intimately ground and passed through a rotary kiln where they are calcined into a material called clinker. The clinker is reground into a fine powder to which is added small quantities of gypsum to control the setting properties of the cement.

The final product, known as portland cement, contains chemical compounds of silicon dioxide, calcium oxide, iron oxide, aluminum oxide, magnesium oxide and sulfur trioxide. These basic compounds are:

Tricalcium Silicate .... C_3S
Dicalcium Silicate .... C_2S
Tricalcium Aluminate .... C_3A
Tetracalcium Aluminoferrite .... C_4AF

which react to water to form set cement.

Cement that is used in oil wells today is subjected to a range of conditions from shallow wells having temperatures of 80 F or lower, to wells whose depths exceed 20,000 ft where bottomhole temperatures are in excess of 400 F. The use of a single type of cementing material to fit these wide variations of temperature and pressure conditions is impossible; therefore, different types of cements must be compounded to accommodate this wide range in well conditions.

The first property of a cement slurry that should be considered is what is commonly referred to as the pumping or thickening time. A cementing slurry must remain fluid for sufficient time to allow pumping down the casing and up the annular space behind the pipe. A suitable cement should possess an adequate safety factor in case of an unavoidable shut down while pumping the cement slurry.

Secondly, the cement, after having been placed in a well, must set in a reasonable period of time and should develop sufficient strength for continuation of normal drilling operations. The strength required of cement before resuming operations will vary with different operators.

According to work by Farris, the minimum strength required for primary casing cement job is 8 psi tensile strength, or approximately 100 psi compressive strength. Because these strength determinations were obtained under ideal conditions, many operators consider 500 psi compressive strength necessary for resuming rig operations.

Types of Cement

The cements used in oil wells are classified as high early strength, common portland cement and retarded cement. The API Committee on oil well cements has further classified the cements used in wells and suggests the depths to which the various classifications may be used. These depth limits are based on conditions of casing well simulation tests and should be considered only as approximate values.

The API Committee on Standardization of Oil Well Cements recommends six classes of cement for use in oil fields. These are classified as follows:

Class A: Intended for use to 6000-ft depth, when special properties are not required. Similar to ASTM C 150, Type I Cement.

Class B: Intended for use to 6000-ft depth, when sulfate resistance is required. Similar to ASTM C 150, Type II Cement.

Class C: Intended for use to 6000-ft depth, when high early strength is required. Similar to ASTM C 150, Type II Cement.

Class D: Intended for use to 12,000-ft depth, when moderately high temperatures and high pressures are encountered.

Class E: Intended for use to 14,000-ft depth, when high temperatures and high pressures are encountered.

Class F: Intended for use to 16,000-ft depth, when extremely high temperatures and extremely high pressures are encountered.

(See late edition of API Spec. 10 for additional information.)

Testing Of Oil Well Cements

In recent years major advancements have been made in the testing of cements and admixtures used in wells. The development of high pressure testing equipment, such as the Stanolind thickening time tester, makes possible the testing of oil well cements under simulated pressures and temperatures encountered in actual well operations. Such tests give an accurate record of how long a cementing slurry will remain fluid when pumping under these conditions.

Standard procedures for testing cements for various well depths are given in the API RP 10B "Recommended Practice for Testing Oil Well Cements." The casing cementing test schedules of this code are based on actual well data. In the curve in Fig. 16-5, the normal static bottomhole temperature line is calculated from the formula 80 + .015 X depth in feet. This is taken to be the average formation temperature at various depths along the Gulf Coast and is usually found comparable to those of other areas where wells exceed 10,000 ft.

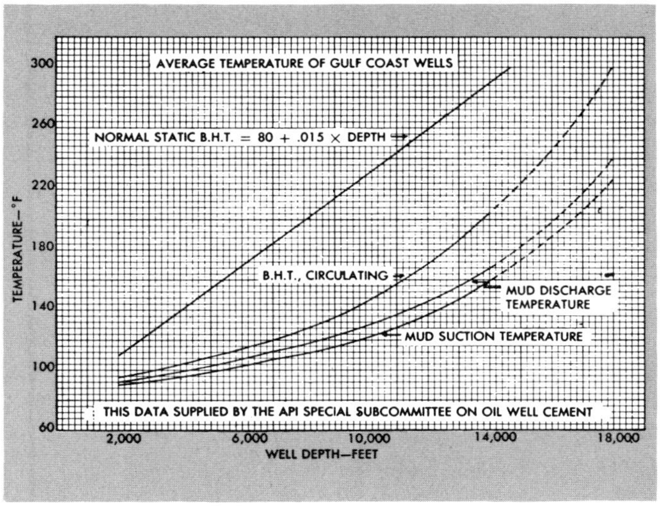

Fig. 16-5. Oil Well Cementing

In many areas, of course, there are exceptions to this average, particularly in shallow wells, but the information given here should be a guide where bottomhole temperatures are not available.

The bottomhole temperature circulating curve was derived from actual measurements made while wells were being circulated. This is the temperature cement slurry should reach during casing cementing at the given depths. The vertical difference between this curve and the static bottomhole temperature curve represents the amount of cooling a well undergoes while circulating.

The mud discharge temperature curve is the average temperature of the mud returns at the surface while circulating. The vertical difference between this and the bottomhole temperature circulating curve represents the cooling a slurry undergoes in returning to the surface from the bottom of the well.

The mud suction temperature curve is the average temperature of mud entering the well through the mud pump. The vertical difference between this and the mud discharge temperature represents the cooling a fluid undergoes while passing through the mud pits during circulation.

These curves represent average measured values to the 14,000-ft depth as shown by the solid lines. The dashed part of the curves at depths below 14,000 ft are merely extrapolations of the data gathered above 14,000 ft.

As an example, from the curve, the average 10,000 ft well should have an actual formation temperature (under static conditions) of 230 F. After circulation has been established and conditions stabilized, the bottomhole circulating temperature will be approximately 144 F and the discharge mud temperature will be approximately 127 F.

These curves indicate that the bottomhole circulating temperature of wells is much closer to the mud discharge temperature than to the actual formation temperature. This is especially true of wells that are less than 12,000 ft deep.

Admixtures To Cement

Often it may be necessary to alter or modify the cements used for a particular job to comply with certain well conditions. Today the different types of additives used in cements for various purposes may be classified thus:

Cement Accelerators:

1. Calcium chloride
2. Sodium chloride

Cement Retarders and Dispersants:

1. Calcium lignosulfonates

Weighting Materials:

1. Barium Sulfate
2. Ferrophosphorus

Lost Circulation Additives:

1. Shredded fibrous materials
2. Cellophane flakes
3. Mica flakes
4. Expanded Perlites
5. Granulated nut shells

Light Weight Additives:

1. Bentonite
2. Pozzolans
3. Expanded Perlites
4. Hydrocarbons

Special Cements:

1. Resin Cement
2. Diesel oil cement
3. Gypsum cements

Handling cement with modern bulk equipment makes it possible to add these materials to dry cement prior to actual cementing operations. There may be occasions, however, where additives such as accelerators are added directly to the mixing water before cementing.

Cement Accelerators

An accelerator is often used to reduce the waiting time on cement. This is particularly true on surface pipe where it is desirable to resume normal drilling operations as soon as possible after cementing. It is common practice to use calcium chloride for this purpose since it is perhaps the most effective accelerator known for oil well cementing slurries and is readily available and economical.

Using 2% calcium chloride based on the weight of the cement or cement mixture will: (a) accelerate the thickening time, (b) accelerate the setting time, (c) increase the early strength, (d) reduce the waiting on cement time. As a rule of thumb, 2% calcium chloride will approximately double the strength of portland cement at temperatures of less than 120 F in 24 hours. The use of greater quantities will not achieve any beneficial results.

When used in approximately the same quantities as calcium chloride, sodium chloride will also cause an acceleration in the setting time. However, sodium chloride is not as effective as calcium chloride and should only be used where calcium chloride is not available. Salt water will have the same effect when the sodium chloride concentration is from 40,000 to 60,000 ppm. A normal concentration of Gulf water will range from 15,000 to 30,000 ppm of sodium chloride.

Cement Retarders And Dispersants

Some more effective cement retarders and dispersants in use today are derivatives of lignin compounds and are generally referred to as calcium lignosulfonates. These are used to retard common portland cement where well conditions are beyond the range normally recommended for this type of cement. In some instances they have been applied to retarded cements to give an extension of the thickening time for use in extremely hot wells.

The amount of retarder will depend, of course, on the bottomhole temperature of the well and can be added either to the cement in a dry form or to the mixing water. These materials are also used with high percentages of gel or bentonite cement to achieve a lower viscosity slurry which is easier to pump and handle while mixing.

Light Weight Additives

Bentonite in Cement. Bentonite has been used as an additive for cements for many years primarily to create lighter weight slurries and reduce water separation.

FIGURE 16-6
PORTLAND CEMENT CONTAINING BENTONITE

Per cent Bentonite	Water require-ments	Slurry weights	Slurry Volume	Compressive strengths 100F	
				24 hours	72 hours
4	7.7	14.1	1.53	1220	2825
8	9.7	13.3	1.82	730	1370
12	13.3	12.6	2.19	395	755

Figure 16-6.

The amount of bentonite will vary from 1% to 25%, depending upon the type of job. Bentonite is a colloidal clay which requires large volumes of water. When used with cement, the additional water is principally responsible in reducing the slurry weights (Fig. 16-6).

High percentages of bentonite in cement will cause a reduction of compressive strengths and thickening time of both portland and retarded cements. According to some authors on cement and concrete technology, the high water ratios required for gel cement lowers the resistance to sulfate attack and increases the permeability of the set cement.

Modified Cements. The term modified cements here applies to those slurries which are composed of common portland cement (Type I) and high percentages of bentonite (5% to 25%) with 0.5% to 1.5% calcium lignosulfonates.

These cementing slurries have properties similar to those of gel cements, except use of calcium lignosulfonate acts as a retarder and dispersing agent, rendering the cement slurry more pumpable under high temperature well conditions. This type slurry, when mixed at a high rate of sheer, has low fluid loss properties as recommended for permanent type well completion practices.

Pozzolans in Cement. Pozzolans are silicious materials, which in themselves possess no cementitious value but in a finely divided form will, in the presence of water, chemically react with lime at ordinary temperatures to form compounds that have cementitious properties.

These pozzolans, when used with portland cement, react with the free lime liberated from the cement and thus become cementitious. Pozzolans have been known and used since the time of the ancient Romans and Greeks and structures built by those people are still standing today in testimony to their resistance to various destructive agents.

In recent years, pozzolans have been used in oil well cementing to create a lighter slurry; improve the workability of the cement; obtain better perforating qualities; reduce cementing costs, and increase the resistance of the cement to attack by sulfate or brine waters.

Generally speaking, pozzolans are mixed with equal proportions of portland cement for use in oil well cementing. These types of materials can be accelerated in the same manner as portland cement in reducing the WOC time or they can be retarded with chemical additives for use in deeper wells.

The early strength of these materials is less than that of normal portland cement, being comparable to gel cements, but will develop strength with time so that the ultimate strength approaches, or may exceed, that of portland cement. These materials are not normally recommended with commercially retarded cements for use in deep wells, as the dilution of the chemical retarder in the cement produces an accelerating effect.

Expanded Perlite in Cement. Perlite is a volcanic material which is mined, screened and expanded by heat to form a cellular product of extremely low weight. It is manufactured for purposes of creating light weight aggregate for cement. It is used with common portland cement or retarded cements, with normally a small percentage of bentonite (2% to 6%) to prevent segregation of the perlite particles from the cement slurry. It reduces slurry weights, increases slurry volumes, and acts as bridging material for porous or fractured formations encountered while cementing.

In field practice it is always advantageous to mix perlite cement with what might appear to be an excessive amount of water because under pressure these materials act as tiny sponges and take up water from the slurry. This volume change under pressure results in a set cement which has a higher density and lower volume than observed in the cement slurry at atmospheric conditions.

Hydrocarbons in Cement. Hydrocarbons have been used to a limited extent with portland cement to reduce the slurry weight beyond the normal range of bentonite or perlite cements. The hydrocarbon is first emulsified with water and an emulsifying agent prior to mixing with cement. With such slurries it is possible to obtain weights in the range of 11 lb/gal.

These types of slurries have a much higher angle of repose than do gel cements and have been used in a few instances on lost circulation jobs where normal weight slurries could not be used.

Hydrocarbons and cement do have disadvantages in that they have low compressive strengths and extra equipment is required to emulsify the fluid prior to mixing. The cost of hydrocarbons, such as kerosene or diesel oil, does increase the cementing costs considerably. These slurries have low fluid loss as measured on the filter press and give high volume yields based on a sack of cement.

Weighting Materials In Cement

Often it is desired to increase density of the cement slurry for use in deep wells to overcome high bottomhole pressures. Barium sulfate with a specific gravity of 4.2 can effectively increase the slurry weights to 18 lb to 19 lb/gal. It is the same type material applied as a weight material for drilling muds.

Other materials having higher specific gravities, such as ferrophosphorous, can be added to increase the density of the slurry somewhat more than barium sulfate. Disadvantages are the cost and lack of availability.

Weight materials do not appreciably affect the thickening time of common cement but in some instances cause an acceleration of retarded cements for use in deep wells.

Lost Circulation Additives

Numerous types of lost circulation additives are used in oil well cementing. These include shredded fibrous materials, cellophane flakes, mica flakes, expanded perlites, granulated nut shells and many others. These agents are used as bridging materials to prevent lost circulation in cavernous formations encountered while cementing.

Some fibrous materials that are commonly added to drilling muds cannot be used in cements because they contain soluble materials which cause a retardation in the setting of the cement. The selection of a proper lost circulation additive will depend on the location of the trouble and the size and type of opening which requires shutting off.

Resins in Cement Resins can be used effectively in high early and portland cements to improve the cementing bond between the formation and to shut off gas or bottomhole waters where other types of cementing slurries have failed. This type material is recommended principally for squeeze work and can be used over a wide range of temperature conditions.

Resin in cement produces slurries that are handled in the same manner as neat cement and in some instances where small volumes are required, the slurries are placed into the well by means of a dump bailer mechanism. In the mixing of resins with cement it is necessary that the resin and water be pre-mixed before they are added to the cement. Frequently they are used for tailing out a cementing job where trouble has been encountered with bottomhole water.

Under pressure, resin cement slurries will compact on the wall of the hole, thus forming a filter cake which reduces the penetration of the fluid from the slurry. The resinous material that has been squeezed into the formation will set and form an effective bond in all types of formations.

The disadvantage of this type of cement is the cost of resin which prohibits its use in large volumes. It is, however, one of the most effective materials known for sealing pipe in salt domes for LPG storage or other operations.

Diesel oil cement. Diesel oil cement is a material used for squeezing off unwanted formation waters. This type of slurry is a mixture of portland cement, a surface active agent and diesel oil, in which the cement will not hydrate.

It is used in the same manner as conventional water slurries but has the advantage that it will not set and harden until in contact with water. This feature gives an unlimited pumping time for spotting and relatively small volumes of slurry are required. The surface active agent in diesel oil cement reduces the volume of diesel oil for a given viscosity slurry.

Such slurries are mixed approximately 15 lb/gal and when squeezed under pressure, produce a high density, high compressive strength material. Water starts to penetrate the slurry as soon as contact is made and continues until the cement has set. The WOC time is governed by the portland cement and the well's temperature conditions.

Gypsum cements. Special gypsum cements with high early strength and controlled setting times have been designed specifically for oil and gas wells. They are used principally for temporary plugs and develop a high strength in a matter of a few hours. Due to their short setting time they are usually placed in a well by means of a bailer rather than pumping.

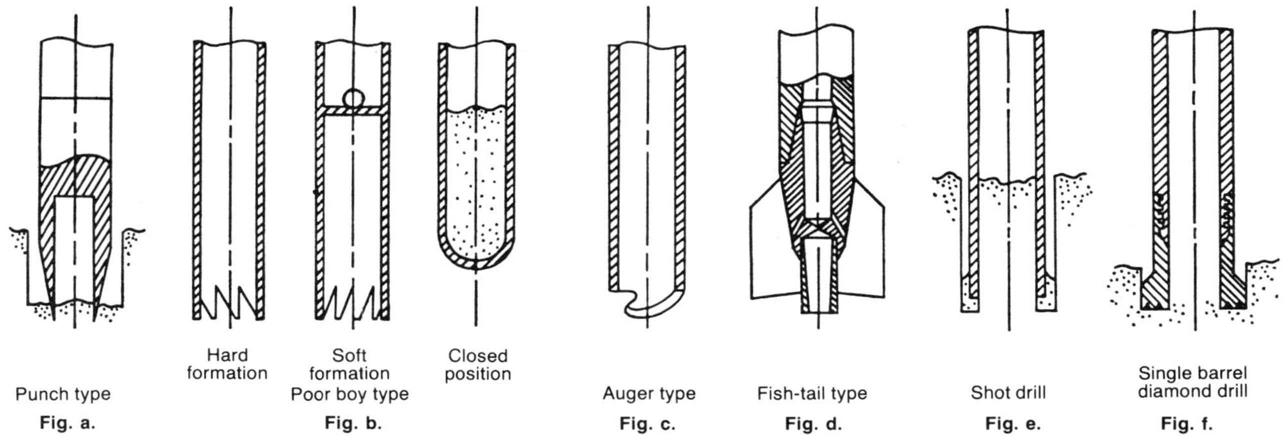

Fig. 17-1.

Chapter 17 Rotary Coring and Formation Testing

One of the most important and reliable sources of information on the lithology of the reservoir comes from cores. In addition to certain other information on porosity, permeability and type of fluid content, which can be obtained from samples, data of considerable value is available from a careful visual examination of the samples.

In cores from fractured reservoirs, an idea of the spacing, extent and number of different directions of fractures can be obtained quite readily. Oriented cores allow direction of the sets of fractures to be determined. Important stratigraphic information may be obtained from a carefully planned coring program. Doubtful pay zones as indicated by the electric log may be verified by sidewall cores.

The general use of rotary coring equipment started about 1921. Before the inception of the rotary core barrel, it was generally believed that a principal advantage of the cable tool system over the rotary method of drilling was that a continuous visual record of the formation was available through the bailed cuttings. Cuttings of hard strata, however, were very fine and often of no value. Oil was usually discovered when it appeared in the bailings or blew over the crown block.

The present rotary core bit is one of the most highly developed tools now in use in the drilling industry. It evolved from early models, some variations of which are in use today. It is believed that a brief scan of the evolutionary development of the coring tools is very important in any discussion concerning the fundamentals of their present day use and application (see Fig. 17-1).

Evolution of Core Bits

1. The punch type, or standard core bit, was once used on rotary rigs by spudding the drillstem. No core catcher was used as the formation was expected to pack in place. A modern variation of this is the sidewall sampler, which punches cores from the sidewall of the well.

2. The poor boy core barrel was made by slotting or shaping the end of a piece of pipe with a hacksaw or a cutting torch. A hole was usually drilled 2 ft or so above the top of the expected core to keep circulation away from the sample and to drain the drillstem when pipe was pulled.

The core barrel was closed by applying weight and rotating rapidly with very little or no pump pressure. This caused the teeth to bend inward, cut off the core, and make a basket to hold it.

3. The auger type or sampler bit was essentially a poor boy core barrel with a lip bent over the bottom. The core was literally "cut" from the bottom.

4. The fishtail type was a short core tube screwed or welded in the center of a fishtail bit. The device filled up when the bit was run and gave a short core. The bore of the entrance portion of the core tube was tapered upwardly to retain the core.

5. The shot drill consisted of a single barrel core tube rotating on a blanket of hard shot that was poured in the hole around the core barrel. The shot imbedded itself in the material of the core tube and was dragged around the bottom of the hole to cut the formation. It was recovered by pumping down nails

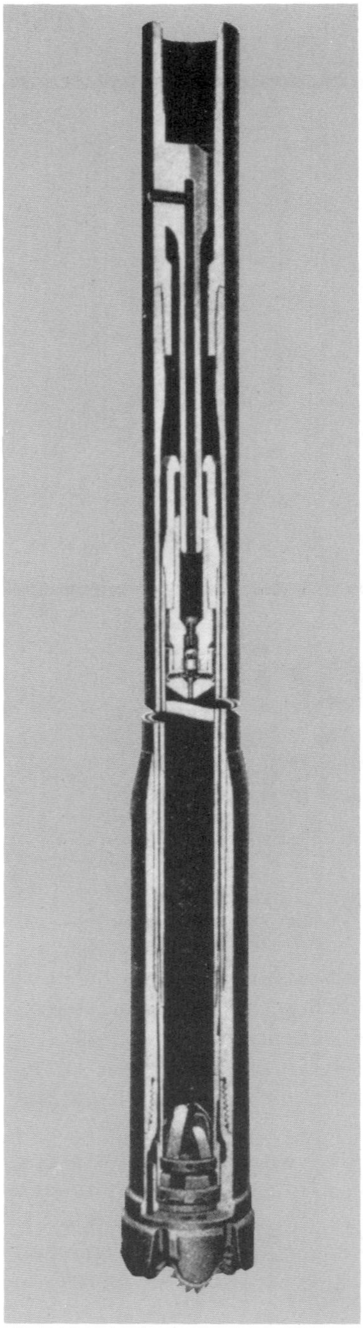

Fig. 17-2a. Conventional core barrel assembly hard formation roller type cutter head.

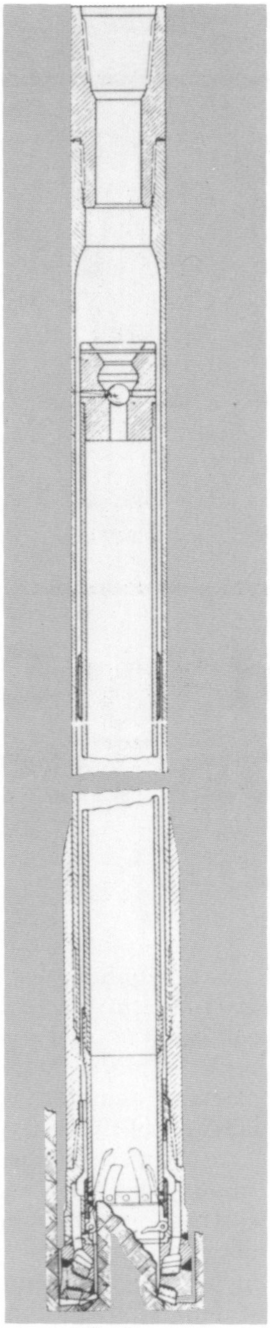

Fig. 17-2b. Conventional assembly, drop-ball valve, short sub hard formation head.

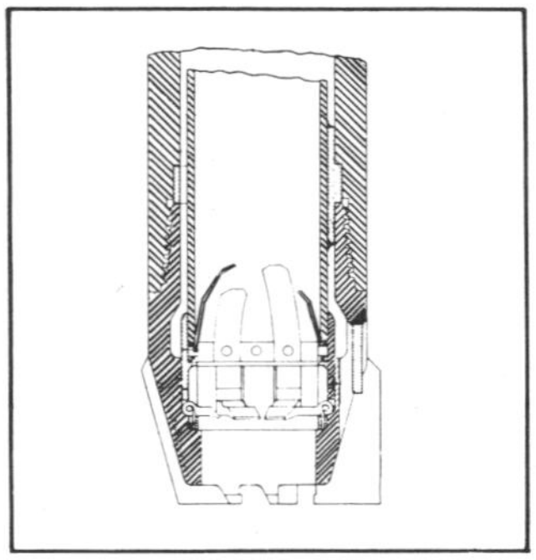

Fig. 17-2c. Conventional core barrel assembly showing soft formation cutter head and core catchers.

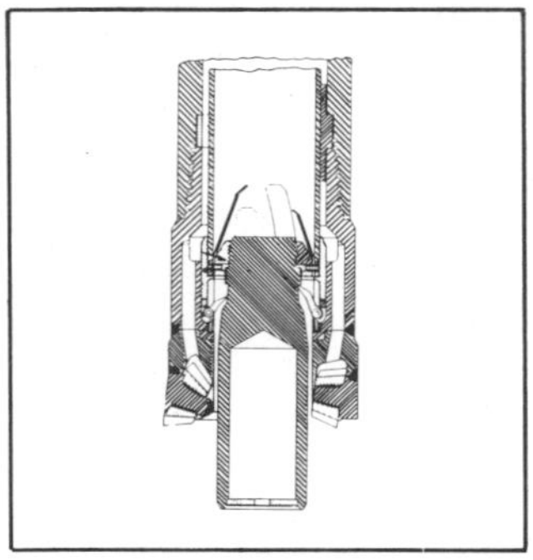

Fig. 17-2d. Conventional core barrel assembly showing core barrel plug in place for going in the hole.

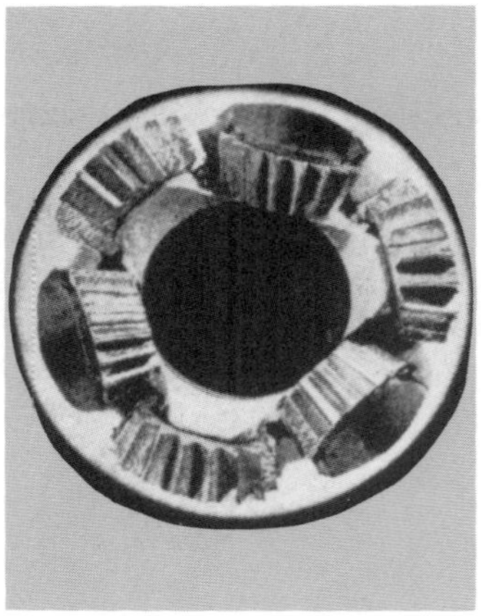

Fig. 17-2e. Hard formation roller type cutter head — bottom view.

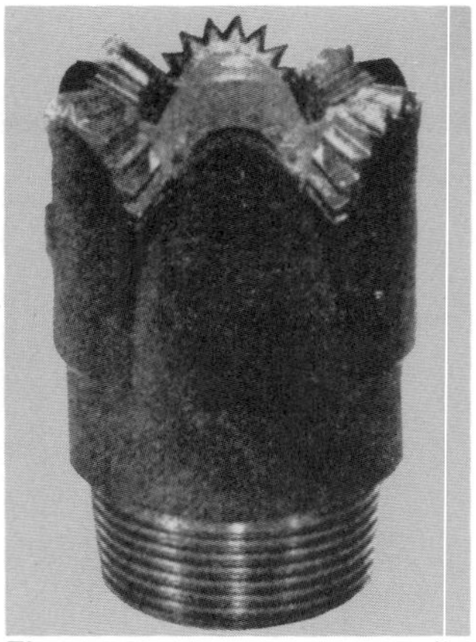

Fig. 17-2f. Hard formation roller type cutter head — side view.

Fig. 17-2g. Hard formation roller type cutter head — bottom view.

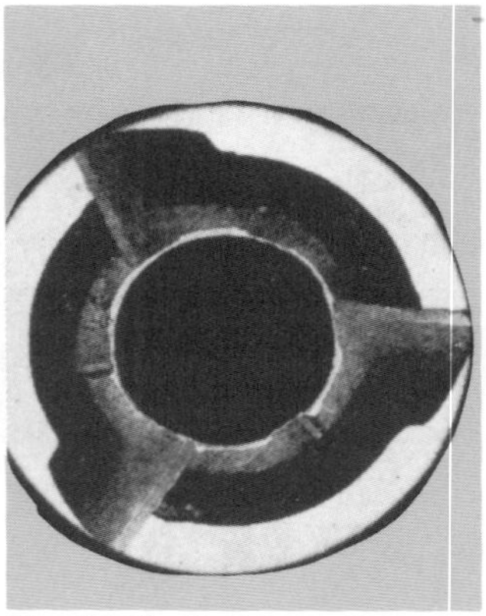

Fig. 17-2h. Soft formation head - bottom view.

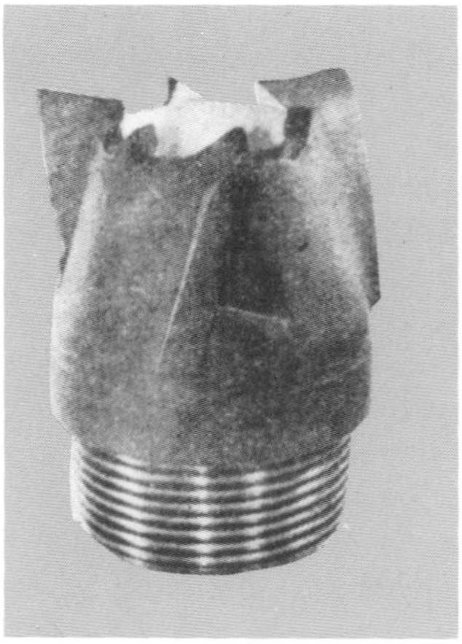

Fig. 17-2h. Soft formation cutter head — side view.

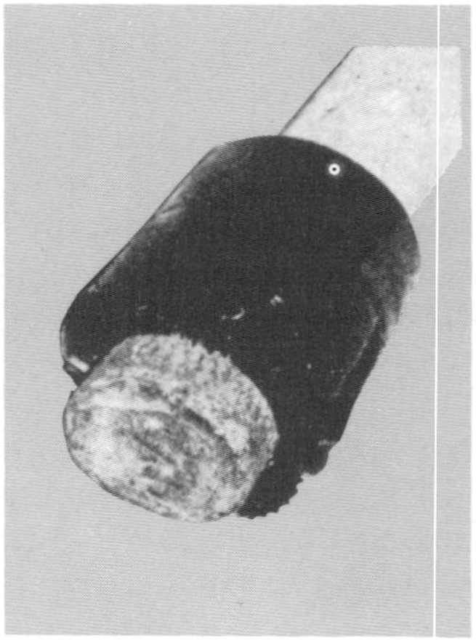

Fig. 17-2j. Hard formation cutter head with core.

and junk to foul the core in the pipe and to hold it as the pipe was being pulled.

6. The single barrel diamond drill in which black diamonds were secured to the bit head to take the place of shot as a cutting means, was a refinement of the shot drill. A slip type core catcher was provided to retain the core.

The first four types listed above were essentially soft formation coring devices, while the latter two were hard formation bits. None of the early types were satisfactory for broken formation coring.

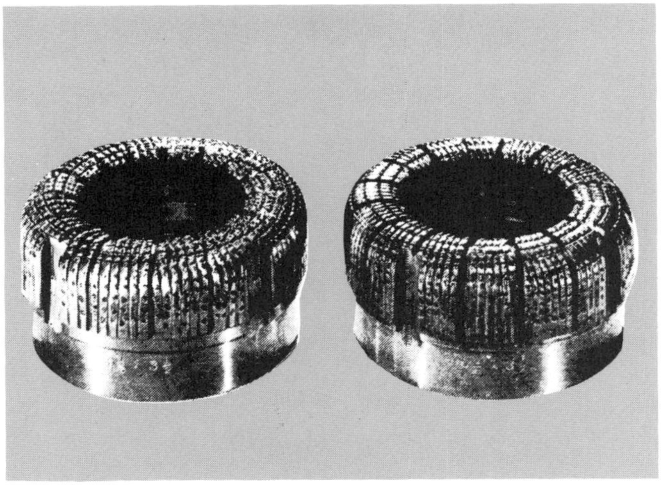

Fig. 17-2k. Conventional hard formation diamond type cutter head.

Core Types and Equipment

There are essentially only two types of cores: those taken as the formation is being drilled, and those taken from the side of the well bore after the zone has been penetrated. Conventionally, cores are classified in accordance with the type of equipment used to obtain them. This classification can be set out as follows:

Conventional Core. This refers to a core obtained by a so-called conventional core barrel, in which the drill string has to be pulled in order to recover the core. A large diameter core (three inches or greater) some 30 ft long, can be obtained by this method. The same core barrel assembly is used in both the hard and soft formation cutter heads (Fig. 17-2).

Diamond Cores. This type core is secured by means of a special diamond core bit. As with the conventional core, large diameters and longer cores (up to 50 ft in length) can be obtained, although the drill string has to be pulled in order to recover this larger core. As the name implies, this type of core recovery is particularly well suited for hard formations.

Regardless of the type of formation, however, recoveries by this method of coring are usually high. The rate of penetration is commonly somewhat less when used in the softer formation, as compared to the hard formation roller cutter core head.

Wireline Retrievable. This type core recovery refers to a core barrel that can be recovered without pulling the drill string. The recovered core is smaller in diameter and shorter in length than that of the conventional type core.

The tool is available in three variations. The first two are provided with a wireline for retrieving the inner barrel. One has the core barrel seat in the cutter head and the inner tube is locked in place and is free to swivel with the core. The other has a core head in the inner barrel which forms the core. This type is locked against rotation and protrudes ahead of the main bit. The third type is held down (in the cutter head) by pump pressure. It is retrieved by reverse circulation or by wireline (Fig. 17-3).

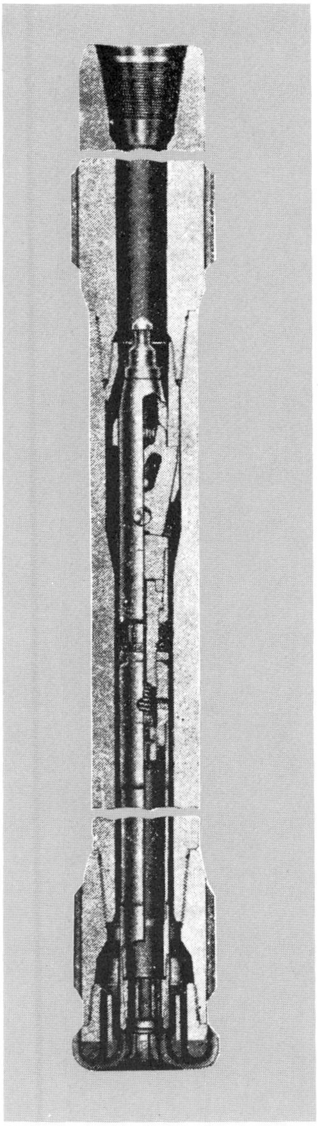

Fig. 17-3a. Wireline retrievable type core diamond head.

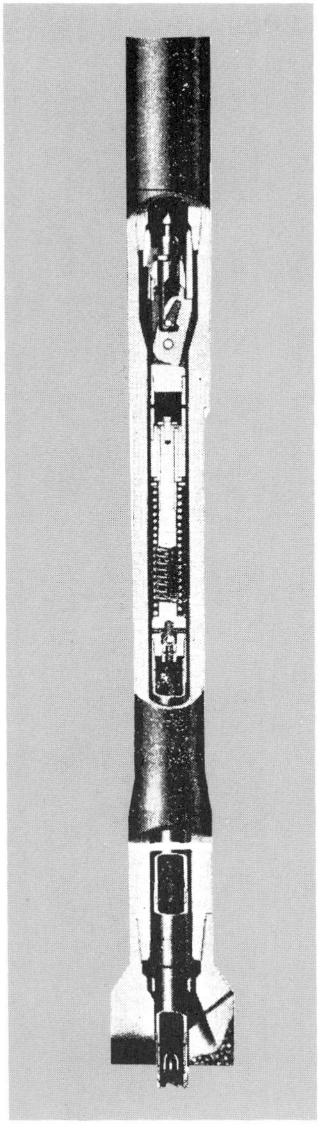

Fig. 17-3b. Wireline retrievable type core barrel assembly.

Sidewall Core. This method of coring allows a formation sample to be obtained after the bit has penetrated the zone. There are two general tools available to perform this technique: the mechanical type sidewall device and the percussion type sidewall device (Figs. 17-4 and 17-5).

It is to be noted that of these modern types of coring equipment, the soft formations use a scraping type bit and the hard formations use the roller type cutters or the diamond bits.

As noted above, both percussion and mechanical type tools are available for sidewall coring, with the mechanical type being most suitable for the harder formations.

Coring Programs

There is no substitute for a carefully planned coring program. The economics of the operation must be

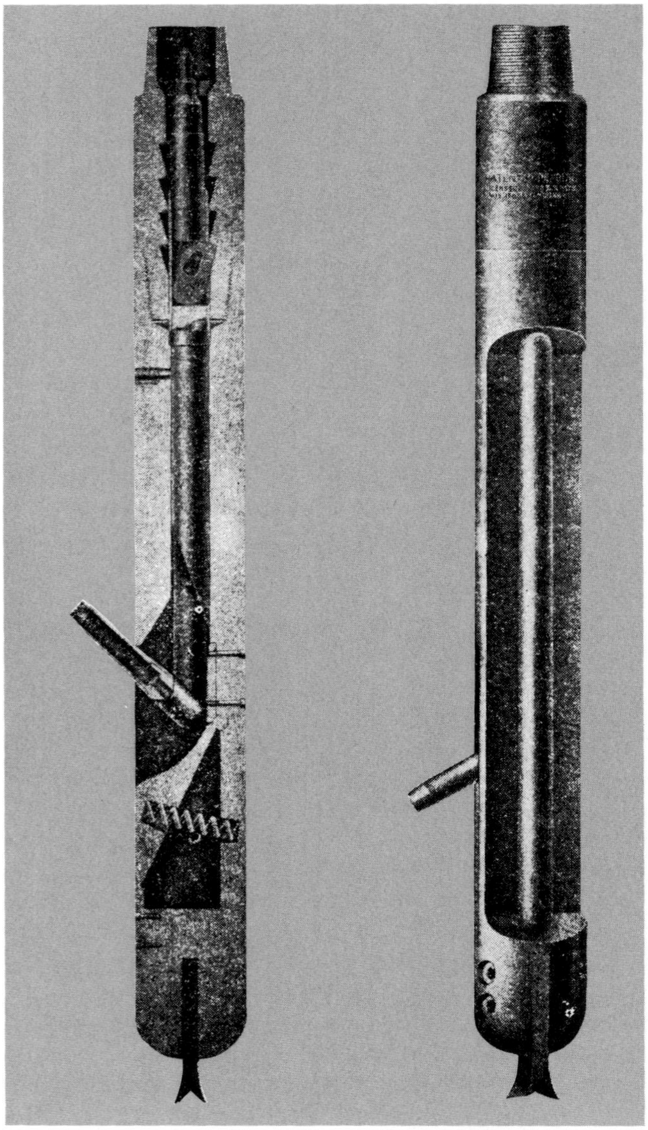

Fig. 17-4. Mechanical type sidewall coring tool assembly.

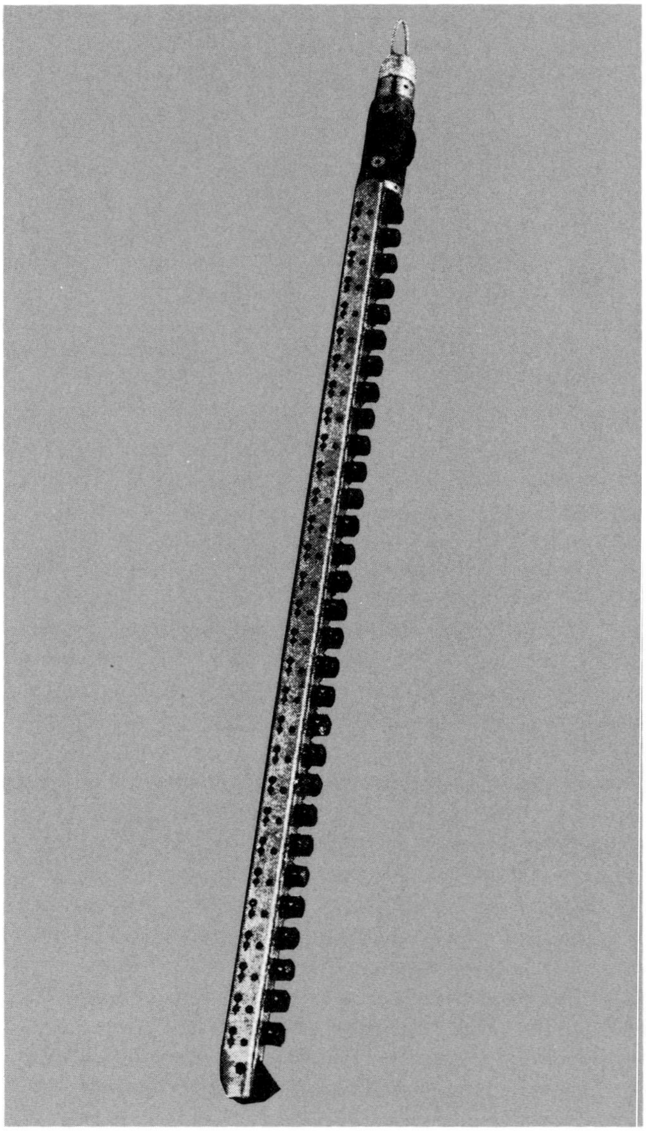

Fig. 17-5. Percussion type sidewall coring tool assembly.

balanced against the information to be obtained and the experience available in the area of interest. The coring program should always be closely coordinated with the drillstem test program.

There are two main factors that should influence the planning of the coring program. First factor is type of well to be drilled, i.e. wildcat, field extension or development; and secondly, the extent of the information available in the local and regional geological picture. Extensive coring on all wildcat wells is obviously economically prohibitive. Depending upon the geological data at hand, however, a carefully planned program can yield very beneficial results.

If, for example, there is little geological data on the area concerned, and it is not a hard rock area, a wireline retrievable core program covering several selected intervals with sidewall cores in the missed zones of interest might yield optimum results for minimum cost.

But if the area is known to be generally favorable for electric log interpretation (such as the normal Gulf Coast province, consisting of the normal salinity Wilcox, the standard Frio, and the moderate depth Miocene) a coring program calling for only sidewall cores may be ample for both wildcat and field extension wells.

In hard rock areas, a program utilizing diamond coring and testing drilling breaks and mud analysis logger shows, would yield maximum return. It is to be emphasized, however, that the diamond method of coring should not necessarily be confined exclusively to the hard rock zones. The record shows that recoveries by this method are substantially higher than by the others. This means that the economics must be considered in the light of the information desired and the recovery experience of other methods in the area in question.

In general, it would seem that only a few wildcat

and field extension wells would call for extensive diamond or conventional core programs, whereas certain selected development wells might make maximum use of this type of coring program.

These remarks are aside from company policy as exercised by some operators. Extensive coring and consequent open hole drillstem testing is widely practiced by certain operators. Others lay out their program on a more selective and conservative basis.

The choice of mechanical or percussion type sidewall coring device will depend on the type of formation and the information desired. The mechanical type is better suited for hard compact zones than the percussion type and will usually recover a large size sample. The percussion type, on the other hand, can recover up to a maximum of 30 samples (100% recovery)on one run.The percussion sampler is excellent for correlation and lithological studies as well as for electric log verification.

Coring Technique

Conventional Cores. Although the remarks to follow apply more specifically to conventional coring, most of them are equally applicable to the other types, except for sidewall. The principal notation is that the retrievable core barrels perform best in the softer formations, although diamond heads are now available for this type of tool.

1. Assemble the bit properly, making sure the catchers are in place right side up and check all threaded joints for tightness. Inspect ball and ball seats to see that they seat properly; replace rubber packers if they are worn or swelled. Doping the core catchers is sometimes helpful to make catchers free to rotate with the core.

2. When going in the hole, care should be taken to avoid striking dog legs or crooks in the hole that might plug the bit water holes or break the cutters or bit blades.

3. After reaching bottom, wash the hole as clean as possible to the formation to be cored. If a core bit plug is used, set the bit on bottom without rotating to shear the rivets.

4. Turn the rotary 20 rpm to 40 rpm and start coring with light weight, gradually applying enough weight to get a definite cutting. Adjust pump speed to the formation being cored; i.e., hard or firm formations and stickly strata need almost as much pump pressure as required by a drilling bit of the same size. Soft, unconsolidated formations are best recovered when the pump is run only enough to prevent sticking. It is very important to feed the weight off smoothly and continuously, so that the bottom is crowded at all times. If it can be avoided, do not make any connections or lift off bottom during a coring operation.

5. Weight on bit. For the hard formation roller type bit, the smaller diameter of conical cutters compared to the larger diameter of a cone used on a regular bit of the same size introduces problems that limit the maximum weight that can be applied safely on a core head. The following factors are involved:

(1) For a given rotary speed the rpm of a core bit cutter at gauge is 1.5 to 2.0 times that of a tricone bit cone. This higher rpm of the cutter operating on a smaller size bearing is conducive to greater bearing wear.

(2) The strength of a cutter head assembly is appreciably less than a tricone in a given size. Excess weight may injure cutters, cutter shafts or shaft welds.

(3) Excess weight will cause higher bearing friction and may result in locked cutters.

Table 17-1 lists the maximum bit weights and rotary rpm for average conditions when using a hard formation roller type core bit. It should be emphasized that local conditions at the well govern, to a large extent, the best operation, and the wisdom and experience of the driller are depended upon to get results.

6. After core is cut, allow the bit to drill off and increase speed of the rotary for a short time. Pick up from bottom slowly with the pump running. Do not pull the bit out of a rat-hole too rapidly; the cutter head may be balled up and tend to swab the rat-hole dry, in which case the core may be pulled through the catchers and lost. Avoid spudding or chasing mud or

TABLE 1. Bit weights and rotary rpm.

Size range in inches	Maximum weight recommended, lb.	Maximum rpm
5¼— 6	10,000	45
6⅛— 7½	12,000	50
7⅝— 8¾	15,000	60
9⅝—11⅞	18,000	60

TABLE 2. General recommendations for maximum core recovery.

Formation	Cutter head	Pump speed	Rotary speed	Weight required on bit	Remarks
Soft shale, sand, gumbo	Soft formation	Fast as necessary to prevent balling up.	25—35	Light weight	Too much weight will "ball up" bit.
Soft with hard streaks-chalk	Soft formation	Fast as necessary to prevent balling up.	25—35	Light to medium	Use light weight in soft formation.
Broken formation, sulphur bearing rock	Hard formation	Medium	25—35	Light weight	Heavy weight and high speed breaks up core.
Anhydrite, chert, hard lime, granite, quartzite	Hard formation	Full	30—40	Heavy weight	On granite, and such rocks, more hole per cutter head can be obtained with slow rotation.

otherwise disturbing the core any more than necessary when coming out of the hole.

7. When core bit comes out of hole, unscrew cutter head. If bit is dressed with working barrel in derrick, do not allow cutter head to drop any distance as inner barrel and core follow cutter head out of working barrel. Raise working barrel to lay down inner barrel. Unlatch elevator and remove cutter head and catchers. Break out stuffing box assembly, shove in core extractor plunger in upper end of inner barrel and screw in core extractor bean. Water is sometimes used to force the core out.

8. General recommendations for maximum core recovery are given in Table 17-2. Local conditions will determine the practice for successful coring.

Wireline Coring

The additional precautions, as set out below, apply specifically to the wireline retrievable coring method:

1. All retrievable barrels and center bit assemblies are sized to pass the proper API drill collar. This should be checked before going in the hole, because many collars in the field are not made to API specifications. Also slightly bent collars and joint shoulders may prevent free passage of the barrel.

2. When going in the hole, if extensive prior clean-out circulation has not been accomplished, it is desirable to go in with the retrievable center bit assembly in place.

3. When the bit is a few feet off bottom, the center bit assembly should be retrieved, circulation maintained for a sufficient period of time to thoroughly clean the hole, and then the core barrel should be inserted. Do not drop either the center bit assembly or the core barrel in dry drill pipe. The recommended practice is to pump the assembly down.

4. When retrieving the core barrel, a set of overshot jars may be needed. Jars are recommended for all wells where coring is conducted at depths in excess of 8000 ft.

5. As an aid to the operator, when going in or coming out of the hole, the hoisting line should be marked with a soft line tie at points denoting a short distance from bottom and top of the well.

6. When drilling in areas marked by the hazard of stuck drill pipe, the pipe should be rotated slowly while going in with the hoisting line.

7. For retrieving the core barrel in certain high pressure areas, a circulating head is desirable because of the blowout hazard due to the swabbing effect of the barrel being removed. Six hundred ft/min (fpm) is maximum recommended withdrawal speed in firm zones and 350 fpm in soft formations. These speeds are recommended as precautions against losing the core.

Sidewall Coring

Mechanical Type. Sidewall coring is both a supplementary tool for conventional coring, and a means for investigation of certain zones that may appear doubtful on the electric log. In the mechanical type sidewall coring operation, the tool is designed to permit rotation and circulation while going in and coming out of the hole. Body of the tool is made up on the drill stem and lowered to the depth opposite the formation to be cored. If circulation is desired before taking the core, a plug may be dropped through the drillstem to close the opening in the main body of the tool. This plug can be removed by a wireline overshot.

After the plug has been removed, the coring assembly is lowered through the drillstem by wireline until it is bottomed in the body of the tool. This assembly is located in position by a positive latching device.

The tool operator lowers the drill string slowly, carefully observing the weight indicator. When the indicator shows weight being taken, the core tube is beginning to bite into the walls of the hole. Dependent upon the formation, the amount of weight taken by the tool will remain fairly constant unless the operator continues to lower until the tool is punched down through the formation.

To prevent this from happening, the distance of safe travel is carefully marked on the pipe above the rotary table. The core tube may be fractured if excessive weight is placed upon it. Generally, a force of 11 points is considered maximum.

After the core has been cut, the drillstem is raised to its original position and the core tube assembly withdrawn through means of a wireline overshot. After core tube assembly has reached the rig floor, the tube proper may be removed by means of specially designed friction tongs.

The core is extracted from the tube by means of a hydraulic jack, a stand and a pusher rod. The length of core is governed by the length of the core tube used, by the character of the formation, and the condition of the hole. Core sizes are possible of 1¼-in. in diameter and up to 10-in. in length, although the recoveries are usually somewhat under this maximum length.

Rate of core recovery depends upon operating depth but experience has shown that cores may average 1 to 2 hour from average depth. It has been operators' experience that any formation that is drillable with a rock bit at the rate of 3 or 4 ft/hour or faster can be cored successfully.

Percussion Type. Principle of the percussion type sidewall sampler is represented by a bullet in the form of a hollow cylinder. The hollow cylinder is fired into the formation by an explosive charge that is ignited electrically. The bullet is attached to the gun mass by two lengths of steel retrieving wire or, in some models, by a flexible steel cable. The bullet containing the core is extracted from the formation by moving the gun in the hole. Only one bullet is fired into the formation at a time.

Shots can be accurately placed with respect to the electric log (SP curve) by recording a second SP curve on an electrode and positioning the gun by direct log correlation. Positioning, firing and retrieving, each bullet takes only a few minutes, so that it is possible to obtain up to 30 cores at different levels, in the period of one to three hours, depending upon the depth of the operation.

Poor Recovery Causes

Poor core recovery may be due to a number of causes, the more common of which are listed below:

1. Trash in mud. Part of bit will be plugged, causing off-center action of the bit head, which will result in a small diameter core or cause all of the core to be lost.
2. Too much weight. Core will be burned or plugged, if soft; broken up, if brittle and hard.
3. A crooked inner barrel will prevent easy entrance of the core, and a crooked outer barrel will cause eccentric action on bottom. Crooked drillstem near the bit will have the same effect.
4. Insufficient weight or intermittent feeding of weight will cause the bit to drill off at times thus allowing soft cores to be washed away or brittle cores to be broken up.
5. Junk in the hole is a very bad condition for good core recovery. Even a small amount can become lodged in the cutter head and destroy the core.
6. Length of core cut may be excessive for the size of the bit used and the formation being cut. In broken or brittle formations, as much core will often be obtained from a cut of 3 ft as can be recovered from a 20-ft cut. Cores from this type formation shatter and wedge themselves so that no further core enters the barrel.
7. Formation changes in the course of cutting a core. Hard cores taken above soft strata will frequently offer sufficient resistance to prevent additional core from entering the core tube.
8. Connections made in the course of cutting the core are sometimes harmful. Each time the core bit is raised from bottom, the core catcher should act. Thus, when returning to bottom, the catcher has caused a bridged condition that may prevent additional core entering. Also, the catcher becomes second-hand, its condition doubtful.
9. Rapid rotation of a core bit is usually harmful. Bit operates more roughly, tending to break up the core. Hard formation roller cutter heads have small diameter cutters that turn several revolutions for each rotation of the drillstem. Rapid rotation causes bearing wear and shortens cutter life.
10. Too rapid withdrawal of bit or core barrel may cause the core to be sucked out of the barrel.

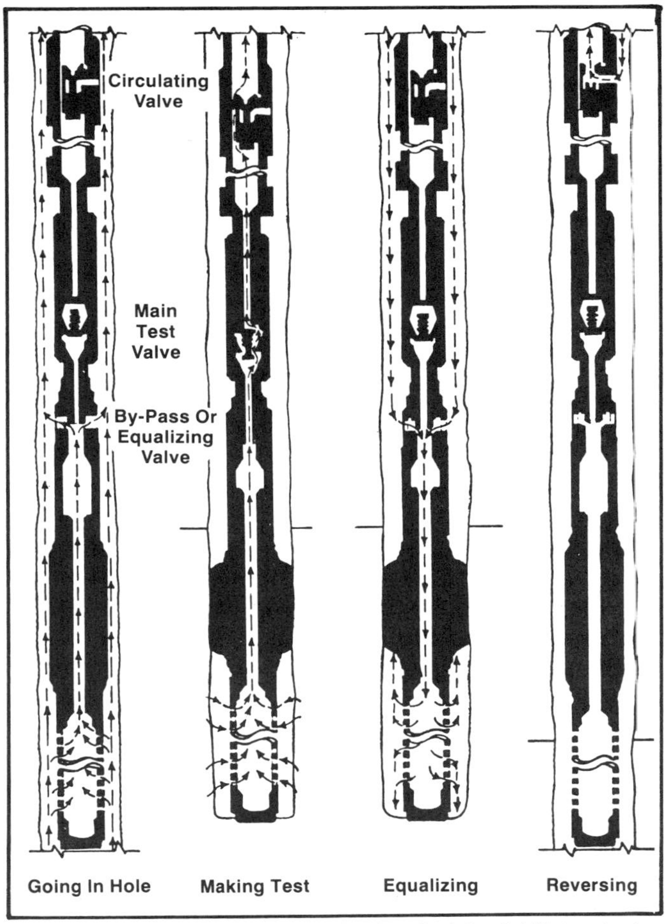

Fig. 17-6. Diagram illustrating the basic principle of drillstem testing. Pressure recording element is usually located in the anchor pipe assembly.

Formation Testing

Despite the recent advances in formation evaluation techniques, there still remain many doubtful pay zones that are penetrated by the drill. The commercial potential of some of these zones can only be evaluated by drillstem test. It should be noted at the outset that a negative test does not always prove the formation to be non-commercial. This fact notwithstanding, the technique of drillstem testing is the most positive method for evaluating the producing potential of a reservoir.

There are several major items of information that can be obtained from a properly conducted drillstem test.

1. The test will evaluate the formation's ability to produce fluid.
2. A sample of the formation fluid may be obtained when the test is made.

3. An indication of the producing capacity of the reservoir is obtained.
4. The shut-in and flowing bottomhole pressures are recorded.
5. The hydrostatic pressure exerted by the mud column is recorded.
6. A formation suspected of taking mud can be checked.
7. The effectiveness of a casing cement job may be checked.
8. A graphic pressure vs time record of the condition of the well bore and well flow performance is recorded for purpose of analyzing results of the test.

The Principle Of Formation Testing

The basic principle of drillstem testing is illustrated in Fig. 17-6. Empty drill pipe, carrying a rubber packer, is extended to block off the annular space between the drill pipe and the wall of the well bore. A valve in the drill stem is then opened to complete the communication between the perforated anchor pipe and the empty drill pipe.

The test zone is thus relieved of the hydrostatic pressure of the mud column and any fluid in the formation is then free to flow up through the drill pipe. After the test period, the equalizing valve is opened in order to equalize the pressure across the packer, so that it may be pulled more readily. The test fluid remaining in the drill stem is then reversed out and the tool is pulled.

Types of Tests

Types of tests and testing tools vary according to the testing conditions. Principal difference is in the packers used and their location with respect to the formation being tested. The different conventional tests are:

1. Single wall packer test.
2. Double wall or straddle packer test.
3. Cone packer test.
4. Wall over-cone packer tests.
5. Casing packer test.

The first four tests listed refer to open-hole tests.

In the single wall packer test (Fig. 17-6), the packer diameter is usually 1-in. less than the bit diameter when used in a full gauge hole. If the packer is to seal in a core hole, the packer diameter is approximately ½-in. less than the core bit diameter. The packer is set by means of weight transferred to the anchor pipe which rests on the bottom of the hole or on a cement plug spotted specifically for test purposes. Special tools are now available that eliminate the anchor pipe in the open hole test.

The double wall or straddle packer test is most commonly used in the full gauge open hole (Fig. 17-7). There are two reasons for this type of test. First, where the depth and pressure make the additional support and seal of a second packer desirable; secondly, where a test zone is to be isolated.

In the latter case, the blank adaptor between the two packers is replaced by a perforated adaptor and the perforated tail pipe is replaced by blank tail pipe. Arrangements may be such that both the pressure between the two packers and the pressure below the bottom packer may be obtained or just the pressure in between the two packers.

The cone packer is used when the core hole is smaller than the hole above (Fig. 17-8). In this case, the anchor pipe does not rest on the bottom of the hole. The seal is obtained by forcing the tapered packer into the upper end of the core hole.

The force for seating the packer is the combination of drill pipe and mud column weight. The shoulder for seating must be in a firm formation.

The wall over cone packer test is used when the core hole shoulder is soft or unconsolidated and the additional seal of the wall packer is desirable (Fig. 17-9).

The casing packer (hook-wall packer) is used for testing through perforations in the casing, for productivity in the open hole below the casing or to test for water shut-off after cementing (Fig. 17-10). As the anchor pipe does not rest on the bottom of the hole, the hookwall packer can be released and reset at any desired level in the casing. Zones may be isolated and tested through the use of straddle packers in this technique.

Testing Tools

Aside from the different packer combinations, there are certain test tools and components that are necessary for the test to be carried to satisfactory completion. Fig. 17-6 illustrates the various components that make up the complete drillstem test tool.

In the conventional test tool, there are three and sometimes four valves, the packer (or packers), the perforated anchor pipe and the pressure recorder. The valves are designated as the by-pass or equalizing valve, the main test valve, the circulating valve and in one arrangement a fourth valve is included.

This is a disc type valve, which is located in the upper part of the tester and serves several purposes. First it insures against a possible contamination of the test fluid. Secondly, leaks in the drill pipe above the test zone can be detected. Finally, it enables an initial shut-in pressure to be obtained. It is opened by dropping a go-devil down the drillstem to shear the metal alloy disc.

It may be placed above or below the water cushion.

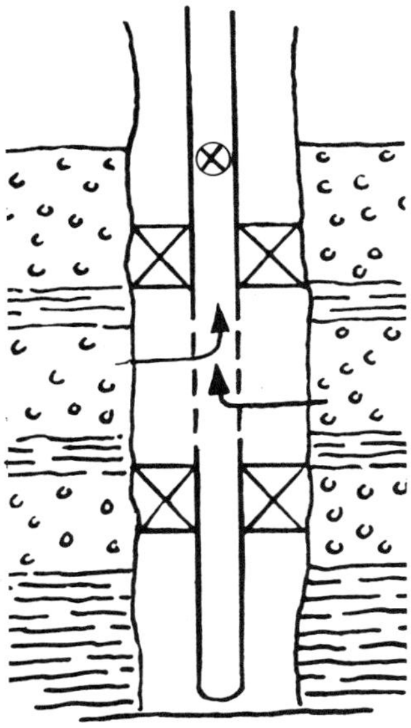

Fig. 17-7. Straddle packer test.

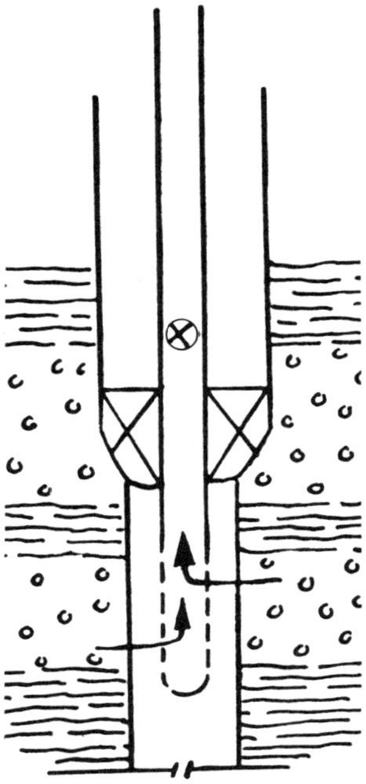

Fig. 17-8. Cone packer test.

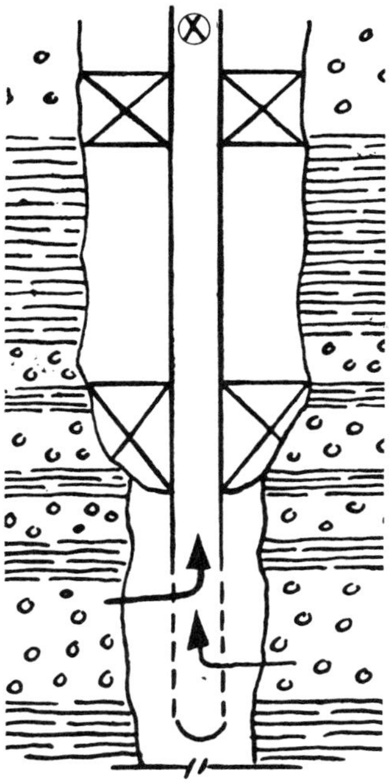

Fig. 17-9. Wall over cone packer test.

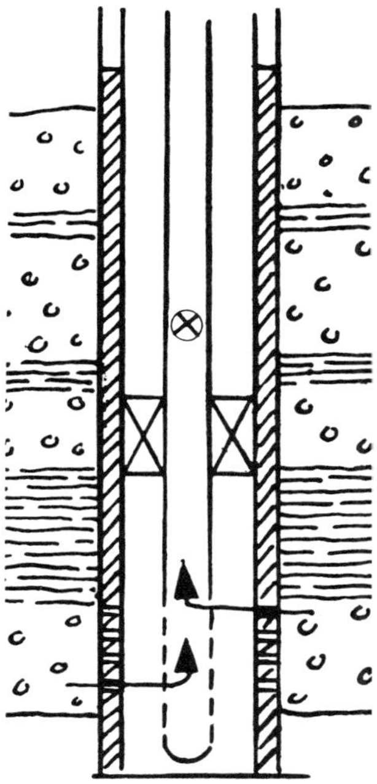

Fig. 17-10. Testing through perforations in the casing.

A heavier go-devil must be used if it is placed below the cushion. One or more pressure recorders are included in the tool make up. Often, a sample chamber is included, if it is desired to obtain a sample of the formation fluid.

The by-pass valve opens a fluid passageway through the bore of the packer to connect the annular space above the packer to the annular space below. This equalizes the fluid pressure across the packer facilitating the seating and unseating operation.

It further reduces the swabbing tendency of the packer when coming out of the hole. A by-pass choke may be used for the purpose of controlling the flow of fluids from the formation during the drillstem test.

The main test valve is one that is opened after the packer has been set. During trips, the tester valve may be locked by a J-slot arrangement. To unlock the valve and open it, weight must be applied to the slip joint, the drill pipe must be rotated slightly and lowered several inches.

The valve may be closed in position again by raising the drill pipe a few inches. This can be accomplished without disturbing the sealed packer or opening the by-pass. A recently developed tool accomplishes the valve opening and closing sequence by means of a hydraulic cylinder, spring and drill pipe weight.

The circulating valve permits circulation of oil and gas samples from the well by either reverse or normal circulation before the tool is removed from the well. The two principal advantages to be gained by use of this circulating valve are (1) the drilling fluid may be reconditioned with the tool still in the hole, (2) a fire hazard is removed, which would exist if combustible fluids were dumped on the drilling rig floor. That is, the complete sample remaining in the drill pipe can be circulated out and the pipe pulled dry.

The pressure gauge records all pressures that enter the perforated anchor pipe during the complete test, from the time the tool enters the surface casing until it returns to the rig floor and is dismantled. Recording is essentially a plot of pressure versus time. Several recording pressure gauges (up to three) may be used depending on type of test.

A safety joint is often used immediately above the packer. This safety joint will permit unscrewing the rest of the apparatus from the packer if the packer becomes stuck in the hole.

Planning The Test Program

It is significant to note that in certain sections of the Gulf Coast open hole drillstem tests are rarely run. The geological province will usually provide an index as to the suitability of the zone for open hole test. Aside from operating policy, there are several reasons for this condition. Certain sections of the Eocene-Wilcox and the deep Frio formation are hard on packers in open hole tests. These sections are tight and have slow pressure build-up characteristics. This means a large pressure differential across the packer for a longer period of time and consequently a high percentage of packer failures.

The down-dip Frio (Corpus Christi area) is characterized by thin sands with many shale breaks. Packer seats are unsuitable in this type of formation and, consequently, the tests usually are of the cased hole type. The three principal objections to open hole testing are: first, the limitations of open hole packers; secondly, the frequent lack of a properly planned program to include complete well preparation prior to testing, and, third, the objection to open hole testing at depth, as the record shows increasing failures with depth.

In certain areas, many companies,as a matter of operating policy, will open hole test only down to a specified depth.

As stated before, however, the suitability for open hole testing will vary with the geological province. The Low Salinity Wilcox and the Rio Grande Valley Frio are very well suited for open hole testing. The Cockfield-Yegua trend also falls in this category. The limestone zones of South and West Texas are commonly open hole tested.

Certain operators make extensive use of rat-hole drilling. Cores will be taken in the rat-hole, the zone tested if core shows are favorable and the rat-hole is then reamed out, with the process repeated until contract depth is reached. This policy will allow rather safe testing in otherwise troublesome areas, but is obviously an expensive procedure. Operators who, as a matter of policy, core extensively will likewise open hole test frequently.

When the work program for a well is planned, selection of a hole diameter favorable to testing should be the first consideration. The best packer seat should be selected that can be found immediately above the producing zone. Generally the harder formations such as sandstone, lime and chert provide the best packer seats. They are firm and do not wash out to oversized diameters during drilling.

A short section test usually gives more conclusive information than a test over a long section.

When long sections are being tested, it should be remembered that the drilling mud in the hole must be produced into the drill pipe before the formation fluid can enter. If the volume of the mud is large, a sizable back pressure may result when the mud is extended in a column up the bore of the drill pipe. Testing long sections with small diameter pipe or tubing should be avoided because of the increased back pressure effect.

Testing Technique

Before discussing specific techniques, it is well to recall a few fundamental factors that are operative during the test procedure.

The packer supports the downward force of the mud in the annulus. This can represent a considerable shock load when the test valve is first opened. The magnitude of the shock is dependent upon the size of the valve opening and the pressure inside of the pipe. The use of chokes and water-cushions in the drill pipe tend to reduce shock loads. One special test tool is specifically designed to ease the shock loads normally present in the conventional testers.

A "dry-test" is the worst mechanical condition encountered in testing. It is evident that the smaller the packer area exposed to the pressure differential, the better chance the packer will have of holding. This is a point in favor of rat-hole testing. On the other hand, the pressure exerted by the mud column helps facilitate the effectiveness of the packer seal.

In the conventional single wall packer, open hole test, the anchor pipe must support the packer from below and carry the combined downward forces of the drill pipe and mud column. Although the friction between the packer and the wall of the well bore helps support a portion of this load, the amount of help is indeterminate.

The drill pipe above the packer is subjected to a collapsing pressure imposed by the hydrostatic pressure of the mud. This consideration should be a principal governing factor in determining the amount of water-cushion to be used in the test.

Hole Conditioning. Proper conditioning of the hole and the drilling fluid prior to running the tester is most essential for successful testing. It has been estimated that fully 50% of mechanical test failures have been due to improper conditioning of the hole.

A full gauge clean hole is important.If the testing tool has to be spudded through tight places, mud cake and debris will be pushed to the bottom of the hole where they will tend to clog the perforated tail pipe when the tester is open.

The drilling mud should have sufficient weight to control the formation pressure and a low viscosity is generally desirable. Before coming out of the hole to take a formation test, circulation should be maintained long enough to clean the hole completely of cuttings, cavings and viscous mud.

Testing Packer Setting. Once the mud and hole are in proper condition, the tester is assembled on the drill pipe and lowered into the well with the by-pass valve open and the tester valve closed (Fig. 17-6). During the trip into the well, the fluid level in the annulus should be observed from time to time to see if any leaks develop in the equipment. The observations must be made while the pipe is stationary. A drop in the fluid level may indicate a leak in the drill pipe or tester.

The annulus should be full before the packer seal is made and before the tester is opened. If a sudden drop in the fluid level occurs when the valve is opened, the tool should be picked up immediately to close the test valve, because the packer has failed to seal and the mud is entering the drill pipe from the annulus above the packer. Failure to quickly close the tool may result in lost control of the well.

Surface Control Equipment. Surface control equipment is attached to the upper end of the drill pipe before the tester valve is opened (Fig. 17-11). This equipment will usually consist of a control head with a conductor pipe leading into a slush pit or a storage tank. A small outlet valve with a rubber hose attached is sometimes used to bubble air or gas through a pail of water, in order to detect light blows.

The control head itself contains two valves arranged in tandem with an outlet between them for attaching the flow line. The kelly may be connected to the control head any time circulation down the drill pipe is required. A control head and chokes in the flow line are essential to control formations which flow under considerable pressure.

Time of Test. The length of the period of test in the open hole often times is limited by company policy. Fear of a stuck packer is the principal reason for short tests. Twenty to thirty minutes appears to be the usual length of time for flow tests in the Gulf Coast area. This time, of course, depends upon the experience that has been gained from the area under test.

In the Mid-Continent, the length of test may run one hour or more. The danger of losing the hole is not nearly so great here as in the Gulf Coast plus the fact that the formations being tested are usually more consolidated and of a lower permeability and hence require a longer buildup time.

No restrictive time limit is placed on the cased hole test.

When the flow has continued for the desired length of time, the tester valve is closed. The first few inches of upward travel of the drill pipe closes the tester valve but does not open the by-pass valve or unseat the packer (Fig. 17-6). The tool is left in this position to allow the pressure below the packer to reach equilibrium with the formation pressure surrounding the well bore. The shut-in pressure is recorded. If the initial shut-in pressure is desired, it may be obtained by use of the disc valve as previously noted.

Trip Out of the Hole. At the completion of the shut-in period, the by-pass valve is open and the packer is unseated. During the trip out of the well with the testing tool it is essential to keep the hole full of mud, otherwise the level in the well may drop to the point of losing control of the formation.

Any fluid that has entered the drill pipe through the tester valve and has not emerged at the surface, is trapped above the valve and is usually brought to the surface with the drill pipe. Emptying the drill pipe of volatile fluids as each stand is unscrewed is a fire hazard and should be conducted with caution.

The circulating valve mentioned earlier offers a method for removing the sample from the drill pipe before the pipe is pulled from the hole. Once the valve opens, it locks to permit circulation in either direction through the drill pipe and the annulus. Thus the sample may be reverse circulated through the drill pipe and gauged as it flows from the pipe.

Circulation of samples from the drill pipe using the circulating valve has certain possible disadvantages to be considered. Pumping out through the annulus leads to a certain amount of liquid mud contamination and gas cutting. Reverse circulation up the drill pipe increases the mud pressure on the exposed formation in the open hole. In most combinations of pipe and hole sizes, the pressure drop in the annulus is less than that in the drill pipe. Reverse circulating may therefore result in breaking down weak formations.

Special Testing Tools. Many efforts have been made to improve the existing conventional testing tools. There is as yet no tool available for testing close to bottom in the open hole that has the ruggedness, simplicity and reliability of the standard testing tools now in use. Certain tools have been developed that have a special application and should be mentioned.

One major oil company has developed a wireline drillstem tester that does not require removing the drill pipe from the hole between the coring operation and the bottomhole formation test.

One of the service companies now offers a retrievable wireline tester for testing in a small core hole ahead of the bit without removing the drill pipe after coring or before the testing operation.

The above tools differ from the conventional wall packer testers by a common feature, in that only a small sample of the formation fluid is obtained. A pressure buildup curve, shut-in pressure and a fluid sample are obtained.

The most widespread application for the above two testing tools is in areas where the wireline core barrel is in common use. The procedure of coring and testing can be repeated any number of times without removing the drill pipe from the well.

A multi-zone open hole tester that does away with the anchor pipe is now offered by one of the service companies. The anchoring device utilizes nine mechanically retractable dogs. The anchoring mechanism does not support the full weight but only enough to allow the packer to take hold.

Pressure Chart Analysis

Proper evaluation of drillstem tests results can only be obtained through the use of pressure chart analysis. Dependence upon the visual observation of the formation fluids is often-times grossly misleading. Many instances are of record in which strong fluid shows proved to be non-commercial. Also many

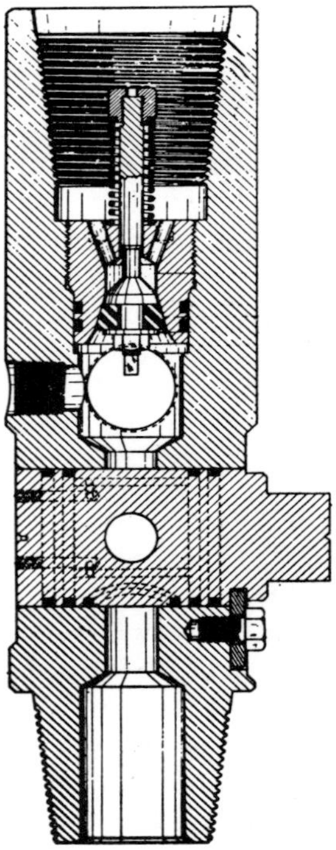

Fig. 17-11. API surface control head assembly.

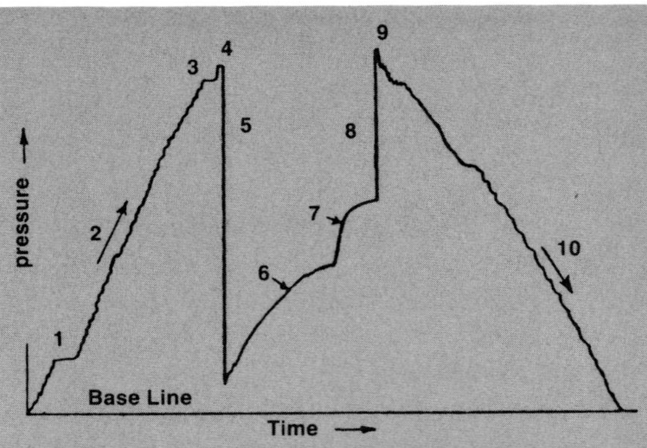

Fig. 17-12. Chart illustrating fundamental pressure recordings (flowing well test).

1. Putting water cushion in drill pipe
2. Running in hole
3. Hydrostatic pressure (weight of mud column)
4. Squeeze created by setting packer
5. Opened tester, releasing pressure below packer
6. Flow period, test zone producing into drill pipe
7. Shut in pressure, tester closed immediately above packer
8. Equalizing hydrostatic pressure below packer
9. Released packer
10. Pulling out of hole

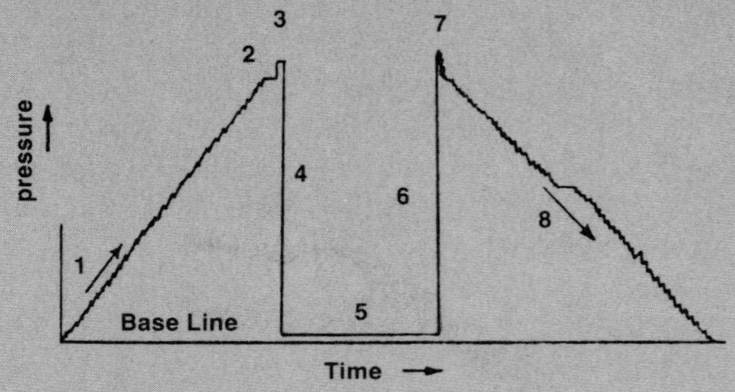

Fig. 17-13. Chart illustrating a "dry" test.

1. Running in hole
2. Hydrostatic pressure (weight of mud column)
3. Squeeze created by setting packer
4. Opened tester, releasing pressure below packer
5. Flow period, test zone open to atmosphere
6. Closed tester and equalizing hyd. pressure below packer
7. Pulled packer loose
8. Pulling out of hole

instances have been noted where apparent dry tests were determined to be commercially productive by analysis of the pressure chart and the analysis was borne out upon completion of the well. Valid results of tests can only be assured through the proper interpretation of these charts. A thorough knowledge of the basic testing tools and the mechanics of the test operation is essential for this analysis.

The chart is no panacea. There is no absolute value of flowing pressure, for example, that will enable a positive decision to be made as to whether or not a given zone will be commercially productive. It is the relative rate of buildup, however, that is the characteristically important criterion.

It is desirable always to have at least two pressure recorders and sometimes three when testing. One is to be placed in the flow stream; a second to be placed immediately below the tester perforations blanked off from the tester flow string. A third recorder is desirable when straddle testing.

Pressure recorders are of various types all of comparable accuracy. The chart is essentially a recording of pressure versus time. The absolute value of pressure, as reflected by the chart, is determined by use of a calibration constant which denotes so many psi/unit deflection away from a base line. For example, in one recording unit, a vertical deflection of 1 in. represents 1,000 psi. Special formulas are used for extrapolation of buildup pressures.

Certain fundamental chart recordings are presented in order to set out the basic problems of chart interpretation. These are followed by more detailed chart recordings wherein most drillstem testing problems are represented and analyzed (Figs. 17-12 and 17-13).

Chapter 18 Straight Hole and Directional Drilling

Crooked holes have worried drillers almost from the start of oilwell drilling. In cable tool operations, crooked holes were readily detected by the way the tools ran or by difficulty in running casing. Crooked cable tool holes usually were offsets or doglegs caused by drilling into steeply dipping formation or by striking a concretion, in which event the hole angle became acute. In Wyoming and other hard rock areas, with steeply dipping formations, spiraled stems and other full hole guides were often used to prevent offsets in the hole.

In rotary drilling operations crooked holes are not so readily apparent and it is probable that when drilling soft shales with little or no dip in the formation there was little tendency for the hole to deviate from vertical. In the twenties, when rotaries were moved into hard rock country, crooked holes became commonplace with reports of wells drilling into the casing of nearby wells. As suspected, surveying instruments confirmed that many wells in the Seminole area of Oklahoma were crooked. The first popular surveying tool was the acid bottle in which a four ounce glass bottle, half full of hydrofluoric acid and encased in a long slim-like cylinder called a bomb, was lowered through the drill string to a point in the drill collar just above the bit. After remaining at rest for about thirty minutes the tool was pulled and the etched line inside the bottle at the acid level revealed the slope of the hole.

When more sophisticated surveying devices were developed, including directional surveying instruments, many were surprised to learn that practically all rotary drilled holes tended to drift updip until direction of the hole stabilized at a slope normal to dip of the formation.

To the practical man engaged in drilling a well, the hole becomes crooked when he is forced to retard drilling as a means of keeping vertical deviations within contract limits. Throughout this book "crooked hole" will carry this same meaning because we are concerned with actual drilling progress and the effect of application of technical knowledge.

Special devices such as ribbed joints of pipe or drill collars and sleeve type or roller type stabilizers have been tried for many years to help drill straight holes. The results of using the various combinations of these devices were uncertain. Since so many factors between wells varied, it was not considered practical to study the wells as a group. Each crooked hole presented a problem to be solved as an individual case. In most cases the only way to drill a "straight hole" was to reduce the weight on the bit and consequently reduce the drilling rate.

Many things have been tried as a means to drill "straight holes" such as changing bit types, deleting teeth from bits, deleting cones, using limber members on the bottom of the drill stem, stabilizing, reducing hole size and then reaming to the desired size, and running light weight and then suddenly increasing weight. With the exception of stabilizing, the effect of any of these methods is questionable and experience has shown that the results cannot be predicted or reliably repeated.

The approach to a crooked hole problem has changed from the old attempt to drill straight holes (small deviation angles) to that of drilling the most economical hole. Larger deviation angles are becoming more acceptable and better control of the hole makes it possible to drill with more weight in the crooked zones. This method has greatly reduced drilling costs. The method, as being used in many places today, was preceded by precise mathematical analysis. The cause of crooked hole and the effect of drill collar size, weight on the bit, stabilizers and related factors have been analyzed mathematically. This information makes it possible to approach the crooked hole as a scientific problem instead of another unplanned experiment.

Through the proper application of the available technical data on crooked holes, drilling costs can be reduced. The amount to be saved will vary with the individual case or locality. When provided with a knowledge of the area in question a practical man now can plan to drill with confidence. The well to be drilled can be classified according to its severity and methods can be employed to combat the situation. Prior planning, utilizing available data, is now possible and is indeed essential to drill economically through crooked zones.

Keep in mind that crooked holes cannot be economically eliminated but a change in prevalent concepts of trouble causing deviation angles will greatly reduce the cost of such holes.

Basic Concepts

In this study it is assumed that the drill collars will lay on the low side of the hole as the hole deviates

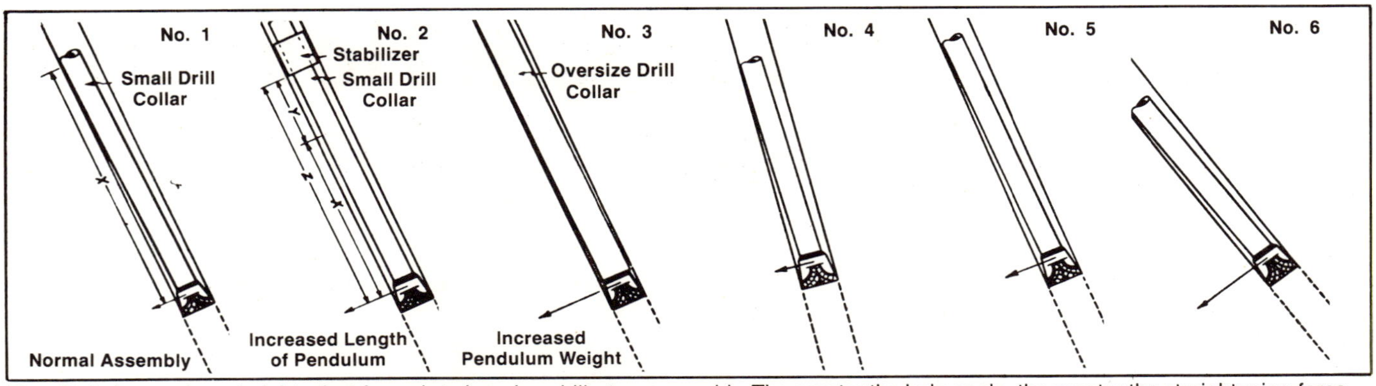

Fig. 18-1. Increasing straightening force by changing drill stem assembly. The greater the hole angle, the greater the straightening force.

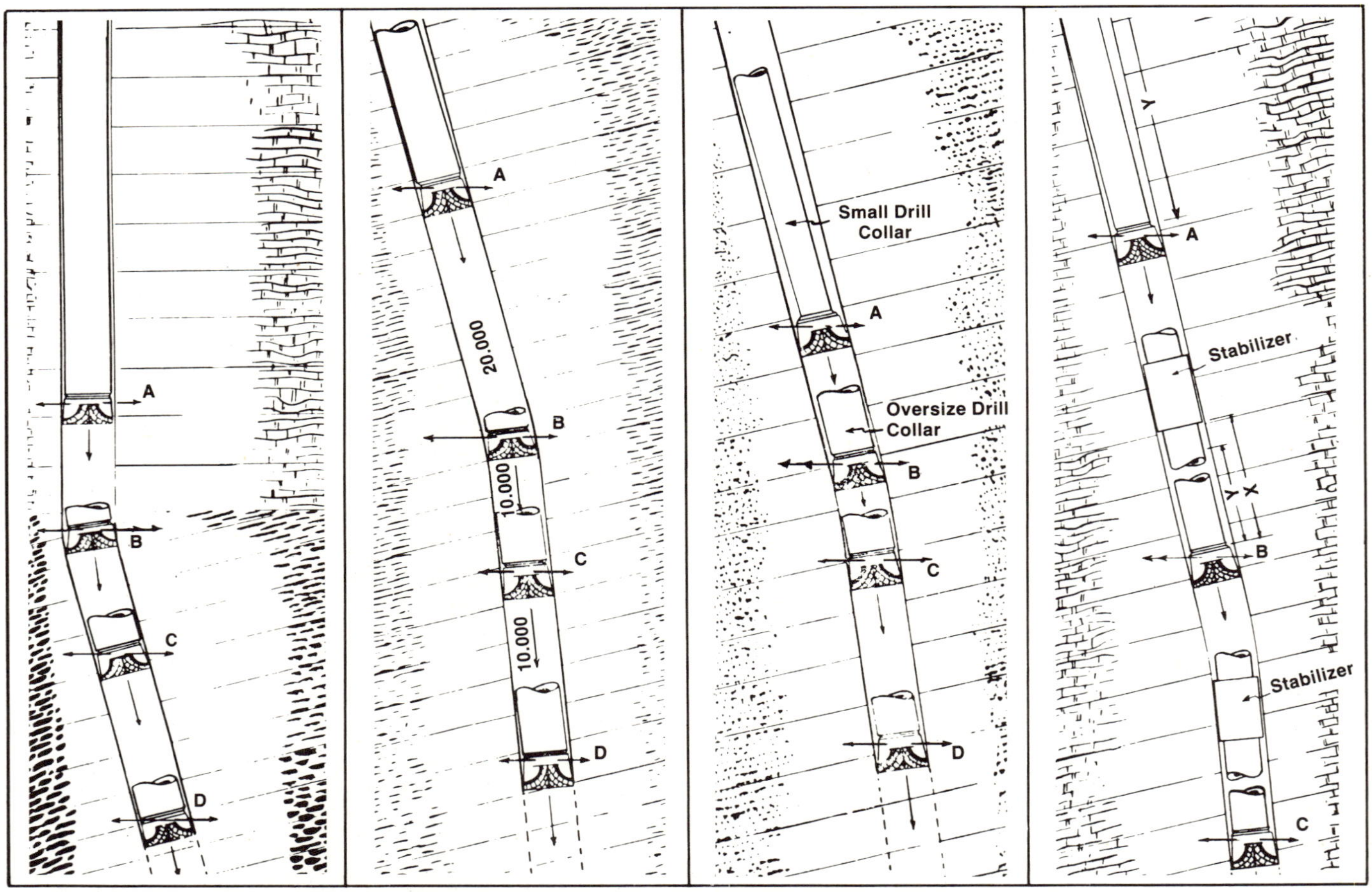

Fig. 18-2. Increase in deviation and establishment of new equilibrium condition after change in formation dip.

Fig. 18-3. Reducing hole deviation by reduction in weight on the bit.

Fig.18-4. Reducing hole deviation by increasing pendulum weight.

Fig. 18-5. Reducing hole deviation by increasing pendulum length.

from vertical. The drill collars are assumed to act as a pendulum from the point of contact with the wall of the hole to the bit; this length, supported only by the bit, tends to swing toward vertical. This tendency applies a lateral force on the bit which causes the bit to cut sideways and acts as a hole straightening force. The amount of this force can be increased by (1) lengthening the pendulum section, (2) increasing the weight of the pendulum section, or (3) by increasing the angle from vertical. These variations are illustrated in Fig. 18-1.

Illustrations are simplified for the purpose of setting forth the principal forces acting on the bit. The forces which determine the hole deviation and drilling progress are (1) the weight on the bit, (2) the reaction of the formation tending to deviate the hole, and (3) the straightening effect of the drill collar "pendulum effect" trying to return to vertical. These forces are shown in Fig. 18-2.

If the forces are in lateral balance, the hole angle will continue as is shown at Point A. A change in deviation must be preceded by some change in the direction or strength of the forces at the bit, as occurs at Point B. These forces can be thrown out of balance by increasing or decreasing the bit weight, by changing drill collar size, by stabilizing or by a change in formation characteristics.

In a near vertical hole, such as shown in Fig. 18-2 to Point A, the bit moved downward because the lateral components of the forces were in balance. When the bit reached Point B the horizontal reaction of the formation tending to push the bit away from vertical was greater than the straightening force of the drill collars and the hole started to deviate from vertical. By the time Point C was reached, the straightening force of the collars had increased as the hole angle increased until the lateral forces were again balanced. The section from Point C to Point D is straight though inclined, which means that a state of equilibrium has been reached and no change in angle will take place until the lateral balance is changed.

A change in hole deviation obtained by a change in weight is illustrated in Fig. 18-3. As weight was decreased lateral forces became unbalanced, the pendulum force of the drill collars was greater than the horizontal reaction of the formation, and the hole started to straighten. At Point C the lateral forces once again balanced and the hole remained straight while drilling progressed to Point D. In this case drilling progress will be retarded as bit weight has been decreased in order to reduce hole angle.

Substituting a larger drill collar will give the same effect. It increases the pendulum force of the drill collars and reduces the hole angle until a state of equilibrium is reached, as shown in Fig. 18-4. At Point C the forces again have become balanced and the hole will continue at this angle. Note that in this case drilling progress is not retarded as the bit weight remains constant, but reduction in angle is accomplished by increasing the pendulum effect of the drill collars.

A similar result can also be obtained by using a stabilizer with small collars to raise the point of contact of the drill collars with the wall of the hole, as shown in Fig. 18-5. The pendulum length is shown as Distance Y without the stabilizer and is increased to Distance X by use of a stabilizer. The increase in the length of the collars acting as a pendulum increases the force tending to return the collars to the vertical position. As the hole is drilled from Point B to Point C the deviation decreases and the lateral balance is again established. Again, drilling progress has not been retarded, as the hole angle has been decreased without reduction of the weight on the bit.

Rapid Changes In Hole Deviation

Production departments of some oil companies, which are producing from wells with large deviation angles, no longer consider hole deviation in itself to be detrimental. The real culprit, from the producer's point of view, is the so-called dogleg or sudden change in hole direction. Slope of the hole, so long as it builds slowly, causes little or no producing difficulty.

Several of the operators in crooked hole country are no longer concerned with deviation or drift as such but insist on a smooth hole without doglegs. It is not uncommon, where natural drift can be reasonably calculated, to shift the surface location to a site which will permit the hole to be drilled rapidly, at low cost and allowed to drift naturally to the target area.

Prevention of doglegs and the assurance of a smooth hole can best be realized by the employment of the "packed hole" technique which may include square drill collars, stabilizers, reamers and various combinations of these tools.

Strict adherence to small deviation angles can penalize the producer in two ways. Forced light weight to maintain small angles raises the cost per foot and the producer must eventually pay this increased cost of drilling. In cases where the hole angle rapidly approaches the contract limit, the producer pays for and sometimes insists on drilling practices which result in a dogleg being formed through a diligent effort to keep the hole straight. Deviation is held within limits only by a drastic reduction of weight.

A misunderstanding as to what actually causes trouble in a producing well has led to a false feeling of security in maintaining small deviation angles. Certainly, deviation angle is important but is by no means the only important factor. Various factors have been under study and new methods to define crooked hole for contract purposes have been proposed.

Fig. 18-6 has been drawn to clarify what happens when abrupt changes in weight are made to control hole deviation. It is apparent from this illustration that even though the desired hole angle was maintained, the well bore was not as trouble free as might be expected. In a hole approaching contract limit, Point A, a large reduction in weight has been made, a dogleg formed, and a smaller deviation reading obtained at Point B. Reduction in hole angle has stopped at this point because the collars touched the wall of the hole at a point close to the bit and the straightening force is lost. The hole will continue to drill at this angle until Point C, when the pendulum length of the collars increases enough to produce a straightening force which will overcome the formation reaction. The hole then gradually straightens until the equilibrium condition is reached.

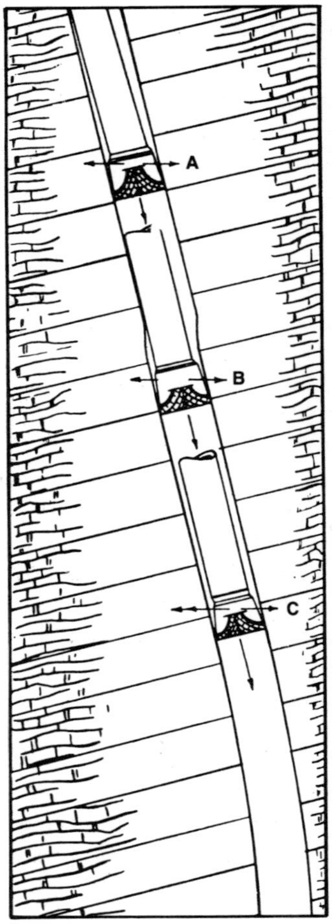

Fig. 18-6. Formation of a dogleg by rapid reduction of weight on bit showing eventual reduction in hole angle.

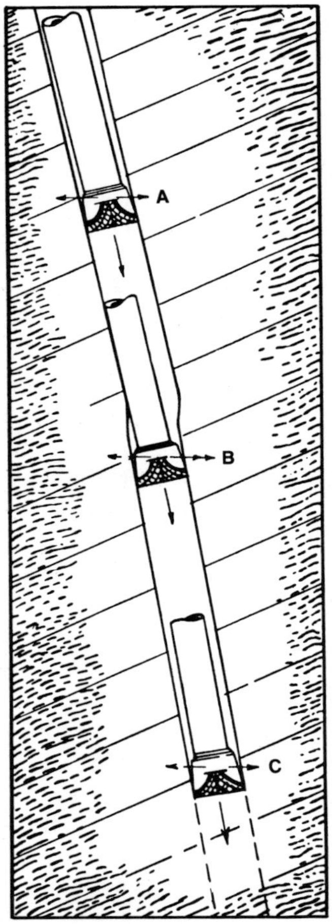

Fig. 18-7. Formation of a dogleg in a more severe crooked hole zone showing slight increase in deviation angle.

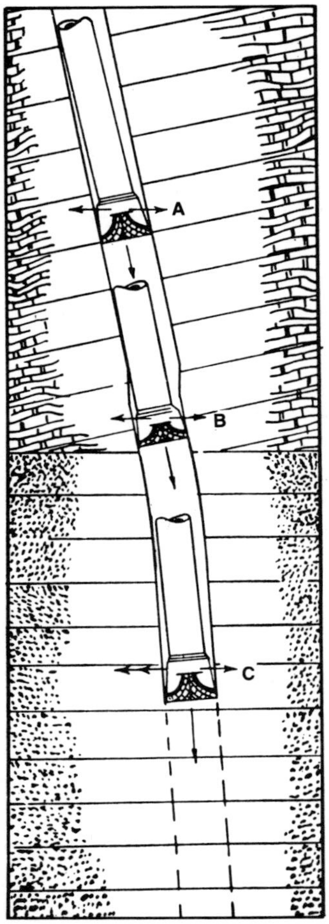

Fig. 18-8. Hole straightening after increase of bit weight near bottom edge of crooked hole zone.

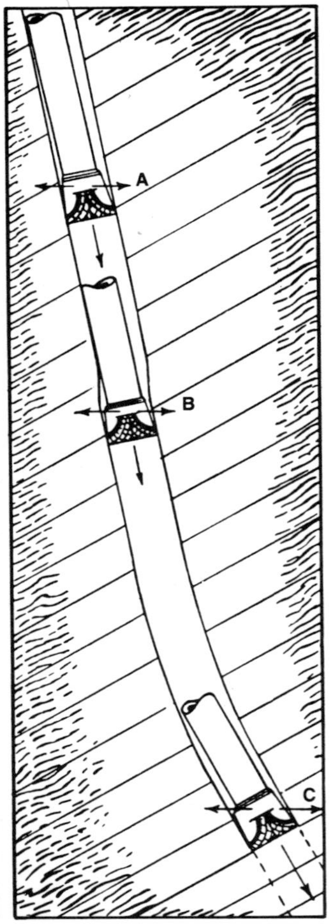

Fig. 18-9. Effect of increased bit weight when a considerable section of crooked hole zone remains to be drilled.

Fig. 18-7 illustrates a variation of this condition in a more severe crooked hole zone. When the dogleg was formed and the drill collars contacted the wall of the hole near the bit, the pendulum force was lost and sufficient formation reaction existed to actually increase the hole deviation slightly. If the hole is permitted to progress, the formation's force will be overcome by the increase in pendulum effect of the collars and a new equilibrium will be established.

The new deviation established will be greater in Fig. 18-7 but in both Fig. 18-6 and Fig. 18-7 better results could have been obtained without the hazard of the dogleg by a gradual reduction of bit weight. Often the tendency is to lighten the weight even further if a deviation survey is made showing the hole angle to be increasing slightly after the dogleg is formed in the case of Fig. 18-7. This practice costs even more money since a reduction in weight will generally slow the drilling rate.

Many field men can remember places where a hole was straightened by fanning bottom and then throwing the weight to it after recording an indication that the hole is beginning to straighten. In several cases which have been examined the situation paralleled that illustrated in Fig. 18-8. After a long period of various experimental operations which carried drilling progress to near the lower edge of a crooked hole zone, light weight operation was tried and the dogleg was formed.

After getting a slight reduction in hole angle the weight was increased to several thousand pounds per inch of bit diameter. Drilling progress immediately increased and when Point B was reached the formation had changed and lateral force of the formation was very much reduced.

The hole angle reduced to near vertical and the hole is said to have been straightened by weight. This was the actual case in several instances examined.

Weight failed to straighten the hole illustrated in Fig. 18-9. This situation was similar to that shown in Fig. 18-8 except that a considerable section of the crooked zone remained and the hole angle went to 9°.

Knowing when the formation will change is apparently the trick to straightening a hole by increasing the weight. This can sometimes be planned and the drilling rate profitably increased while approaching the formation of lesser dip.

Size of the dogleg and the rate of change of hole deviation will be much less for large drill collars as they will touch the hole wall sooner as the bit cuts sideways, as shown in the illustrations.

Solution Of Actual Drilling Problems

Many cases where theoretical data has been confirmed by actual results are available and a few have been selected to illustrate certain types of problems and how they were solved.

Learn all possible about the area in preparation for drilling crooked hole; the depth, thickness and severity of crooked zones. After this has been done the best equipment for the job can be selected, the program planned and the hole can be drilled as economically as possible for the conditions which exist.

Example 1: Two rigs were moved into a new field where limited information was available. The records available did not show any severe crooked hole sections so no preventive measures were taken. Slight crooked hole was encountered wih Rig A, while Rig D found the same zone was severe and difficult to drill. Fig. 18-10 was drawn to show depth versus hole angle for Rig A and the other two wells which had been drilled. This graph was made only to show the depth at which deviation changes could be expected and ignores all other factors.

It was noted that in each of these three wells the deviation angle increased at 5200 ft and in each case returned to near vertical at 6100 ft. The well Rig D was drilling was plotted next to the other curves and it too showed an increase in angle at 5200 ft. It was then reasonable to assume that it would start to lose angle as it approached 6100 ft. The weight was gradually increased from 5000 lb to 20,000 lb in the lower portion of the crooked hole section and the drilling rate was increased from 2.56 ft to 13.2 ft/hour.

By predicting where the hole would straighten, heavy weight could be run through a zone when normally light weight would have been carried. Starting the heavy weight program at the maximum deviation of 3° as indicated by the broken line in Fig. 18-10, Well D, the hole had returned to 2° at the end of the first bit run and continued to reduce angle to 1° as drilling progressed.

These first two wells (A and D) required an average 62 bits, 60 days and 720 rotating hours. With proper planning and accumulated knowledge of the area wells are now drilled with 40 bits, 42 days and 535 rotating hours, a saving of 22 bits and 18 days.

From the Woods and Lubinski charts it was deter-

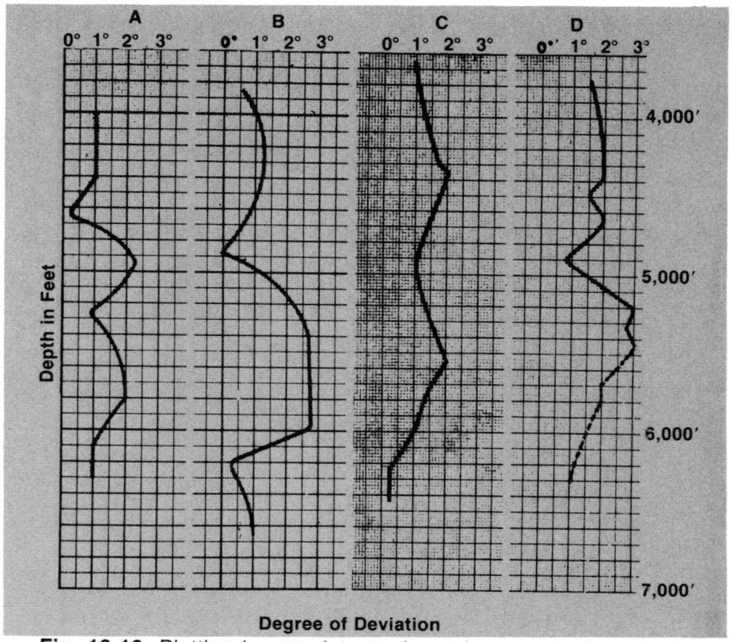

Fig. 18-10. Plotting known data to determine extent of crooked zones.

mined that 8¾-in. hole using 8-in. drill collars would be more economical to drill than the 7⅞-in. hole, 6-in. drill collar combination used on the first two wells. Calculated on a basis of 3° maximum deviation, 15,000 lb was carried with the 6-in. collars in 7⅞-in. hole and 34,000 lb could be carried with 8-in. collars in 8¾-in. hole at the same deviation. Actual drilling in this field with 35,000-lb 8-in. collars and 8¾-in. hole produced a maximum deviation of 2¾° which was reached after most of the section had been drilled. The 35,000-lb weight was used to drill the entire section and the hole was free of doglegs.

In this area electric logs and sample descriptions have been used to correlate the wells and when crooked zones are approached in drilling, reamers are placed just above the bit to retard the rate at which the hole gains angle. It is accepted that the hole will be allowed to approach the maximum contract angle since the most economical drilling rate will be at the maximum weight usable to establish equilibrium just under the contract limit. To retard the rate at which the hole gains angle permits more hole to be made under heavy weight conditions before weight must be reduced to establish equilibrium near the contract limit.

No set rule can be made to say when reamers should be used or stabilizers moved in the string. Proper use of these tools comes from field analysis experience. An excellent knowledge of the crooked zone formation is necessary to evaluate results properly. The presence or absence of reamers can have a profound effect on drilling angle and rate change of this angle.

Three crooked zones exist in each hole drilled in

TABLE 18-1.

Well	Hole size	Collar size	Weight 1000's	Maximum deviation	No. bits through zone	Days through zone
A	7¾″	5¾″	22	4″	63	66
B	8⅝″	5¾″	18	3″	58	
Calculated					4 Dia	45
C	8¾″	8″	47	4″	?	30
Actual						
C	8¾″	8″	40	3″	42	32

TABLE 18-2.

Depth from	to	Feet drilled	Bits used	Days	Feet per bit	Cost/ft @ $40/hr Rig time
1091	2838	1747	40	24	42.15	19.50
2838	2696	857	13	8	65.5	12.20
3696	4715	1019	23	17	44.3	20.15

this area and similar solutions have been applied to each zone. The savings mentioned are cumulative for all zones and not for the one zone shown in Fig. 18-10.

Example 2: Two wells drilled in Gaines County, Texas, showed a marked tendency for the wells to deviate at approximately 6300 ft. The most severe section came at approximately 7800 ft but the entire interval from 6300 ft to 8400 ft had to be drilled with light weight. An analysis was made of the hole size, weight on bit, drill collar size and hole deviation through the section in question. The Lubinski and Woods curves were utilized and an estimate made as to how much faster the well could be drilled with oversized collars.

Table 18-1 presents a tabulated account of what was done:

Using oversize collars and drilling at maximum angle saved 13 days and 22 bits. The most rapid rate of penetration will be obtained when the hole angle is near the contract limit. To reduce the weight and straighten the hole only results in slower penetration and greater cost per foot. Why drill with light weight and maintain 1° or 2° when the drilling rate can be increased while maintaining and controlling deviation at 3° or even 4°? Once the equilibrium angle is obtained the hole is under control; that is, to increase weight slightly increases the hole angle and conversely, to decrease weight decreases the hole angle.

Contractors actually did establish equilibrium in the first two wells listed in Table 18-1. They continually reduced weight and found that they had to maintain a large angle and comparatively heavy weight to drill the hard formation. Several runs were made with the same weight and the hole remained at the same angle which means the hole was in equilibrium. The small collars offered a small straightening force and consequently the weight had to be reduced. The Lubinski and Woods charts offered a means of calculating the effect of using larger drill collars and increasing the pendulum effect. The use of larger drill collars permitted more weight on the bit while keeping the hole in lateral equilibrium at the same deviation angle.

Example 3: A well in the Rocky Mountains was being drilled with 6-in. drill collars in a 9-in. hole. The formation was known to dip 30°. At a depth of only a few hundred feet the hole was off to 4° and the weight had been reduced to 4000 lb. Every known bit type and construction was used, including two cutter, three cutter, four cutter and diamond bits. In 24 days 1747 ft were drilled, using 40 bits, including diamond bits. Four degree deviation was still held with 8000 lb weight. It was calculated from the Woods and Lubinski charts that with a 5° deviation and 8-in. drill collars, 17,000 lb could be run. Permission was obtained to try this procedure.

Two 8-in. collars were run immediately above the bit with 6-in. collars completing the string, and 17,000 lb was placed on the bit. Eight days later 857 ft had been drilled with 13 bits and the hole was 4¼°. At this point the operator chose to try to straighten up the hole. The 8-in. collars were removed and light weight operation resumed. The final result was the establishment of a 5° to 5½° deviation angle. 23 bits had been used to drill 1019 ft. Bit usage had been higher due to the practice of reducing the hole size and then reaming to full size.

Table 18-2 is a presentation of the progress on this well:

Example 4: Stabilizers are being used to increase the straightening force of the drill collars by increasing the distance between the bit and the point at which the drill collar contacts the side of the hole. The distance the stabilizer is placed above the bit depends upon the hole size and the collar size. When trying to obtain the maximum straightening effect with 6¼-in. collars in a 7 ⅞-in. hole the stabilizer was found to be very effective at 68 ft above the bit. Unpublished theoretical data shows the optimum distance above the bit to be 73 ft.

While drilling a crooked 7 ⅞-in. hole with 6 ¼-in. drill collars the hole was in lateral equilibrium with 15,000 lb of weight at 3°. A stabilizer was placed in the string and the same weight carried on the bit. The hole now stabilized at 2°. A weight of 20,000 lb was placed on the bit and the hole stabilized at 3°. Several changes were made in the weight and stabilizer arrangement and it was found that 5000 lb additional weight could be carried on the bit when properly using the stabilizer.

Example 5: All theoretical work must of necessity assume uniform bedding planes and formation dip. No mathematical formula can be used to evaluate the forces expected by fractured formation. Since a great amount of crooked hole is caused by fractured formation, it is necessary to assume an equivalent uniform case.

For example, if a fractured zone is encountered

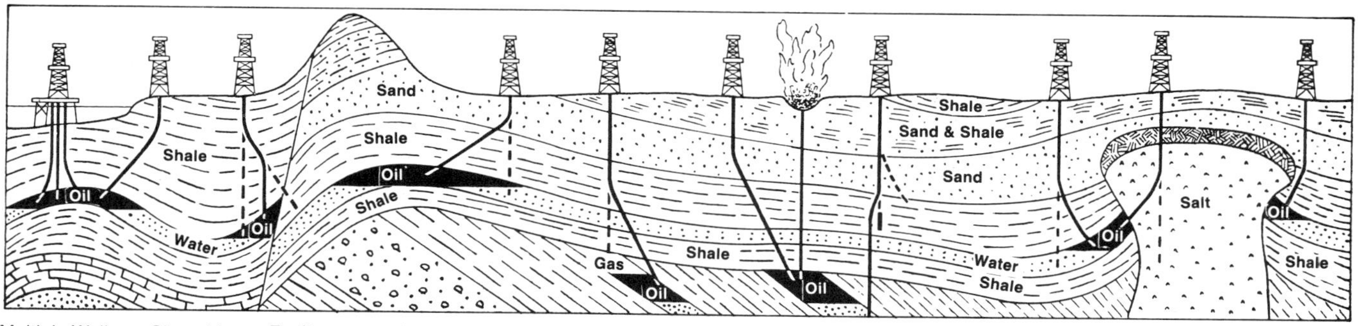

Fig. 18-11.

while drilling and the hole reaches a state of lateral equilibrium at 5° with 8-in. collars in an 8 ¾-in. hole with 15,000 lb.of weight, predictions can be made as to what results will be obtained if certain changes are made. Treat the information the same way as any other set of field data and enter in the charts. Care should be exercised when working with a hole of this nature since rapid changes in direction are possible.

On one such well as mentioned above, the hole deviation changed 2° in about 25 ft. The sudden change, although the total deviation angle was small, indicated that a severe crooked hole condition could be expected. Weight was reduced to minimize the rate of change and even with 8-in. collars in 8 ¾-in. hole the deviation angle reached 8° in about 150 ft. Several bits were run and the hole stabilized the same as in other crooked holes.

The operator decided to core and it was noted that the lower 2 ft of the core was free of the cracks and secondary deposition which marked the upper portion. After reaming down the core hole more weight was placed on the bit and as expected, the hole started to straighten.

By carefully reducing weight and controlling the rate of change as much as was practical the hole was permitted to reach a comparatively high deviation angle for West Texas. This was necessary in order to carry enough weight to drill the hard formation.

The hole was drilled more than 1000 ft beyond the severe crooked zone with no evidence of dragging.

Summary

Various methods of drilling crooked holes, all based on straightening the hole, are questionable in their effectiveness and certainly can prove to be expensive practices. A planned scientific approach to the crooked hole problem through the use of Lubinski and Woods charts can result in lower drilling costs in most cases.

Even greater reduction in well costs can be realized by employment of the packed hole technique and allowing the hole to drift naturally, from a properly chosen surface location, to the target area.

Directional Drilling

Controlled directional drilling is the art of drilling a well to a pre-determined underground target or location. Concurrently, this is the fundamental of all modern drilling — whether it be a bore hole held as close as possible to the vertical or a deliberately planned deviation from the vertical. In either of these categories, the mechanical principles of application are basically the same.

In earlier days, directional drilling, pioneered by John Eastman in California, was used primarily as a remedial operation either to side-track junk tools, bring the bore hole back to vertical or applied to the drilling of relief wells. Today's concepts and methods of oil recovery have so broadened its scope as to separate it from the original objectives and give it a clearly defined field of its own. Where the target is located immediately below the rig, the industry now refers to it as straight-hole drilling and,where necessary, directional methods are applied as a corrective measure.

In its own field, directional drilling is now defined as a preplanned operation to drill a bore hole to an underground target whose location is a given lateral distance from the vertical.

Contemporary economics of oil recovery have dictated the progressive development and continuous refinement of controlled directional drilling. Today, we find it applied in addition to its remedial field, in six other production situations that make it economically feasible and desirable.

These basic applications are outlined below (Fig. 18-11 and 18-15).

1. Offshore drilling in which an optimum number of wells are drilled directionally from the same platform or artificial island. In the majority of instances, only one vertical well may be drilled from any one platform. It is obvious that if only one vertical well

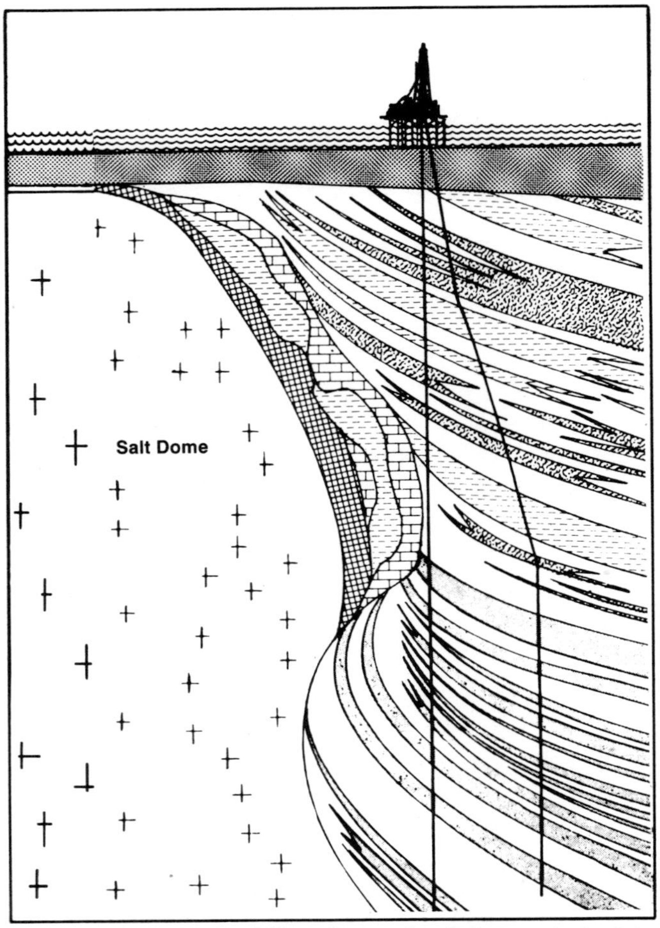

Fig. 18-15. Through the drilling of several wells from a single platform directional drilling is applied to reduce location costs. Typical use shown above is being used to develop saltdome type production in the Gulf of Mexico.

were possible per platform, drilling and production costs would be excessively high. Directional drilling has helped to cut these platform and production costs greatly through the drilling of a number of wells from the one location.

A variation of this application is the T-head, a man-made island from which a number of directionally drilled holes are employed for optimum formation penetration.

2. Onshore drilling for underwater formations has been very successfully used on the West Coast. In the Huntington Beach, California, area some 460 wells have been directionally drilled with borehole deviations as high as 70° from the vertical. Lateral distances of up to two miles have been attained.

3. Inaccessible location of producing formation dictates remote rig location and directional drilling. Production may be located under plants, surface tanks, etc., not feasible to move.

Into this category also falls the problem of rig location versus production where the latter may be under a river bed, mountains, harbors, roads, etc. In some cases when difficulty is found in obtaining rig locations, multiple wells are drilled from one location in a manner similar to that used in offshore applications.

4. Salt dome exploration has given directional drilling a fertile field of application. The two most prominent types of domes, the deep-seated and piercement, have been thoroughly explored and where possible, brought into maximum production by carefully planned and executed controlled directional drilling programs.

5. In formation tilt and fault control, directional drilling is employed to facilitate more economical drilling programs in the former by following the natural drift of formation or well deflection to bottom the hole at the desired subsurface location.

In fault control, the well may be deflected across or parallel to the fault to better production.

6. Stratigraphic trap exploration finds directional drilling extensively used to explore the oil-bearing formation when the original vertical well bored into water or gas. Well is then plugged back and deflected in an attempt to bottom out in the oil strata.

To complete the list, mention again should be made of its remedial or corrective applications: side-tracking lost tools, straightening crooked holes, etc., and drilling relief wells to control a blow-out.

These techniques reflect the lowest and highest precision uses of directional drilling. Side-tracking and straightening operations usually enjoy rather slack tolerances both with respect to angle and direction. Short distances and limited time of operation reduce need for stringent controls. On the other hand, the drilling of relief wells requires extensive control both of deviation and direction. Since the relief well must bottom as near the bore of the burning well as possible, considerable planning and accurately controlled drilling must be utilized throughout the drilling process. Due to relatively large distances, close tolerances are required.

Types Of Directionally Drilled Holes

In introducing the discussion on types of deviated bore holes, the choice of a particular type to fit a given drilling problem involves many complex factors. In a preplanned, deflected bore hole, a carefully conceived and executed drilling program based on geological information, knowledge of and the selection of the right tools, mud and casing programs and coordination of target point with casing program and rig location may insure success of the venture.

Actual experience in recent years has patterned controlled directional drilling to three types. Practically all deflection-drilled holes will fit one of these types or a combination thereof (Fig. 18-12).

Type I is planned so that the initial deflection angle is obtained at a shallow depth and from that point on, the angle is maintained as a straight line to the target zone. In practice, the hole is spudded to a depth in excess of 100 ft and the hole below this point deviated to the proper angle and direction. Once the angle and direction is obtained, surface casing is set through the deviated section and cemented to act as a bearing for subsequent drilling operations.

Type II also sets the initial deflection angle near the surface. After the angle is set, drilling continues on this line until horizontal or lateral distance from the vertical is attained. The hole is then returned to vertical and drilled until objective depth is reached.

Surface casing is set and cemented as in Type I— after the desired deflection angle is reached. Intermediate casing is set and cemented after the hole has been returned to vertical.

Type III starts its deviation well below the surface. The angle is then maintained to target point.

Only generalization can be applied where the drilling or production problem is to be correlated to type of hole configuration. As previously stated, a number of complex factors govern either the selection of a pattern type or the successful completion of a directional drilling program. Thus, many of the problems encountered often require adaptation of a pattern or combinations so as to make the problems and treatments quite individual.

Generally speaking, Type I is employed in moderate depth drilling in areas where intermediate casing is not required and where oil-bearing strata is a single horizon. Type I is less expensive than Types II or III for several reasons and is possibly the simplest to complete.

Deflection is made at a shallow depth and thus trip time for whip-stocking, surveying, etc., is appreciably less than for other types. The curved portion of the bore hole can be readily cased also for prevention of key seats.

The deflection angle is considerably less, too. To its possible disadvantage is the fact that directional-drilling operations must be maintained through a large part of the drilling time. Difficulty in accurate logging interpretation has been encountered also. This is chargeable generally to the borehole angle.

Type II is employed on deep wells in areas where gas troubles, salt water flows, etc., dictate the setting of intermediate casing. Also, it permits more accurate bottomhole spacing in multiple pay areas. Deflection angle may be set in surface zones where drilling is fast and round-trip costs can be held to minimum. When hole is returned to vertical and intermediate casing set, directional operating crew can be released and straight-hole drilling practices followed to completion of hole.

To some extent, it does eliminate logging interpretation problems encountered in Type I and does allow a vertical hole into the producing horizon. But, it also imposes high angles and a double reversal in the drill string. Possibilities of doglegs and key seats are increased if casing is not set in straightening out portions of hole.

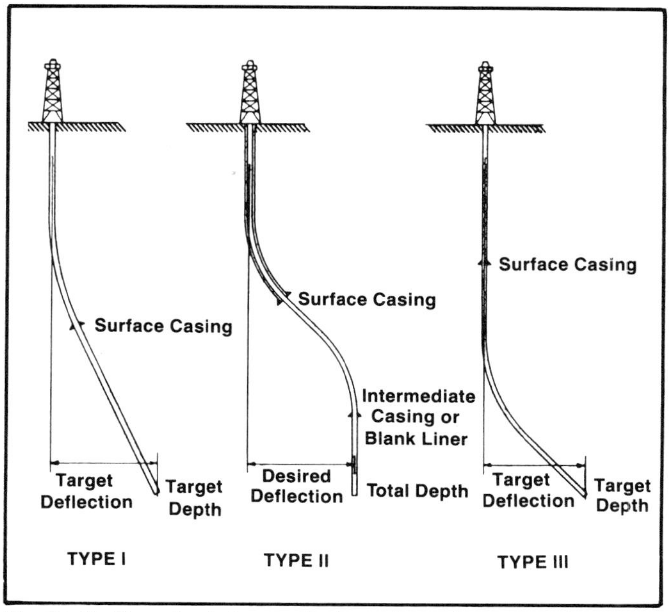

Fig. 18-12. Three basic types of hole configuration are employed in directional drilling programs. Type selected for any one well program is dependent on formation and well conditions.

Type III is used in multi-pay fields where multiple completions are not desired. Deflection angles may be relatively high and lateral distances from vertical line to desired penetration point relatively shorter than the other types.

Typical application would be exploring a stratigraphic trap or obtaining additional geological data on a dry hole.

Since deflection operations take place deep in the hole, trip time for such operations are high and the deflected part of the hole is not normally protected by casing.

Mechanics Of Directional Drilling

While the whipstock is the symbol of directional drilling, major drilling activity is accomplished with conventional drilling tools. Whipstock operations are used to initiate deflection or make major corrections in the direction of the hole. Conventional techniques and principles are employed to maintain or control angle and direction during the drilling operations.

The objective of conventional drilling operation is to maintain the direction of the project hole along a predetermined course with the most favorable drilling rates permissible. Since drilling rate is a function of weight, rotary speed and circulation rate, all of which also affect directional tendencies of the bit, conditions must be compromised to permit optimum directional control with acceptable drilling rates.

Directional Tendencies Of Drilling Strings

The tendency of the bit to drift is a function of the drilling string characteristics, formation dip and direction, weight applied to the bit and to some extent the rotary speed.

In conventional drilling, weight applied to the bit has the same reaction on the lower part of the drill collars as end loading on a column. When little or no weight is applied, the drill collars hang straight. As weight is applied, the column flexes until the drill collars contact the wall of the hole. Additional weight results in this point of contact traveling down the hole closer to the bit. The angle formed by this point of contact and the clearance between the drill collars and hole approximates the change in direction or deflection that would accompany the next increment of hole drilled. From this it follows that the higher the bit weight, the greater the tendency of the hole to be deflected.

In vertical or near vertical holes, the direction and magnitude of this deflection is unpredictable. In high angle hole, the point of contact will in the majority of cases be on the lower side of the hole with the result that the change in deflection is additive and tends to increase hole deflection. Increasing drill collar sizes, moderate bit weights and higher rotary speeds are used to reduce this tendency. Small drill collars, high weights, etc., are often used where it is desired to increase drift tendencies.

Effect Of Formation Dip

There is definite indication that the dip of formations has a pronounced effect on the direction of drift. While each area has its own characteristics, certain general tendencies have been observed. Where alternately hard and soft formations are being drilled with a well stabilized sharp bit and weights high enough to cause drill collar flexure, travel will tend toward a course perpendicular to the bedding plane of the formations.

Where the opposite bit conditions exist, i.e., dull bit, low bit weight and stability, the reverse effect seems to be true and the bit will tend to slide or track into the softer formation and in the down slope direction. This latter condition is particularly prevalent where the angle of incidence of the bore hole with bedding plane is high.

Where thick formations are encountered or where the degree of hardness between formations is small, the effect of dip on the direction of drilling is negligible.

Well programs can be planned to take advantage of these characteristics where data is available. These tendencies are also utilized in controlling the progress of the well bore. High bit weights are used often to increase deflection. A limber drilling string, reduced bit weight and a sharp coarse tooth bit with an acute edge angle are used in many areas to reduce the deflection angle when the drift of the hole exceeds the bedding plane angle of the formation.

Use Of Reamers And Stabilizers

In recent years reamers and stabilizers have been added to the line of useful tools for directional drilling. In effect they may be used to minimize or exaggerate as needed the characteristics of a conventional drill collar string. Placed near the bit, these tools act as the point of contact or fulcrum to increase the angle of the bit with respect to the bore of the hole. Properly placed in the string they may be used to maintain the angle of the hole or under some conditions utilized to decrease the hole angle.

The point of tangency as well as the proper placement of reamers and stabilizers is dependent upon the OD of drill collars, diameter of the hole, rotary speed and weight on the bit. (The Woods and Lubinski charts presented at the Mid-Continent District Meeting, API, March 1954 and 1955 are excellent aids in selecting stabilizer points.)

Stabilizers are used in three types of operations in directional drilling.

1. They are used as a fulcrum by placing the stabilizer or stabilizers just above the bit to produce high angles.
2. They are placed at an optimum point above the natural point of tangency to produce a hole straightening effect or reduce hole angle.
3. Stabilizers are placed accurately in the drill collar string to maintain hole angle during the drilling operation.

In this latter condition one operator recommends the use of one stabilizer for every two drill collars where the hole angle is less than 30° and one stabilizer per joint where the angle exceeds 30°.

Theory Of Whipstocks

As previously stated the role of the whipstock in directional-drilling operations is primarily that of correcting the course of the hole or the initiation of large deflections in the drilling program. Simple in design, its effective application approaches the complexity that is associated with other drilling tools.

Essentially the whipstock is a wedge which deflects the bit from its previous course. However, the hole drilled and the ultimate deflection and direction obtained in subsequent drilling operations is not that of the wedge angle alone. In practice three vectors or angles are required to determine the ultimate change in direction that will be obtained with any one whipstock setting. These are: the angle of the face of the whipstock, the roll-off angle which is a measure of the tendency of the bit spin off the face of the tool and the follow up angle which is caused by the drill string tending to flex at the point of change and the enlargement of the hole in the whipstock section. This latter angle may take effect in the first 50 ft or in some cases over several hundred feet of hole.

While the angle of taper of the face of the whipstock can be readily measured, the other two angles are functions of the individual tool and formation characteristics. Normally they may be reasonably defined in the various areas, such that their effect can be used in making settings.

Roll-off generally affects the direction taken by the bit but does not alter the angle. As a result of this it is applied as a correction in orienting the whipstock face. Follow-up has little or no effect upon direction but normally adds to the deflection angle provided directly by the whipstock. This effect is projected over the subsequently drilled hole and is affected by the reamer and drilling assemblies used. In practice the total angle that will be delivered by the one whipstock setting is used in determining the setting.

Graphic methods are used to determine proper whipstock settings. The overall deflection angle is used in the solution and consists of face angle and follow up angles. The roll-off angle is applied as a correction.

In Fig. 18-13, A represents the angle and direction of the hole and C the desired conditions for further drilling. Closure of the triangle AOC shows that a setting of S73E would be required to obtain this change and a deflection of 5° would be desirable. Assuming that the roll-off angle is one degree to the right, the setting would be made with the whipstock facing S74E to allow for this tendency.

Similar constructions may be used to determine multiple whipstock settings and in the solution of other directional problems.

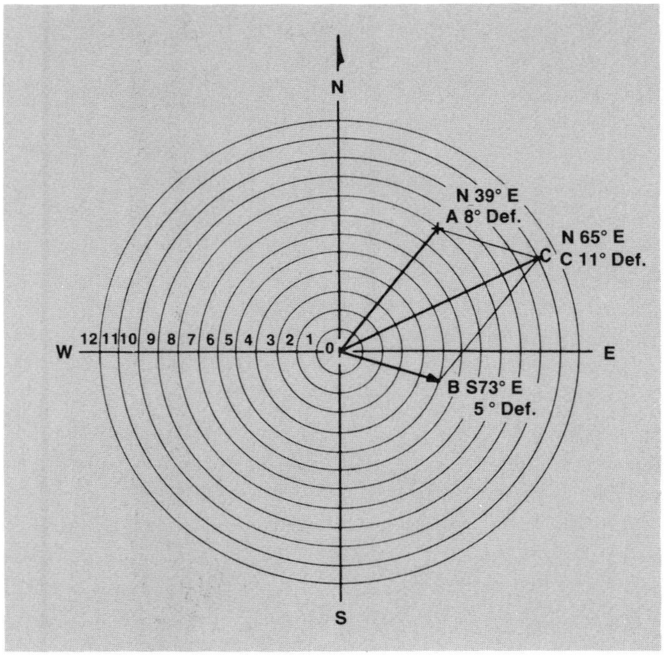

Fig. 18-13. Graphic techniques and plotting are used to determine whipstock settings and chart the projection of the hole. As shown, whipstock setting is obtained readily by closure of triangle AOC.

Tools And Their Application

The basic controlled directional drilling tool kit consists of the retrievable whipstock, pilot reamer, non-magnetic tool joint and the single shot directional survey instrument. Special adapters and subs used in conjunction with these tools complete the kit.

Whipstock

The whipstock is a tapered or wedge-shaped steel casting with a groove on the wedge side which will deflect the bit to the desired angle. At the top of the device is a ring providing a guide through which the drill pipe revolves and by which the whipstock is retrieved. The bit used is smaller than the gauge of the bore hole. For example, 8⅝-in. bits are the largest that can be used in 10⅝-in. to 14¾-in. holes (Fig. 18-14).

In actual practice, the whipstock is selected for the desired deflection angle (maximum permitted being about 4°), bolted to a special sub just above the drill bit and run to the bottom of the hole. The desired direction is then set by orienting with instruments or a magnetic sub. Some operators have used a method of setting with tools on the surface using a transit. The latter method is now considered the least desirable and is inaccurate below 5000 ft but was used extensively in earlier years.

After the whipstock is set in the hole, the bottomhole orientation tool is run and set in a special sub just above the whipstock assembly. The instrument measures the vertical angle or low side of the hole, the position of the face of the whipstock relative to the low side of the hole, and the horizontal angular differences between tool face and the face of a magnet in the orienting sub.

The orienting tool is then withdrawn, readings taken and correlated to determine the correct rota-

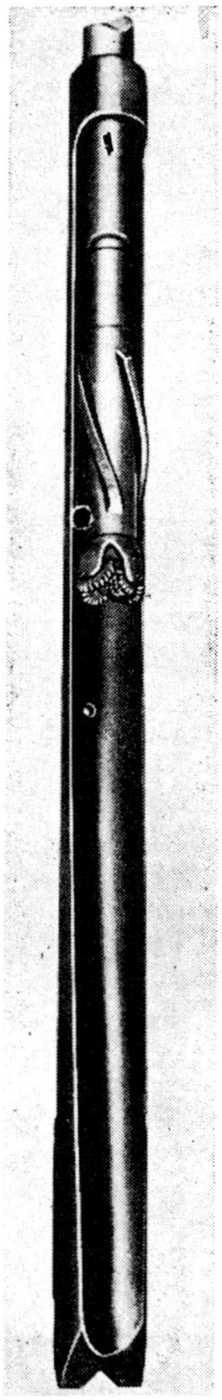

Fig. 18-14. Trade mark of the directionally drilled hole, the retrievable whipstock shown is used almost exclusively in modern directional operations.

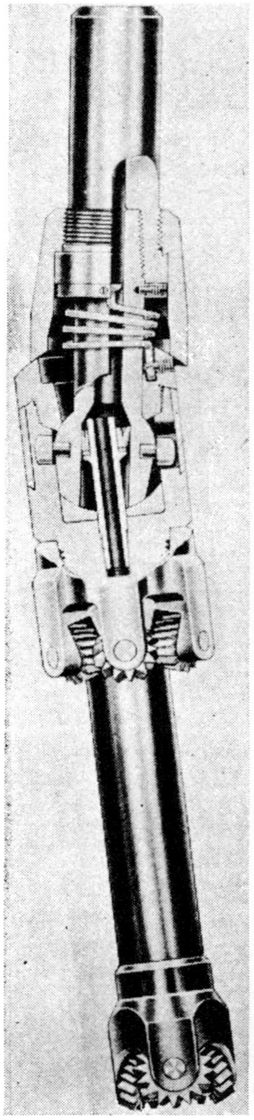

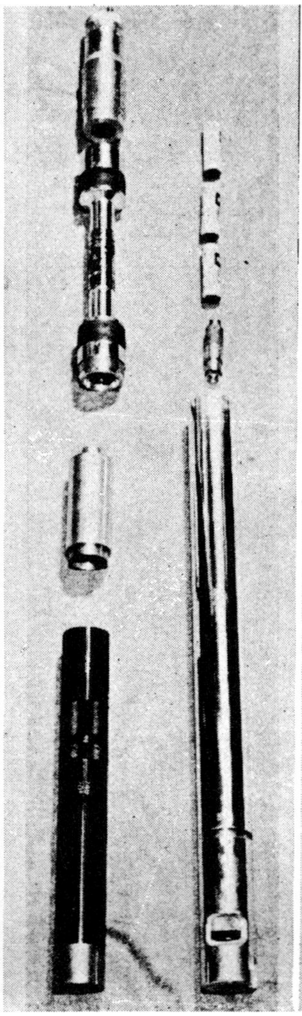

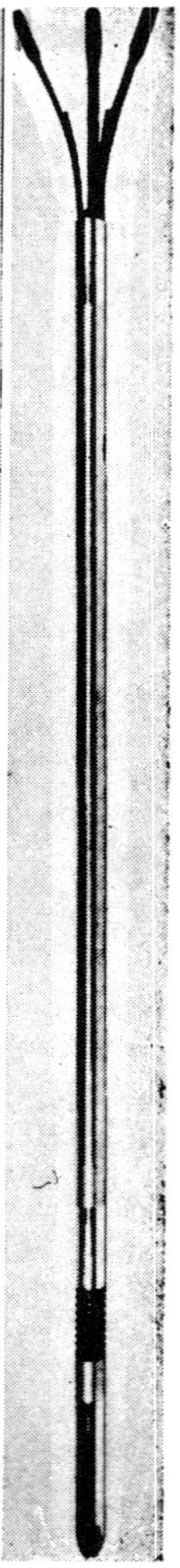

Fig. 18-16. Full gauge hole deflectors have found only limited use in deflecting wells due to difficulty of controlling direction. This assembly is used to follow-up whipstock settings where a considerable increase in deflection is desired.

Fig. 18-17. Components of the single shot survey instrument. Units include camera, plumb bob, and magnetic compass.

Fig. 18-18. Single shot survey tool made up into a go-devil with a feather type head. Unit lands in non-magnetic collar and is retrieved when the drill string is pulled.

tion of the whipstock assembly for desired alignment. The instrument is then rerun for a final check on the whipstock.

After checking and pulling orienting tool, weight is applied on drill string to shear the pin holding bit to whipstock and the small diameter bit then drills a rat hole to deflection angle for a distance of 10 ft to 20 ft.

The whipstock assembly is then withdrawn and the rat hole reamed to full gauge with a piloted reamer. Upon completion of the reamer run, a single shot survey instrument is dropped into the string which lands in a non-magnetic collar located in the reamer string. This is done to check the new deflection and direction.

If found satisfactory, a conventional drill string is run to drill ahead. If unsatisfactory, whipstock operations are repeated until the correct direction and drift is obtained.

Alternate Modern Method

At least two downhole drilling machines are now being used in controlled directional drilling. One is turbo-drill and the other is the dyna-drill. Both are driven by drilling fluid and require no rotation of the drill string. The dyna-drill is a positive displacement, low speed motor, which is actually a pump run in reverse. The turbo-drill, used in directional drilling, is a short multi-stage mud turbine which operates at a relatively high speed.

Because rotation of the drill string is not required it is possible to initiate a deflection by placing a bent sub above the mud motor (of either type) and orienting the drill string to the desired direction. Surveys are taken in the usual manner and corrective measures are taken as required.

Surveying

The non-magnetic drill collar and the single-shot survey instrument form a single component in the directional kit. Since direction is obtained by magnetic means, the non-magnetic collar or sub is used as the landing sub for the survey instrument. It is normally run just above the bit to provide the most accurate information on deviation.

Non-magnetic collars have the same diameters as a conventional steel collar but are normally manufactured from K-Monel which has a magnetic permeability roughly equivalent to that of air. Used in 18 to 30-ft sections, it permits the operator to obtain a true magnetic compass bearing at the bottom of the hole. K-Monel drill collars have mechanical properties comparable to SAE 3140 and 4140 steels which are presently used in conventional drill collars.

The single-shot survey tool makes a record of the angle of inclination and a compass reading showing the direction of the angle. Instruments are dropped into the string for seating in non-magnetic collars or subs, or may be run on wireline. Other techniques include open-hole, surveying, dropping through core bits, etc.

The tool consists of a pendulum or other method for indicating the vertical, a magnetic compass, timer, light source and photographic section. Most popular practice is to run instruments in a go-devil assembly which is dropped into the string. This unit sets in the non-magnetic collar and is retrieved when the drill pipe is pulled. Single shot instruments are constructed to have ranges 0° to 10°, 0° to 20° and 15° to 90°.

Concept Of The Project

A controlled directional-drilling project is so concerned with a number of factors as to very nearly be a unique undertaking each time one is planned and developed. In conceiving a project seven basic factors must be carefully correlated. These are:

1. Accurate as possible geology including, if possible, operator's personal experience and knowledge of area to be drilled and known drilling hazards.
2. Casing program.
3. Surface location of rig.
4. The target or bottomhole location accurately correlated to items 2 and 3.
5. If more than one well is to be drilled, spacing requirements are to be considered.
6. The mechanical job of drilling, itself, including careful selection of tools and their operating capacities.
7. Mud programs.

Due to the greater wall forces involved and the longer drilling time, hole conditions should be maintained as best possible. Mud and casing programs are individual bore hole and geological requirements. Detailed discussion of such is outside the scope of this treatise.

However, a number of operators indicate that an extra-tough filter cake forming mud with rapid cake replacement is mandatory to the program. Oil emulsion muds have found favor in reducing drillpipe friction, particularly where high deflection angles are required.

Directional drilling is normally more expensive and time-consuming than conventional vertical. However, under some circumstances directional wells have been found less expensive and are being applied for this purpose. With the increasing use of controlled directional drilling, new and better tools are being developed. Improved methods will undoubtedly result in reducing the cost of this type drilling.

Hole troubles are more frequent. Drill string sticking seems to be the greatest hazard with stuck pipe occurring about twice as often. Penetration is slower (some average 20% less than in vertical wells).

The selection of kickoff points is dependent both upon the hole type and formations present. The preferable points at which to commence deviation of the hole should be at a juncture where formations are relatively thick and more or less uniform in drillability. Moderate to medium hard formations yield best results. Soft shales and poorly cemented sands prove difficult and enhance opportunities for key seating troubles and poor whipstock drilling and reaming.

Highly tilted and faulted horizons containing alternately hard and soft streaks are to be avoided until desired angle of deviation and hole direction is attained.

Chapter 19
Blowout Prevention — Well Pressure Control

An oilwell blowout is an uncontrolled flow of well fluids and/or formation fluids. A threatened blowout is called a kick, hence blowouts result from uncontrolled kicks. A kick, in turn, is caused by an under balance of mud column pressure which permits entrance into the well bore of formation fluids, either oil, gas, salt water or any combination of such. These extraneous fluids, as they enter the well bore, displace an equal volume of drilling fluid but may not otherwise become evident until circulated to near the surface, at which time the expanding gas causes the mud returns to surge or kick.

In addition to the personal hazards involved, the physical and economic waste resulting from blowouts continues to be enormous. It is ironic that many of the most important oil and gas discoveries, during the earlier years of this century, were ushered in by spectacular blowouts which clearly placed the particular well in the gusher class. Apparently there was little concern about the physical waste involved and none at all for potential reservoir damage. The size of the blowout seemed to be a sort of status symbol.

The Lucas gusher at Spindletop, near Beaumont, Texas, which blew in January 10, 1901, was probably the most highly publicized oil well blowout that has ever occurred. It is believed to have launched the Petroleum Age, now better known as the era of an abundant supply of low cost energy. The Lucas well was drilled only to a depth of 1020 ft when it started flowing at an estimated rate of 50,000 b/d of oil. There were no blowout preventers or other flow controls available at that time so the well continued to flow for nine days before the pressure declined sufficiently to permit installation of a valve on the casing and the well to be shut in.

In subsequent years, many spectacular blowouts occurred in such areas as the Tampico Basin of Mexico, the San Joaquin Valley of California and at Conroe, Texas, where, during the early development of the field, a total of twelve wells were reported to have blown out. During the second and third quarters of this century progress has been made in equipment and methods for the control of well pressures. The discussion which follows reasonably reports on the state of the art as it exists.

Basic Considerations

In the hydraulic rotary drilling system, formation fluids, under pressure, are normally confined to their respective reservoirs by the hydrostatic pressure of the mud column.

The hydrostatic pressure created by a column of fluid is equal to density in pounds per gallon (lb/gal) multiplied by .052 times the vertical height of the column in feet. The multiplier, i.e., .052 is a number worth remembering.

Abnormal pore pressures, i.e., pressures in excess of a normal salt water gradient; .465 psi per foot of vertical depth, are found occasionally wherever deep wells are drilled.

If the mud column becomes too short or the density too low, formation fluids will enter the well bore.

This is called a kick. The volume of fluid in the mud system will be increased at the same rate as formation fluids enter the hole. A kick, encountered while drilling, is indicated at the surface almost immediately by an increase in flow rate of mud returns and, soon thereafter, by an increase in volume of pit mud.

If there are no instruments present to monitor increase in mud return rate or to gauge minor gain in pit mud volume, there are three other indications of probable kick development that should be considered.

(1) A significant increase in drilling rate. As mentioned earlier in this chapter, rate of penetration improves with reduction in over-balance of bottomhole pressure. If an abnormal pressure formation is encountered which creates possibly an underbalance of BHP, the rate of penetration, other factors remaining the same, will surely increase drastically.

(2) Any considerable drop in pump pressure, at same speed, or an unexpected increase in pump speed, is indicative of a decrease in hydrostatic pressure in the annulus due to incursion of low density fluids.

(3) Appearance of salt water, with some gas and/or oil in the mud returns.

Kicks which occur while tripping, as they so often

do, are not readily apparent at the surface but can be detected indirectly by close surveillance of mud volume required for hole fill as drillstring is pulled. Unless the hole fill volume consistently at least equals the volume of steel removed, a kick is indicated.

If a kick continues undetected, hence uncontrolled, a blowout is almost certain to ensue.

Successful well kick control depends on:

(a) Early detection of the kick.
(b) Prompt arrest of kick development.
(c) Proper analysis of the problem.
(d) Removal of kick fluid from the hole.
(e) Changeover to mud of proper density.

Steps (d) and (e) above may be performed either concurrently (known as the wait and weight method) or consecutively (known as the driller's method). In either event, both (d) and (e) must be accomplished while maintaining at all times a bottomhole mud pressure slightly greater than formation pressure.

In deep wells, particularly where abnormal formation pressures exist, there often is little margin between pore pressure gradient and fracture pressure gradient. Over-domination of pore pressure is likely to fracture the formation at some level in the uncased portion of the hole, causing loss of circulation thus greatly compounding the problem.

Successful kick control requires that the well be cased adequately and that it be equipped with surface controls and instruments sufficient to allow early detection of a kick, contain the maximum probable surface pressure and to monitor the entire killing operation.

Instruments And Hardware

An instrument known as a flow rate indicator (illustrated and described in Chapter 11) is an essential device for detecting a kick at the earliest possible moment. The pit level indicator (see Chapter 11) also is a valuable device in confirming the occurrence of a kick and in determining kick fluid volume.

Mud pressure gauges to indicate drill string pressure and casing pressure are essential. Mud pump stroke counters facilitate carrying out the planned pumping schedule of the kick removal operation. The benefits of these various instruments can be enhanced by use of recording devices in addition to the indicators.

The essential hardware needs for well pressure control are:

(a) A cemented casing string of adequate strength, set to a depth sufficient to avert fracture of the formation by pressures likely to occur in a killing operation.

(b) A casing head suitably situated for supporting the blowout preventer (BOP) stack.

(c) A blowout preventer stack consisting of the several items considered appropriate for the existing situation.

(d) Closing unit and control console to provide the hydraulic pressure required to actuate the preventers.

(e) A choke manifold which includes at least one adjustable choke.

(f) Kelly cocks to control back flow through kelly, swivel, etc.

(g) Inside blowout preventer that can be made up into drill string if kick occurs while tripping.

Figures 19-2 through 19-8 illustrate the various pieces of hardware mentioned above. API Bulletin D-13 illustrates and describes numerous BOP stack arrangements. Fig. 19-1 illustrates one such arrangement that is considered quite popular.

Replacing Kick Fluid With Weighted Mud

If kick is detected while tripping, install inside BOP, close annular preventer, shut in well and strip drill string to near bottom.

While each stand is being lowered, bleed annulus through choke as necessary to avoid excess pressure.

Fill pipe periodically but do not break circulation until pipe is near bottom, then install kelly and break circulation through choke.

Thereafter proceed as if kick had occurred while drilling.

If kick is first suspected while drilling, withdraw the kelly and shut off the pump.

Unless mud flow from annulus stops within 15 seconds, close BOP and shut in well completely.

Make certain drill string is full of mud, then, with well completely shut in, allow a reasonable period of time, possibly 15 minutes,for pressures to stabilize.

Observe and record on worksheet the shut-in pressure of drill string and of annulus.

Calculate mud weight required to slightly overbalance formation pressure. Note: The drill string and the casing annulus are regarded as two legs of a U tube. With kick fluid in the annulus, shut-in casing pressure will always be greater than that of the drill string full of drilling mud of normal density. The shut-in drill string pressure, plus the hydrostatic pressure of the mud column therein, equals bottomhole formation pressure. The procedure next described applies only to the wait and weight plan.

Wait And Weight Plan Of Kick Control

Keep the well shut in while conditioning all pit mud to kill mud density.

During the shut-in period, while mixing mud, the

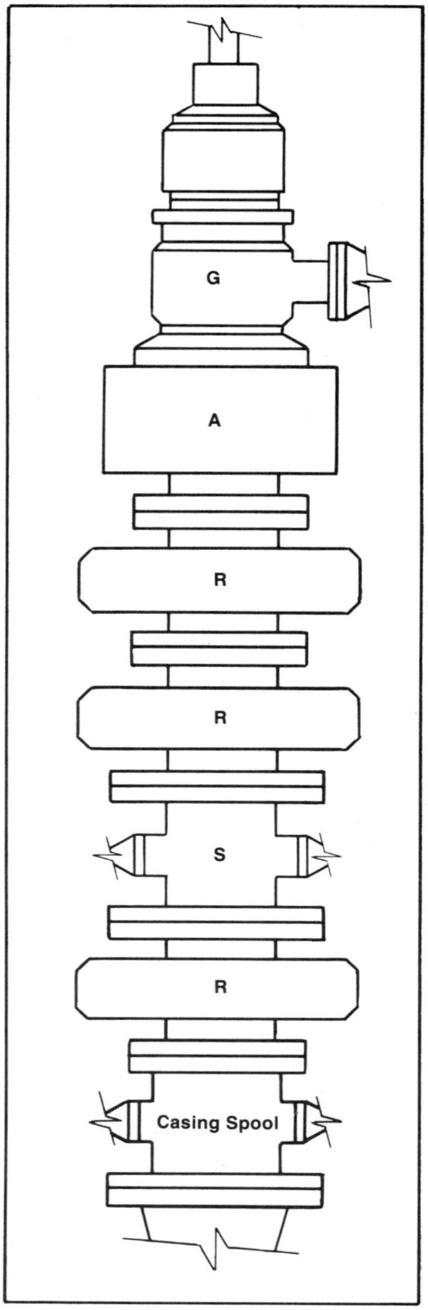

Fig. 19-1. Blowout preventer stack arrangement RSRRA.

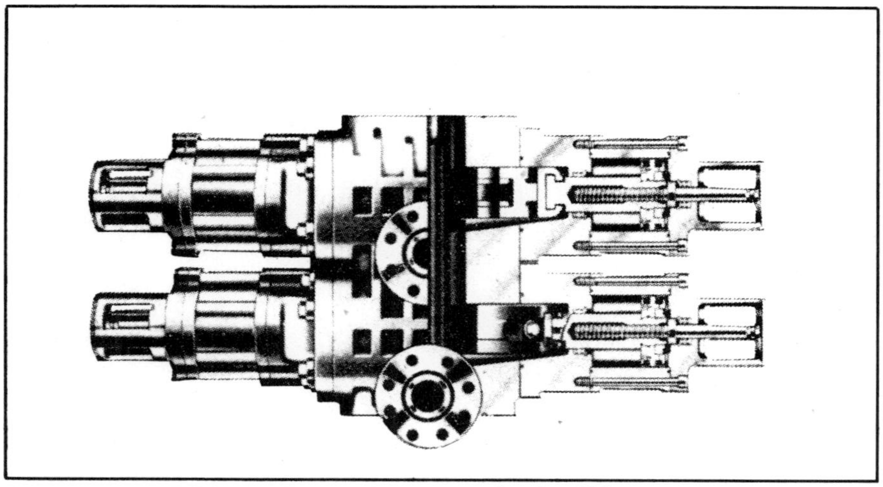

Fig. 19-3. Type LWS Double Blowout Preventer—Sectional View.

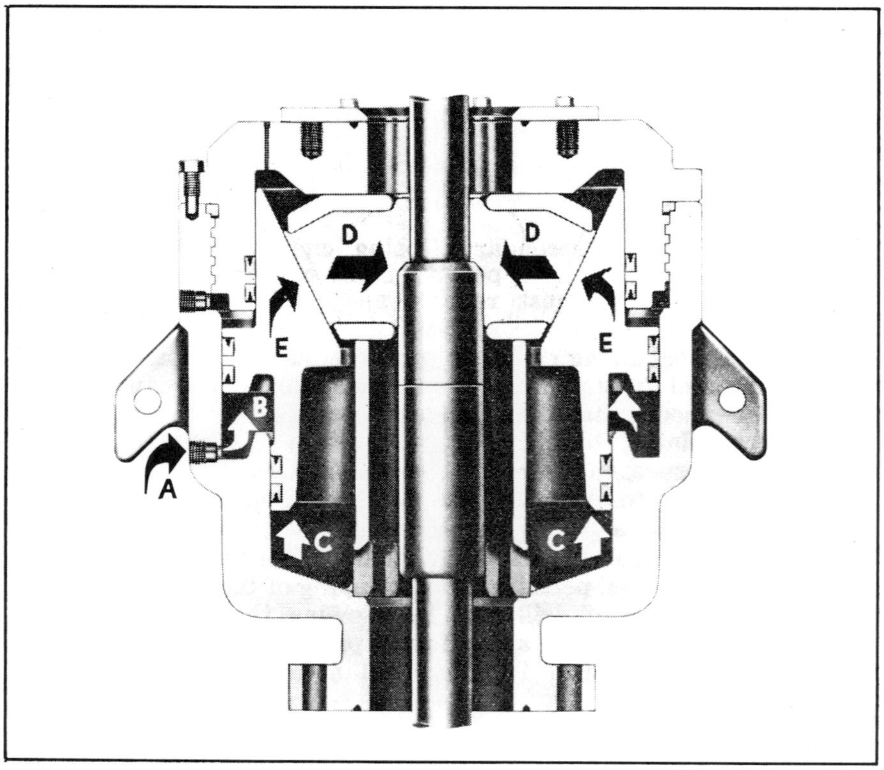

Fig. 19-4. Hydril Type GK annular BOP.

drill string pressure should be observed at regular intervals. If the drill string contains an effective back pressure valve, it will be necessary to run the pump a few strokes in order to obtain a true drill string pressure. Stop pump if or when casing pressure increases as a result of the few strokes of the pump.

If, during shut-in period, the true drill string pressure increases significantly, indicating possible pressure inversion, due to rising gas bubble in annulus, bleed annulus through choke until drill string pressure declines to normal shut-in pressure and thereafter hold at less than 200 psi above original shut-in pressure.

Determine and record on worksheet the capacity of each rig pump, in gallons per crankshaft revolution and revolutions per barrel, assuming a volumetric efficiency of 95%.

Calculate and record on worksheet the interior volume of drill string and the volume of the annulus.

Prepare pumping schedule for the removal of kick fluid and the replacement of old drilling fluid with newly weighted mud.

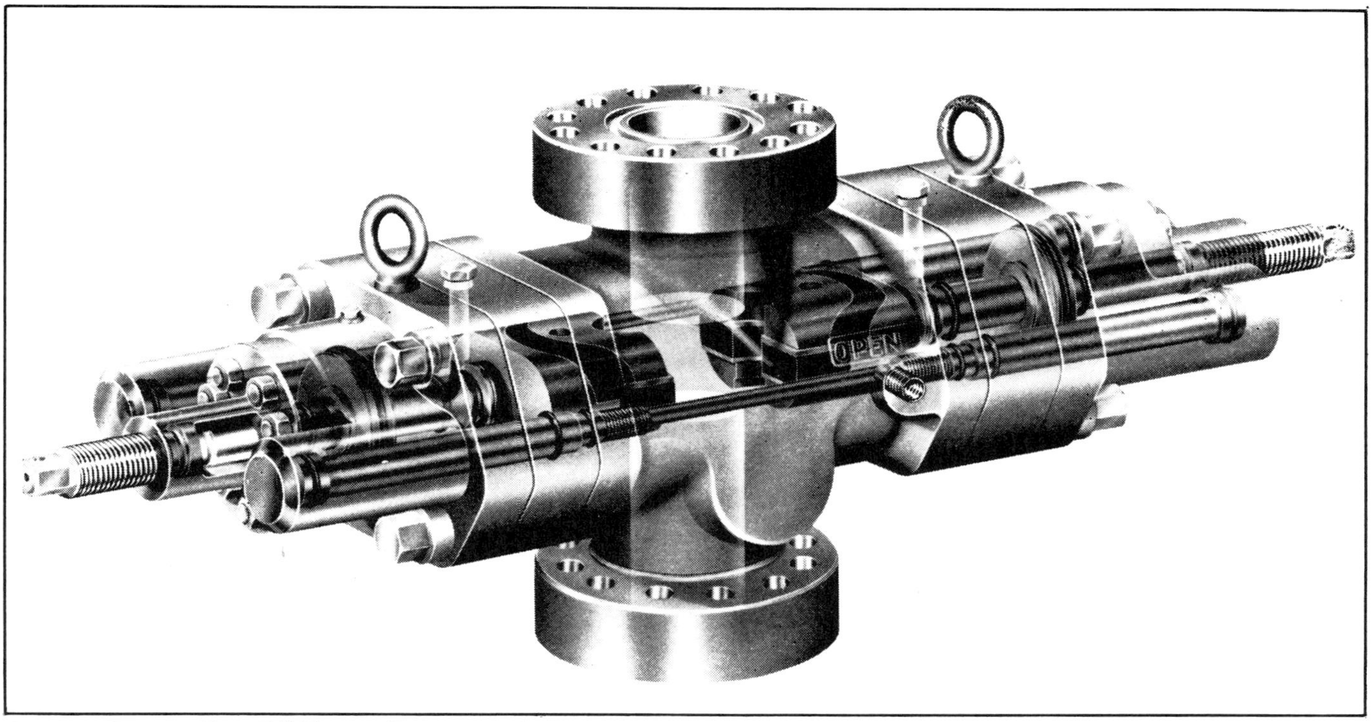

Fig. 19-2. Cameron Type "U" Blowout Preventers.

Fig. 19-5. Koomey S series accumulator unit.

Note: Two basically different plans have been used successfully in this pumping operation, either of which makes possible the interchange of drilling fluids while maintaining a bottomhole pressure slightly above formation pressure. One such plan controls and monitors the casing pressures, relative to volume of new mud pumped, calculated to preserve a slight over-balance of formation pressure. This plan has some advantages, such as more direct control of scheduled pressures; the pumping rate may vary or be changed at will; and the casing back pressures are all predetermined. The disadvantages are volume

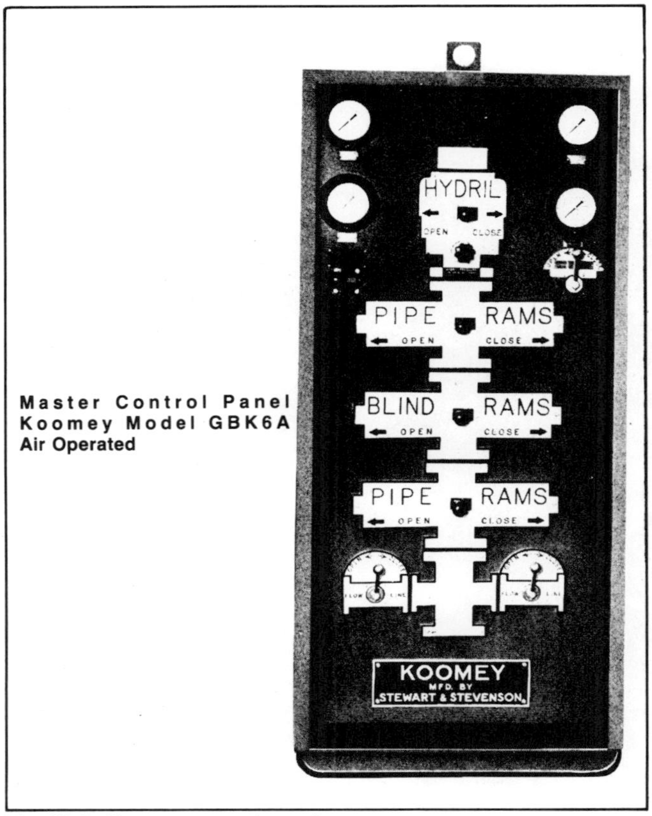

Fig. 19-6. Koomey control panel.

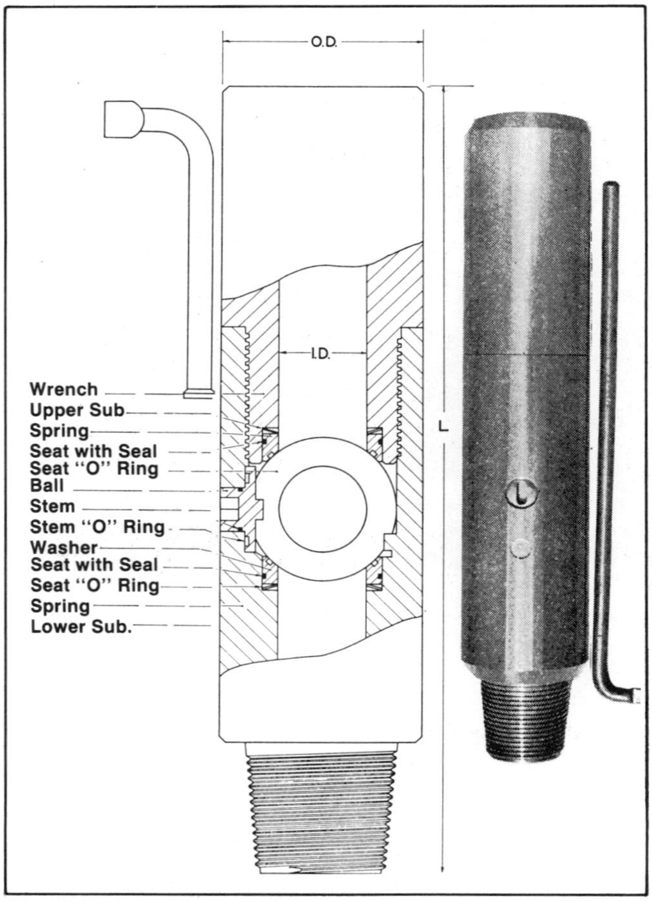

Fig. 19-8. Omsco lower kelly valve.

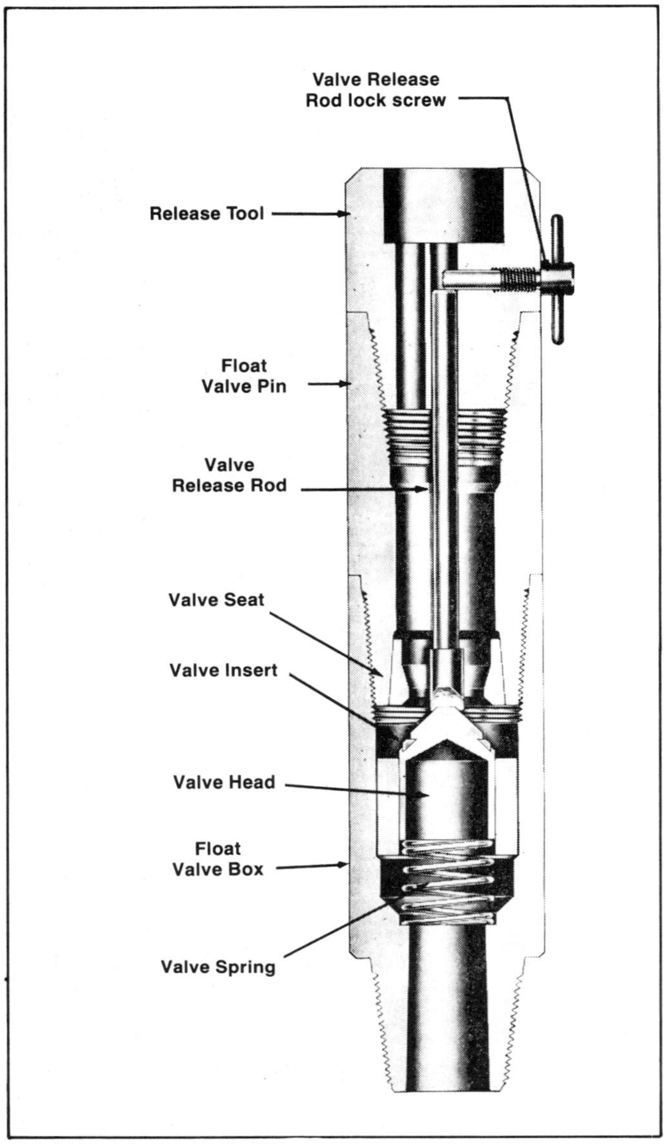

Fig. 19-7. Gray inside BOP.

and composition of kick fluid must be ascertained; requires time consuming calculations; and more complex in theory and harder to explain to crewmen.

The other plan disregards casing pressures unless they become dangerously high for the rating of the wellhead and casing installed. The plan is known as the constant rate, circulating pressure method. The pumping rate is maintained constant while the drill string surface pressure is regulated to calculated values by adjustment of chokes on the casing.

The advantages of this method are its simplicity; volume and composition of kick fluid is not critical; and requires a minimum amount of paperwork. This is the method used in the following example.

The hypothetical well was drilling at a depth of 13,500 ft when kick was first detected. 9⅝-in. casing was set at 11,900 ft; 5-in. OD drill pipe and 8⅝-in. bit were being used. Pre-kick free circulating pressure was observed to be 2400 psi at 50 rpm of pump No. 1.

TABLE 19-1. PUMPING SCHEDULE TO KILL WELL, USING PUMP NO. ______

	NEW MUD PUMPED		STATIC	CONSTANT RATE CIRCULATING DP PRESSURE			
	BARRELS	CUM. REVS.	DP Pr	20 RPM	30 RPM	40 RPM	______ RPM
	5	28	700	1085	1565	2240	
	80	440	467	862	1353	2043	
	160	880	233	638	1141	1847	
D.P. VOL	240	1320	0+	415+	930+	1660+	
	342	1881	0+	415+	930+	1660+	
	444	2442	0+	415+	930+	1660+	
	546	3003	0+	415+	930+	1660+	
AN VOL	648	3564	0+	415+	930+	1660+	
	728	4004	0+	415+	930+	1660+	
	808	4444	0+	415+	930+	1660+	
TOT VOL	888	4884	0+	415+	930+	1660+	

Drilling mud weight was 13.1 lb/gal. Shut-in casing pressure was 1200 psi and shut-in drill string (static) pressure was 700 psi. An increase of 1.0 lb/gal, to 14.1 lb/gal was calculated to balance formation pressure. $700 \div 13{,}500 \div .052 + 13.1 = 14.1$

Internal volume of drill string was calculated to be 240 bbl. The annulus volume was calculated to be 648 bbl. Pump No. 1, with 6½-in. liners has a rated displacement of 8.04 gallons per revolution or 5.50 revolutions per barrel, assuming a volumetric efficiency of 95%.

Based on the observed pre-kick free circulating pressure, the calculated net pressures at reduced speeds of pump No.1, with old mud (13.1 lb/gal) were found to be 385 psi at 20 rpm, 865 psi at 30 rpm and 1540 psi at 40 rpm, whereas the calculated net circulating pressures with new mud (14.1 lb/gal) were found to be 415 psi at 20 rpm, 930 psi at 30 rpm and 1660 psi at 40 rpm.

The foregoing calculations were based on the theory that friction head or pressure loss, in the system, varies directly with fluid density and in proportion to the square of the flow rate.

The next step is the preparation of a pumping schedule (Table 19-1). The initial circulating pressures indicated are the sum of two basic components; i.e., drill string static pressure, due to under-balance of hydrostatic head, and the friction head, called net circulating pressure. As the drill string is filled with new (heavier) mud the static pressure declines, in this case from 700 psi to zero, whereas the net circulating pressure increases, in this case at 30 rpm, from 865 psi to 930 psi, thus resulting in an initial circulating pressure (at 30 rpm) of 1565 psi and a final circulating pressure, when drill string is first filled with new mud, of 930 psi.

The decline in static pressure and also in the circulating pressure is uniform and proportional to the volume of new mud pumped. After the drill string has been filled with new mud the pumping rate and the circulating pressure are to be held constant. The friction head in the annulus will increase only slightly due to increase in mud weight, therefore this minor effect is disregarded.

In the actual constant rate pumping operation drill string pressures are to be controlled entirely by adjustment of choke (or chokes) on the casing outlet.

Drill string static pressures may be monitored from time to time by stopping pump and closing casing outlet for a brief period. If it is observed that static pressure disagrees wih those shown on prepared schedule, the chokes should be adjusted to revise circulating pressure either up or down. A zero reading of static pressure is ambiguous, therefore even after drill string has been filled with new mud, the static pressure should always be maintained slightly positive.

As indicated on the pumping schedule (Table 19-1) several check points are provided. It must be remembered that the critical pressure to be held constant is the bottomhole pressure, being the sum of mud column pressure and static drill string pressure. Exact mud density, the pumping rate and the calculated net circulating pressure are all subject to human error, therefore it is suggested after pumping the first few barrels of new mud to get pump speed and chokes adjusted, that static drill string pressure be monitored by stopping the pump, closing casing outlet and observing static pressure.

The difference between circulating pressure and static pressure is the initial net circulating pressure for *old* mud at the preselected pumping rate.

The initial circulating pressure also may be determined directly while pumping the first few barrels at the preselected rate by holding casing pressure constant and equal to last observed shut-in pressure, by choke adjustment. This method is particularly useful if you do not have current pre-kick circulating pressures at normal pumping rate and at a useful reduced rate.

If the initial net circulating pressure thus determined disagrees significantly from that employed in preparation of the pumping schedule, the scheduled circulating pressures should be adjusted accordingly. The pumping schedule provides for three additional check points while filling the drill string with new mud and seven more check points while removing the remaining contaminated annulus fluid and the old drilling mud.

It is very important to maintain constant pumping rates and not to adjust circulating pressures by changing rate slightly, but rather that circulating pressures be regulated only by choke adjustments. It may be advantageous to change from one pumping rate to another, like from 30 rpm to 20 or 40 rpm, but if this is done the corresponding circulating pressures must be maintained.

When the hole is filled with new mud, the well should be dead but just barely. It is advisable to continue circulating, through open choke, while conditioning all mud in the system to a density which will provide desired margin of over-balance. When this is done, normal drilling operations can be resumed.

Driller's Plan Of Kick Control

The so-called Driller's method of kick control does not differ in main objectives from the wait and weight plan but does differ in procedure.

In this plan the contaminated fluid in the annulus is displaced with *uncontaminated old mud,* using the constant rate/constant pressure technique to maintain a slight over-balance of formation pressure. The circulating pressures at constant pumping rate are controlled soley by choke adjustment. The *static* drill string pressure must be maintained constant, throughout this phase of the operation at a value equal to or slightly above the original shut-in pressure.

The initial constant rate *circulating* pressure is most often determined by adjusting choke to maintain constant casing pressure while pumping the first few barrels of old mud at pre-selected constant rate. The circulating pressure then observed must thereafter be maintained constant throughout the first circulation at constant pumping rate. The static drill string pressure can be monitored at any time by stopping the pump and closing the casing outlet.

If static drill string pressure differs from that chosen to slightly over-balance formation pressure, the circulating pressure may be corrected by altering choke adjustment.

When all contaminated fluid has been circulated from the well and uncontaminated mud reaches the surface, pumping may be stopped and the well shut-in. At this point, with the well completely shut-in, casing pressure and drill string pressure should be equal or nearly so. If casing pressure is significantly higher than drill string pressure, it would suggest the probability that additional kick fluid had entered the bore hole, in which event it would be advisable to continue pumping until a complete interchange of the well's total mud volume is achieved.

The well may then remain shut in until heavier mud of calculated density is prepared. Thereafter the second circulation can be started at the same constant pumping rate as used in the first circulation. While the entire drill string volume is being pumped, the chokes must be adjusted to maintain constant *casing* pressure at same value as the last observed shut-in pressure.

When the drill string is full of new mud and pumping is being continued at the same constant rate, the *drill string pressure* then observed shall become the controlling factor. It must be held constant by adjusting chokes on the casing until new mud reaches the surface. The second circulation is thus completed and the casing pressure should have declined to zero with open choke. When pump is shut down the drill string pressure also should be zero. If pressures are not reduced to zero, the required mud density should be recalculated and a third circulation with heavier mud should follow.

Considering again the U tube analogy, one must remember that drill string hydrostatic pressure plus circulating pressure, *minus* friction pressure is always equal to annulus hydrostatic pressure plus casing pressure *plus* friction pressure in the annulus.

The annulus area normally is much greater than the interior drill string cross sectional area, therefore a slower flow rate in the annulus produces a significantly lower friction pressure than is produced in the drill string with turbulent flow.

As indicated previously, the Driller's method of killing a kick requires essentially two complete circulations, the first of which displaces all annulus fluid with *old mud* of uniform density and no variation in constant rate circulating pressure is required.

The casing pressure during this cycle of operation will change considerably with the highest casing pressure occurring when the gas bubble, if any, first reaches the choke. As a matter of fact, with all methods of kick fluid removal the highest pressure at any level in the annulus occurs when the bubble first reaches that level. If kick fluid is being displaced with low density mud, these annulus pressures become higher than would occur with weighted mud rising in the annulus.

The higher annulus pressures associated with the Driller's method may become of serious concern if inner casing string has a low pressure rating or if it is set at a shallow depth where fracture pressure of the formation is less than the probable annulus pressure that will occur when the gas bubble reaches the shoe.

The second circulation starts with both legs of the

WELL KILLING WORK SHEET
DRILLER'S METHOD

1. RECORDED INFORMATION

A. Casing size: 9 5/8 Depth: 11,900 ft

B. Rated casing burst: 9,710 psi

Max. Allowable Pressure Set by Operator 9,000 psi (Use lower figure)

C. Circulating Pressure (P_n) 2,400 psi

Pumping Rate 50 SPM (Normal Drilling Circulating Rate)

D. Reduced Circulating Press. (P_r) 865 psi

Rate: 30 SPM

2. STOP PUMP AND CLOSE WELL COMPLETELY ALLOW PRESSURES TO STABILIZE

Do not let Casing Pressure exceed lowest of 1-B. If pressure builds to this valve circulate at highest allowed casing pressure and use low choke pressure method to kill the well.

A. Drill Pipe Pressure (P_{dp}) 700 psi

B. Annulus Pressure (P_a) 1,000 psi

C. Mud Density (in DP) (W_1) 13.1 ppg

D. Drill String Length (D) 13,500 feet

E. Pit Gain — barrels

3. SET CIRCULATING RATE AND PRESSURES TO CLEAN WELL

A. Start pump and simultaneously open choke. Pressure on the choke initially should be the shut-in casing pressure and this pressure should be maintained while the pump speed comes up to the desired strokes per minute.

B. Adjust choke to obtain P_a psi and to obtain drill pipe pressure of $P_r + P_{dp}$ psi. Use adjusted reduced circulating pressure P_r and rate of pumping from 1-D.

C. Record: Circulating DP Pressure 1565 psi

Rate: 30 SPM

D. Maintain pump rate constant at the selected reduced speed and maintain CONSTANT DRILL PIPE PRESSURE. If drill pipe pressure increases, open choke; if it decreases, close the choke slightly.

E. When well is free of gas, oil, or saltwater, stop pump and close well. At this time the annulus and drill pipe pressure should be same as original closed-in DP pressure, P_{dp}.

F. Record new annulus pressure (P_a) 700 psi

4. MUD DENSITY TO KILL WELL

A. The Mud Density Increase needed is calculated from the Recorded Information of Step 2.

$$\Delta W = \frac{20 \times P_{dp}}{D} = \frac{20 \times (700)}{(43{,}500)} = 1.0 \text{ ppg}$$

B. Required Mud Density:

$$W_r = W_1 + \Delta W = 13.1 + 1.0 \text{ ppg}$$
$$= 14.1 \text{ ppg}$$

5. INCREASE SURFACE MUD SYSTEM TO REQUIRED DENSITY.

(If mud weighting can be done in separate pit it should be started at Step 3.)

6. SET CIRCULATING RATE AND PRESSURES TO KILL WELL

A. Start pump and simultaneously open choke. Pressure on the choke initially should be the shut-in casing pressure 3-F and this pressure should be maintained while the pump speed comes up to the desired strokes per minute.

B. Adjust choke to hold the new P_a and hold constant until the drill pipe is full of the required density mud.

C. After drill pipe is full of the required density mud, record drill pipe pressure and hold pump rate and DRILL PIPE PRESSURE CONSTANT by varying choke size until the annulus is filled with new mud.

D. When required weight reaches surface, choke pressure, if any, is bled off. Stop circulating and check for flow.

Figure 19-9. Driller's Method Work Sheet

U tube full of unweighted old mud of uniform density. The sole purpose of the second circulation is to displace all old mud with new mud of prescribed density. The second circulation is accomplished in two stages. While filling the drill string with new weighted mud the constant rate circulating pressure will decline while adjusting choke to maintain constant *casing pressure* equal to the last observed shut-in pressure. The friction pressure in the annulus provides a safe over-balance of formation pressure.

When drill string is full of new weighted mud, observe circulating pressure and hold both circulating pressure and rate constant. As new mud rises in annulus the casing pressure will decline until the hole is full of new mud at which time the chokes should all be open and the casing pressure should be reduced to zero. Shut off the pump and if no flow continues, the well is dead and normal drilling operations can be resumed.

Due to its simplicity and minor shortcomings, the Driller's method has been chosen as standard procedure by the International Association of Drilling Contractors (IADC). The subject of Blowout Prevention and Kick Control is fully discussed in their home study course, Unit III-lesson 3, prepared by Petroleum Extension Service, University of Texas.

Figure 19-9 is a reproduction of Well Killing Worksheet taken from Unit III-lesson 3. The same basic data as used in the wait and weight example for killing a kick in a hypothetical well is used also in completion of the Driller's method worksheet.

It has been realized for many years that the drilling industry must be permitted to use a common kick killing procedure that is acceptable to operators and drillers alike. It now appears that the Driller's method will best meet this criteria. It is the simplest to explain to crewmen and quite easy to execute. After all, drilling crews most often are alone when a blow-out threatens and the plan they can understand should be used.

Chapter 20
Air And Gas Drilling

Air and gas drilling, which is more accurately described as rotary pneumatic drilling, is the utilization of air or gas for the circulating fluid with conventional rotary tools. This drilling method has been successfully employed after cable tool and rotary hydraulic methods proved uneconomical or impractical for drilling under certain conditions.

Characteristics of Rotary Hydraulic Drilling Method

The rotary hydraulic method provides the most economical drilling for the conditions found in the majority of wells drilled today. The characteristics of the rotary hydraulic method responsible for this superiority over cable tool drilling are:

1. Higher penetration rates in soft and medium formations.
2. Primary control of well pressure.
3. Better support for sloughing formations.
4. Means of directional drilling.

Under some conditions, however, rotary drilling is used to disadvantage because of the following characteristics:

1. The possibility of damaging formations due to the drilling fluid and particle penetration.
2. The susceptibility to loss of circulation.
3. The tendency to contaminate cores and cutting samples.
4. The possibility of overlooking productive formations.

Characteristics of Rotary Pneumatic Drilling Method

The use of air or gas for the circulating fluid minimizes the four disadvantages listed above for rotary hydraulic drilling; however, two of the advantages are nullified as well: the primary control of well pressure and the support for sloughing formations. In addition, air or gas drilling has in several cases shown an inability to cope with formation water and in the case of gas drilling or drilling in with air, a considerable fire hazard is introduced. Thus, the important characteristics of rotary pneumatic as compared with rotary hydraulic drilling are:

1. Less possibility of damaging formations by drilling fluid and particle penetration.
2. Less susceptibility to loss of circulation.
3. Less tendency to contaminate cores and cutting samples.
4. Less possibility of overlooking productive formations.
5. Higher rates of penetration.
6. Longer bit life.
7. Less control of formation pressure.
8. Less support for sloughing formations.
9. Less ability to cope with formation water.
10. Higher fire hazard.

Applications of Air and Gas Drilling Methods

The scope of application of air and gas drilling is limited to certain conditions according to present day concepts. However, in the future the field may be expanded as more experience is accumulated.

Rotary pneumatic drilling has been used to a limited extent in the past but its recent success started around 1951 — just 50 years after the first rotary hydraulic rig drilled at Spindletop. At present air or gas drilling appears to offer advantages where formations do not tend to slough nor contain appreciable amounts of water and one or more of the following conditions exist:

1. Producing formations are susceptible to injury from drilling fluid or particle penetration.
2. Severe loss of circulation exists.
3. Drilling fluid is expensive or non-available.

San Juan Basin, New Mexico

Gas drilling has proven to be successful in this area where the formation is conducive to loss of circulation and contamination by the wetting action of drilling muds. The Mesa Verde formation of the San Juan Basin consists of approximately 800 ft of alternating sands and shales topped at about 4700 ft to 5000 ft.

Conventional rotary drilling was first tried for penetrating the producing zone and found to require 10 to 20 days and 8 to 10 bits. In addition the plugging action of loss of circulation materials on the producing formation definitely eliminated drilling with mud. Cable tools were then tried and often required

5 to 8 weeks to drill in the 800 ft of formation with practically continuous fishing jobs and waste of gas from the hole.

Gas drilling was then tried and drilling in cut to an average of 4½ days and 3 to 4 bits with a corresponding reduction in drilling costs and and overall saving of gas. The 6¼-in. and 6⅛-in. hole is drilled with an average gas volume of 2.2MMcf/day and an average stand pipe pressure of 416 psig. Operations successfully conducted with gas circulation in this area also include coring and directional drilling.

West Texas Area

Air drilling has been used in an attempt to improve completions in the Spraberry field of West Texas. The Spraberry formations have a very low order permeability and mud cake is detrimental to the productivity of these wells. Air drilling has improved this condition. The 6⅛-in. holes are drilled and cleaned out using 1 to 1.5MMcf/day at pressures around 250 psig.

Gas drilling has been successful in drilling below the surface casing from about 1500 ft to just above the pay zone at around 9500 ft in Val Verde County. An average of 1.5MMcf/day gas about 125 psig was used to drill the 8¾-in. hole. Compared with a mud drilled hole, gas drilling saved 17 days, 34 bits and 34 trips in drilling from 1510 ft to 5321 ft.

South Mountain, California

Air drilling has been applied successfully to combat severe loss of circulation coupled with limited water supply for drilling the surface hole in the South Mountain and Oat Mountain fields of California.

At South Mountain the surface formations consist of highly fractured shales and broken volcanic formations to depths of 1200 ft. Conventional rotary methods proved expensive because of high mud and loss of circulation materials costs. Cable tool drilling consumed excessive time and presented difficulties with crooked holes and fishing jobs.

Similar conditions exist at Oat Mountain where fractured shale and some sands compose surface formations and crooked hole and loss of circulation were experienced primarily above 1200 ft.

In both areas insufficient water is available to drill without returns. Air drilling eliminated expense of loss of circulation materials and loss of rig time suffered with conventional rotary and cable tool drilling, and in addition produced increased drilling rates on the order of 40% to more than 200%, increased bit life from 36% to more than 300%, and reduced drilling costs from 30% to 50%.

The 12¼-in. and 15-in. surface holes drilled required air volumes of 3000 cfm to 5000 cfm at pressures of 70 psig to 110 psig. While drilling with air, fishing operations (including washing over and side tracking, as well as setting whipstocks and directional drilling) have been accomplished without difficulty.

Geophysical Shot Hole Drilling

Air drilling has found application in drilling shot holes for seismic surveys. Conventionally, shot holes are drilled with rotary tools and water or mud for circulation to depths up to 400 ft. In remote areas water may have to be trucked long distances and loss of circulation or drilling without returns is expensive. Air volume of 300 cfm to 400 cfm is sufficient for drilling the 4¼-in. or 4¾-in. hole with circulating pressures ranging from 25 psig to 100 psig.

Results so far with air in shot holes have not consistently shown the economic advantages found in other applications mentioned. Advantages may become more apparent as crews gain experience with this method of drilling.

Mechanics of Air and Gas Drilling

Experience with air and gas drilling has established that higher penetration rates and longer bit life are possible, compared with conventional rotary hydraulic drilling. This is believed to be due to:

1. Cleaner hole bottom due to rapid removal of cuttings by extremely high velocity around the bit.
2. Absence of mud cake, no chip-hold-down effect.
3. Low viscosity and solids content of the gaseous drilling fluid.
4. The cooling effect of large volumes of expanding and highly turbulent air or gas.

Functions performed by gaseous drilling fluid are removal of cuttings and cooling of the bit. Volume of gas required for drilling is governed by the particle lifting capacity that is proportional to the density and square of the velocity of the rising gas column. Relationship between gas density, gas velocity, particle shape and particle size may be expressed as follows:

$$V = 2.67\left(\frac{d_{pn}}{p_t}\right)^{1/2}$$

This describes the slip velocity of a sphere in gas that varies considerably from the shape of rock cutting; however, in the highly turbulent velocity of the annulus, particle shape becomes less important and the few measurements taken so far indicate this to be satisfactory criteria for the design of gas circulating systems.

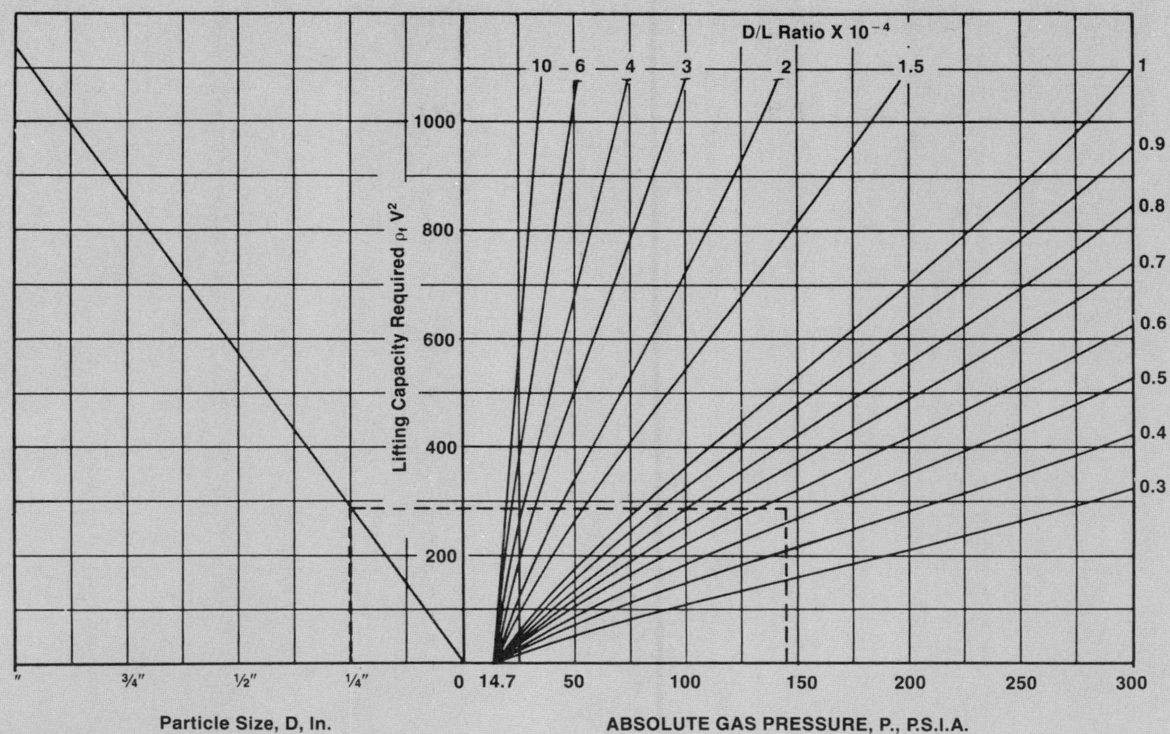

Fig 20-1. Bottomhole pressure determination.

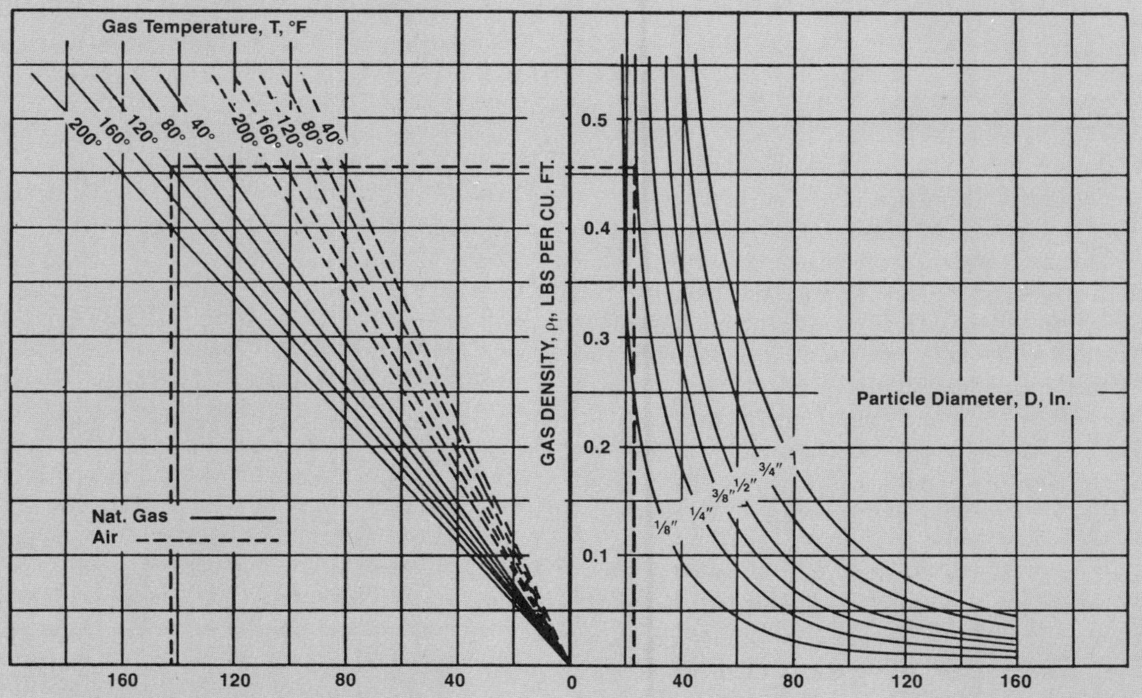

Fig. 20-2. Gas and air lifting capacity determination.

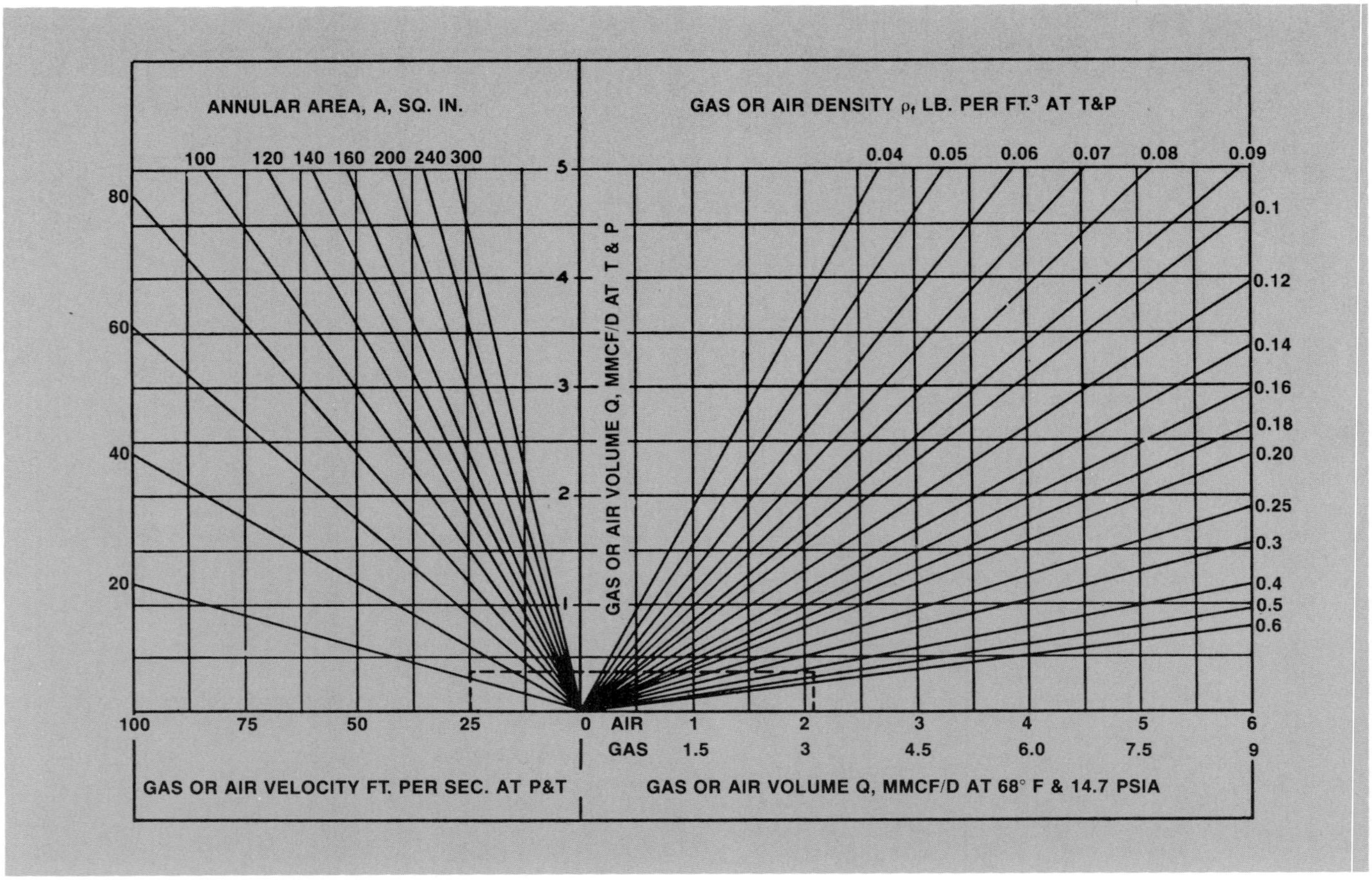

Fig. 20-3. Gas and air volume determination.

Air and Gas Lifting Capacity Determination

Lifting capacity determination may be made graphically by referring to Figs. 20-1 and 20-2. In order to determine the required lifting capacity, it is first necessary to find the pressure at the bottom of the hole. Fig. 20-1 gives a graphical solution for bottomhole pressure for a given particle size lifting capacity, hole depth and drill pipe diameter. Fig. 20-2 graphically solves the gas density and velocity required to lift a given particle size at the given depth.

Lower velocities are capable of lifting smaller cuttings and higher velocities will lift large cuttings. In practice, high lifting capacity (over ¾-in.) requires excessive power; low lifting capacity (under ⅛-in.) results in regrinding and slow penetration rates.

Gas volume required to produce the above determined density and velocity may be graphically solved by reference to Fig. 20-3.

Air volume is found knowing velocity and annular area of hole. This volume is corrected to standard conditions of temperature and pressure. If surface conditions vary appreciably from standard atmospheric conditions, additional compressor capacitor must be provided to supply this quantity of gas.

Example. Assume the following: depth of hole 4000 ft, hole size 6¼-in., drill pipe size 3½-in., temperature at bottomhole 120 deg F. Find volume of free gas required to lift ¼-in. particle.

TABLE 20-1. Equivalent diameter to depth ratio, $D/L \times 10^{-4}$.

Hole Dh size	4¼	6¼	10⅝	10⅝	10⅝	12¼	12¼
Pipe Dp size	2⅜	3½	3½	4½	6⅝	4½	6⅝
Depth,							
500	3.1	4.6	11.8	10.2	6.7	12.9	9.4
1000	1.6	2.3	5.9	5.1	3.3	6.5	4.7
2000	0.79	1.2	3.0	2.6	1.7	3.2	2.3
3000	0.52	0.77	2.0	1.7	1.1	2.2	1.6
4000	0.39	0.58	1.5	1.3	0.83	1.6	1.2
5000	0.31	0.46	1.2	1.0	0.67	1.3	0.94
6000	0.26	0.38	0.98	0.85	0.56	1.1	0.78

Basis: Hydraulic radius of annulus $= \frac{(D_h - D_p)}{4}$

Equivalent diameter of the annulus, $D = D_h - D_p$

TABLE 20-2. Calculated free air and gas volume required for ⅜-in. particle lifting capacity.

Pipe size	2⅜	2⅜	3½	3½	4½	4½	6⅝	6⅝
Circ. fluid	Air	Gas	Air	Gas	Air	Gas	Air	Gas
Hole size	Volume mmcp per day							
4¼	0.9	1.2						
6¼			1.8	2.1				
10⅝			4.6	5.6	4.3	5.3	3.9	4.5
12¼					5.6	6.8	5.2	6.0

TABLE 20-3. Pressure drop in typical gas and air circulating systems.

Description	Length ft	Size in	Pressure Drop, psi		
Hole size			4¼	6¼	12¼
Drill pipe size			2⅝	3½	6⅝
Depth			400	5000	1000
Circulating fluid			Air	Gas	Air
Particle size lifting capacity			¼	¼	⅜
Surface manifold	50	1½	3.2		
	150	4		0.7	
	150	6			1.0
Standpipe	18	2	0.2		
	60	4½		0.1	
	60	4½			1.0
Rotary hose	20	1½	1.5		
	55	3		0.9	
	55	3½			4.0
Swivel		1½	1.5		
		2		1.0	
		2½			4.0
Kelly	30	1½	2.2		
	60	2¾		1.1	
	60	3½			4.5
Drill pipe	340	2⅜	13.6		
	4700	3½		210	
	840	6⅝			6.5
Drill collars	60	1½	8.3		
	240	2¼		28	
	160	3½			35
Bit			4.0	15	15
Annulus			12	160	7
Exhaust	100	4	0.6		
	300	7		0.9	
	300	10			0.8
Total pressure drop			46.1	417.7	78.8
Free gas volume MMMcf per day			0.4	3.5	3.9
Theoretical adiabatic compressor horsepower			21	610	450

Step 1. Refer to Table 20-1 to determine the equivalent diameter to depth ratio, D/L, which for 6¼-in. hole, 3½-in. drill pipe and 4000-ft. hole depth is $0.58 \times 10\text{-}4$.

Step 2. On Fig. 20-1 enter chart at ¼-in. particle size, find intersection with D/L ratio of 0.58 and read bottomhole pressure of 145 psia.

Step 3. On Fig. 20-2 enter chart at gas pressure of 145 psia, find intersection with bottomhole temperature of 120 deg F, project to intersection with ¼-in. particle diameter. (Note: Density of gas under these conditions of 0.46 lb per cu ft). Project vertically to read gas velocity of 25 ft per second. This is the required velocity of 0.46 lb per cu ft gas which will float a ¼-in. particle.

Step 4. To determine required compressor capacity to provide above determined amount of gas, enter Fig. 20-3 at gas velocity of 25 ft per second; find intersection with annular area of the 6¼-in. hole and 3½-in. drill pipe of 22 sq in.; project horizontal to intersection with gas density of 0.46 lb per cu ft and read free gas volume of 3,100,000 cu ft/day at standard conditions of 68 deg F and 14.7 psia.

If the example were desired for air drilling, the amount of air may be determined in the same manner and read on the same air scale in Fig. 20-3.

Equipment for Air and Gas Drilling

Drill Pipe. Hole size and pipe size are of prime importance as the gas or air volume, and to a large degree, pressure is determined by the annular area and size of drillstem.

If the drill pipe is selected as large as feasible in keeping with the fishing hazard, then gas volume requirement will be low, and the pressure drop will also be kept to a minimum.

On the other hand, if small drill pipe is selected with respect to the hole, comparatively large volume of gas is required and the pressure drop will be high. Table 20-2 shows the calculated gas volume necessary to provide a ⅜-in. particle lifting capacity in various hole and pipe size combinations.

Drill Collars. Much of the pressure drop occurs in drill collars and it is therefore advantageous to select large bore collars when gas or air drilling is contemplated. Table 20-3 shows the magnitude and distribution of pressure drop in three typical air and gas circulating systems.

Bits. Conventional roller rock bits have been used for most part in gas and air drilling operations. Circulation nozzles have been enlarged in some cases where large volumes were being circulated, although smaller nozzles have been tried in others to ascertain if jetting action is beneficial as in hydraulic drilling.

Lack of conclusive results seems to indicate that there is insufficient mass in the gaseous nozzle stream to accomplish any appreciable cutting action. Therefore, it would be advantageous to keep the pressure drop across the bit low by enlarging the circulation nozzles.

Surface Piping. Pressure drop throughout the circulation system should be minimized where practical to maintain the compressor power reasonably low. Any reduction in pressure drop in the system will reduce the compressor fuel consumption or allow the circulation of larger volumes of gas. Compressor manifold, stand pipe, rotary hose and kelly should therefore be of ample size in keeping with the volume of gas or air being circulated.

Compressors. Compressor requirement is dictated by: (1) the hole and drill pipe size, which determine gas volume, and (2) the circulating system, which determines circulating pressure.

In air drilled surface holes or shot holes, pressures can be maintained below 110 psig which is as high as one stage of compressors can produce. For drilling in two stages of compressors are necessary unless, as in the San Juan Basin field, gas is available at an adequate pressure. Table 20-4 shows the gas and air volumes, pressures and bottomhole conditions for several typical drilling examples.

Kelly Packer. Packing off the annulus at the surface and diverting gas and cuttings into the exhaust line has been accomplished by rotating kelly packers.

Little pressure exists at this point with conventional direction circulation (less than 1 psig); however, with reverse circulation the full circulating

TABLE 20-4. Typical gas and air drilling conditions.

Circulating fluid	Hole size in.	Drill pipe size in.	Hole depth ft.	Circulating rate MMMcfpd	Circulating pressure psig	Gas density bottom hole pcf	Gas velocity bottom hole fps	Lifting capacity bottom hole in.
Air	4½	2⅜	200	0.4	35	0.12	60	0.38
Air	6¼	3½	7000	1.5	280	0.56	16	0.12
Gas	6¼	3½	6000	3.0	405	0.60	20	0.21
Air	12¼	6⅝	800	5.0	90	0.14	57	0.39

TABLE 20-5. Comparison of conventional vs reverse circulation.

Hole size in.	Depth ft.	Drill pipe size in.	Particle size in.	Circulation direction	Circulating fluid	Flow rate MMMcfpd	Compressor horsepower
6¼	5000	3½	¼	Conventional	Gas	3.5	610
6¼	5000	3½	¼	Reverse	Gas	0.8	60
12¼	1000	6⅝	⅜	Conventional	Air	3.9	450
12¼	1000	6⅝	⅜	Reverse	Air	1.3	72

pressure of perhaps several hundred psi is exerted on the packer. A typical gas and air drilling riser is illustrated in Fig. 20-4.

Exhaust Pipe. It is important that gas and cuttings are discharged a considerable distance from the rig floor. Exhaust lines are run 300 ft more or less on down-wind side of rig and in case of gas drilling, the exhaust is flared.

Size of exhaust line should be selected to provide a higher lifting capacity than in the annulus to prevent a build up of cuttings in this line.

Air and Gas Drilling Techniques

Circulation of gas or air is controlled by the stand pipe valve on rig floor, as shown in Fig. 20-5. Closing this valve stops circulation and compressors automatically unload. Connections are made by stripping the kelly and single through the packer in place.

It is necessary to circulate on bottom for 5 or 10 minutes before starting a round trip in order to clean the hole. Even so, there is a fill up of several feet of cuttings composed of larger particles traveling slowly or suspended in the hole.

Only small cuttings travel near gas velocity, where larger cuttings float until reground to lifting size and then move up the hole. There is little danger of these cuttings sticking the pipe as they become quite fluid when circulating gas. It is possible to circulate the bit through cuttings without rotation.

Loss of circulation reduces the velocity in the annulus, and there is a reduction in the size of cuttings in the returns. Low pressure in the annulus prevents loss of circulation with gas or air from being serious.

Fig. 20-4. GAS DRILLING RISER.

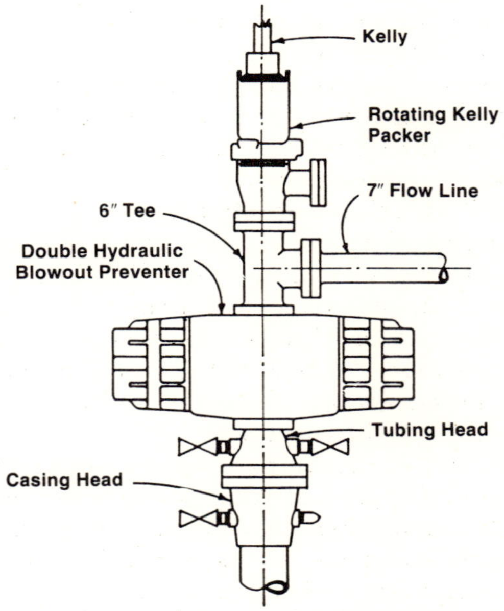

Fig. 20-5. GAS DRILLING HOOKUP.

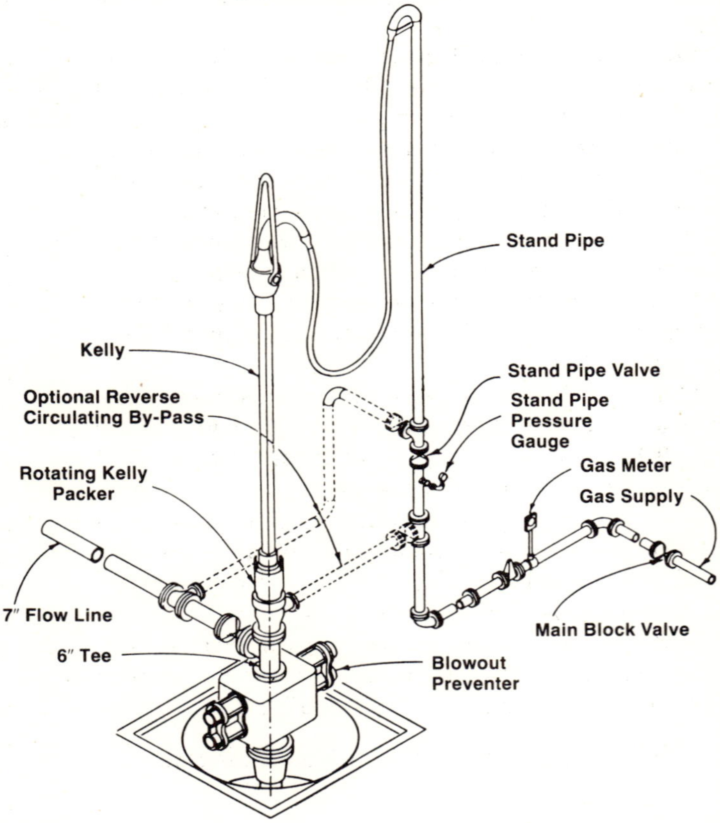

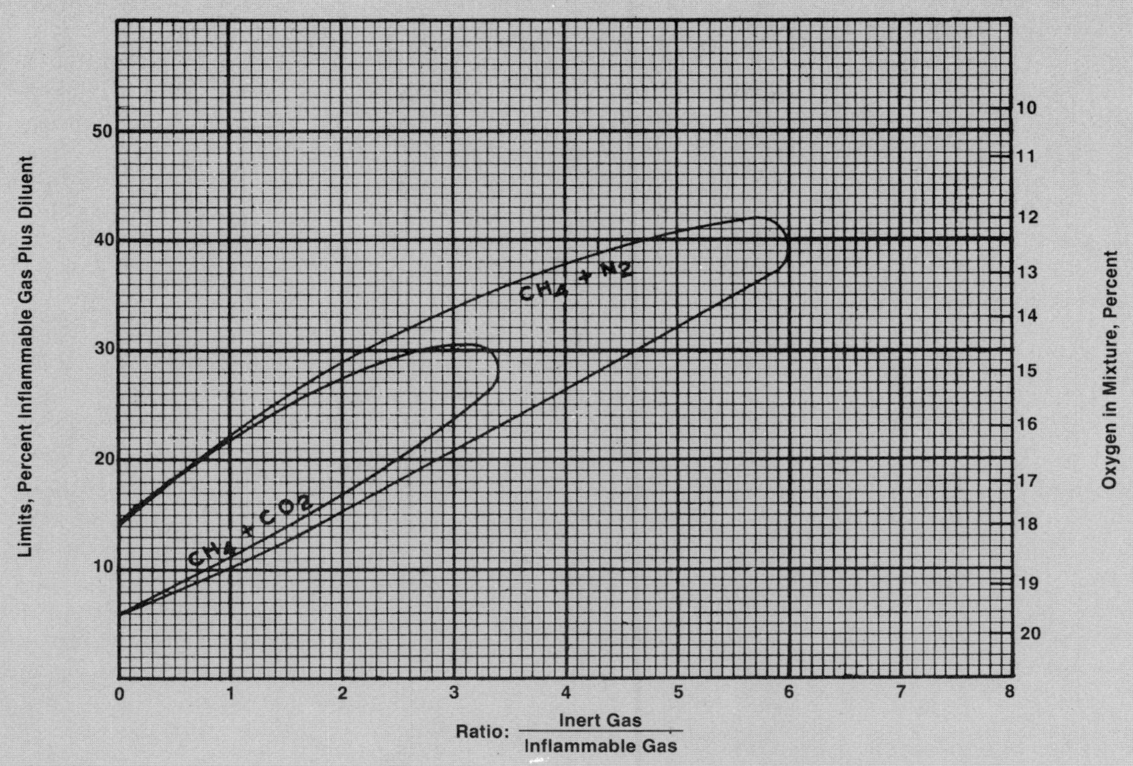

Fig. 20-6. Limits of inflammability of methane when mixed with various proportions of air, nitrogen and carbon dioxide.

Sloughing hole can stop gas or air drilling. Symptoms are a reduction in size of return particle and increased amount of fill in. The enlarged hole reduces gas velocity and hence lifting capacity. A means to combat a sloughing hole has not been devised.

Weight on bit is commonly increased with gas or air circulation and rotary speed reduced. This can be accomplished as there is some evidence of less tendency to make crooked holes with gas drilling than with hydraulic drilling.

The appearance of the exhaust is indicative of several conditions. Returns are seen within seconds after circulation is begun; dust in the exhaust indicates cuttings are being lifted and an abrupt stoppage of dust indicates moisture in the hole.

Reverse circulation has been tried on several occasions with varied results. This method cannot be used in loss of circulation formation as the full circulating pressure is exerted on the annulus. It has, however, been successful for drilling in where increased lifting capacity was desired.

Reverse circulation provides the same lifting capacity with reduced volume and pressure, or a higher lifting capacity with the same quantity of gas as with conventional circulation. Table 20-5 shows the quantity of circulating gas and compressor power required to provide equivalent lifting capacity in two hole and pipe size combinations with both conventional and reverse circulation.

Operations other than drilling have been successfully accomplished with gas or air for circulation. Fishing jobs, including washing over and sidetracking, have been conducted, directional drilling, including milling windows in casing and setting whipstocks, has been executed, and casing has been run and cemented without difficulty. Coring has also been accomplished with good results.

Fire Hazard

The few fires experienced on gas drilling operations have not occurred while drilling, but rather when the pipe was out of the hole or while running tubing, and these fires have been traced to carelessness or forgetfulness in use of matches on the part of the drilling crews.

Precautions ordinarily taken on gas drilling operations are: checking the piping for leaks; installation of adequate blowout preventers; use of a 300-ft flow line; injection of water in the engine exhaust; provision of adequate fire-fighting equipment; provision of proper electrical systems, and education of the drilling crews.

Possibility of using air to drill into gas producing zones has been given some thought. It is possible to mix sufficient inert gas with air to preclude an explosive mixture. Fig. 20-6 shows the limits of inflammability of methane when mixed with various proportions of air and nitrogen or carbon dioxide. It can be seen that if the percent of oxygen in the mixture is kept below 12% there can be no explosive mixture.

Conclusion

A. *Advantages.* Gas and air drilling methods, compared to rotary hydraulic methods:

1. Less possibility of damaging formations by drilling fluid and particle penetration.
2. Less susceptibility to loss of circulation.
3. Less tendency to contaminate cores and cutting samples.
4. Less possibility of overlooking productive formations.
5. Higher rates of penetration.
6. Longer bit life.

B. *Disadvantages:*

1. Apparent inability to cope with formation water.
2. Less support for sloughing formations.
3. Less control of formation pressures.
4. Higher fire hazard.

According to present day concepts gas and air drilling has economic application where formations do not tend to slough nor contain appreciable amounts of water, and one or more of the following conditions exist:

1. Producing formations are susceptible to injury from drilling fluid or particle penetration.
2. Severe loss of circulation exists.
3. Drilling fluid is expensive or non-available.

Gas-and-air drilling has been successfully utilized in the San Juan Basin of New Mexico and the Spraberry field of West Texas to prevent damage to the formation from drilling fluid and particle penetration.

Air drilling has been successful in combating loss of circulation coupled with inadequate or expensive water supply, for drilling surface holes in the South Mountain and Oat Mountain fields in California, and for shot hole drilling in geophysical work.

Chapter 21
Well Completion

After drilling to projected total depth, running the open hole logs and possibly taking side wall cores or making other such tests for hydrocarbons, the well owner must make a decision, based on cumulative information, whether to complete at total depth, plug back and complete at a lesser depth or to plug and abandon.

For the purpose of this discussion, we shall presume the decision is to complete at total depth. The several operations therefore which likely will follow are:

1. Recondition hole for casing.
2. Running and cementing of production casing.
3. Perforating for production.
4. Possibly take sand control measures.
5. Testing for productivity.
6. Well stimulation.
7. Squeeze cementing.
8. Tubing installation.

Strictly speaking, some of the completion operations named above are not to be considered Drilling Technology but because the rotary rig and crews are retained to perform some of them and to assist third party service companies in the performance of others, it seems appropriate to discuss them.

Reconditioning Hole for Casing

As a prelude to running the production casing it usually is advisable to: (1) run drill string with bit back to bottom, (2) circulate out all cuttings and/or carvings which may have accumulated, and (3) recondition the mud to desired composition and consistency.

If prospective producing zones, of good permeability, have been exposed to long periods of mud circulation after having been drilled through, it is most likely that a tough wall cake (consisting of cuttings, shale carvings and other solids) will have been deposited on the wall of the hole and packed, by pounding of the drill string, into a mass that cannot be effectively removed by any type of scratcher now available to the industry.

Before laying down drill pipe, it is suggested that such intervals be subjected to the reaming action of redrilling with full size bit to remove as much of the wall cake as possible. Wall cake thickness can reasonably be determined by a caliper log.

Running and Cementing Production Casing

Presumably the drilling prognosis for subject well includes: (1) a detailed design of the casing string, specifying size, grade and weight per foot, for the several sections of hole and (2) the prescribed casing, in proper order for running, placed on the pipe rack.

Since it is important that the cement in the annulus—between casing and wall of the hole—makes an effective bond to both casing and formation, several precautions are necessary.

It has been demonstrated that the bond between casing and cement is influenced by the surface condition on the outer wall of the pipe. Mill varnish inhibits the bonding properties more than a clean sand-blasted surface and much more than a weathered or rusty surface.

An even better bond can be attained if the casing, for use through a critical section of the hole, is prepared by sandblasting and by applying a bonded layer of angular sand grains to the exterior surface of the casing.

All production casing strings should include a float shoe and a float collar (or baffle collar) placed some distance above the float shoe. The float shoe serves a dual purpose of guide shoe and back pressure valve in addition to its nominal function of providing buoyancy to the casing string, thus lessening the hook load on surface equipment.

The float collar (or baffle collar with hole) serves to stop the two cementing plugs. The spacing between float shoe and float collar usually is varied somewhat in relation to pipe size and length of the casing string.

A good rule of thumb for small casing is to space such collar a distance above the shoe equal approximately to 1% of the casing string length and never less than one full joint for larger diameter casing.

Casing centralizers (Fig. 21-1) are used at preselected depths so as to center the casing in the hole and thus provide a more uniform annular cementing space for the cementing operation.

Centralizers minimize the hazards of channeling cement along the side of casing having the larger annular space. Also, when used in conjunction with wall scratchers they serve to ensure consistent scratching efficiency over the entire circumference of the hole.

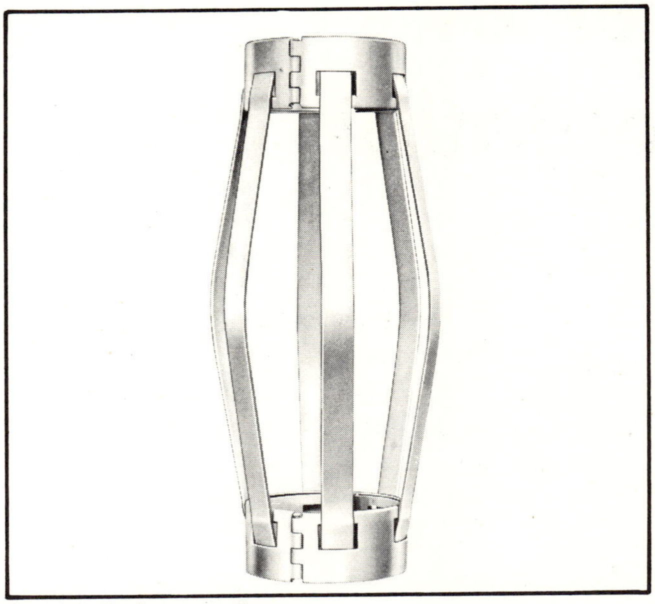

Fig 21-1. Casing centralizers.

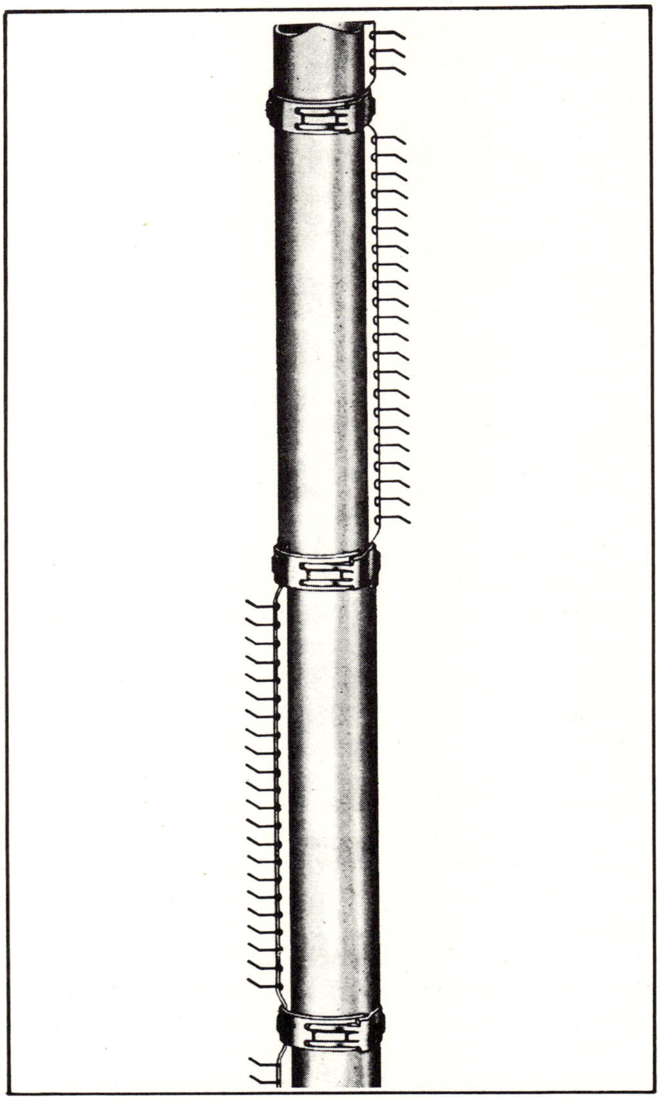

Fig. 21-3. Wall scratchers, rotating type.

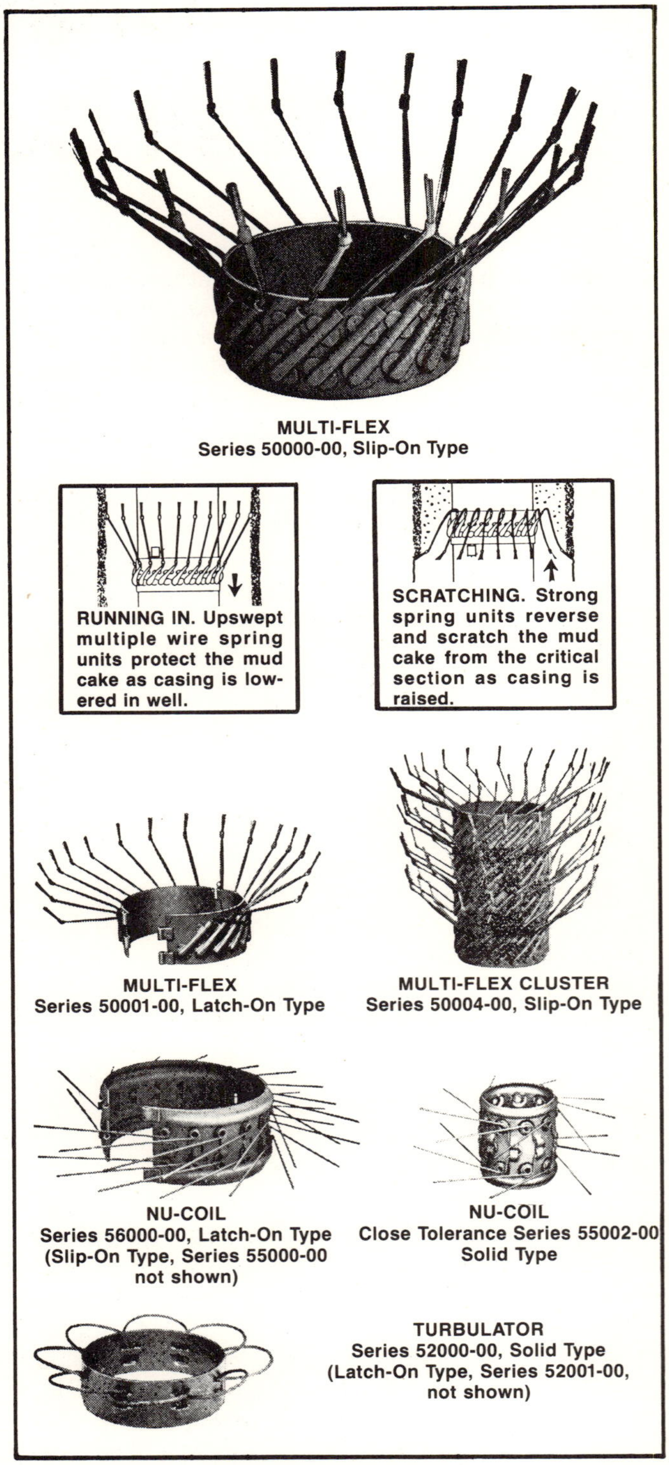

Fig. 21-2. Wall scratchers, reciprocating type.

Wall scratchers (Figs. 21-2 and 21-3), attached to the casing string at preselected intervals, are used to remove filter cake from the face of prospective producing formations so as to assist in obtaining a better bond of cement to formation wall.

Wall scratchers are available in two general types:

reciprocating and rotating. When using either type it is general practice to provide sufficient scratchers to cover slightly more than the prospective productive interval.

The scratchers are attached rigidly to the casing by welding and are actuated either by reciprocation or rotation of the casing after installing same in the well. Normal circulation of drilling fluid is maintained throughout the scratching operation for removal of the debris.

There is little choice between the two types but neither type currently available can be expected thoroughly to remove the tough wall cake previously mentioned. The actual running of casing has been discussed.

Cementing the production casing string is one of the most critical operations in the whole drilling and completing process.

There is no substitute for a successful primary cement job! There are many pitfalls in this important undertaking. Bad cement jobs are less common now than formerly but they still do occur. The cement slurry must be tailored to fit the peculiar conditions found in each well.

The basic portland cement, developed for the construction industry, can be modified in composition and ground to fit a wide range of temperature needs and setting times.

Various additives are used to control density and fluidity. Cement slurry for use in extreme temperatures has been successfully compounded, so there is very little reason to have a bad job due to the slurry.

The more common failures result from poor slurry placement practices such as ineffective cementation around the lower few joints of casing. This is especially bad if formation fluids are present in the poorly cemented interval or if the casing is to be worked through while drilling deeper. Drilling any considerable distance through casing with the lower few joints swinging about in the hole is likely to result in parting of casing and that's bad.

The common causes of ineffective cementation of the bottom joints are either over displacement or contamination of the cement slurry last to go around the shoe. Some operators prefer not to use a bottom plug (or wiper plug) in the cementing operation. This practice is hazardous since the cement slurry will drop through the mud column to some extent leaving a film of mud/cement mixture adhering to the inner wall of the casing which the top plug wipes off only to accumulate atop the cement column. In a long string of small diameter casing a very thin film of contaminated cement, when wiped off by the top plug, results in a volume sufficient to fill several joints of casing.

This problem can be alleviated, but not eliminated entirely, by the diligent practice of using a bottom plug. Unfortunately the bottom plug is not so perfect a wiper as the top plug and it is for this reason that the float collar should be spaced a considerable distance above the shoe so that no contaminated cement gets past the shoe. The cement left inside the casing, even after hardening, can be drilled out easily if the space is required.

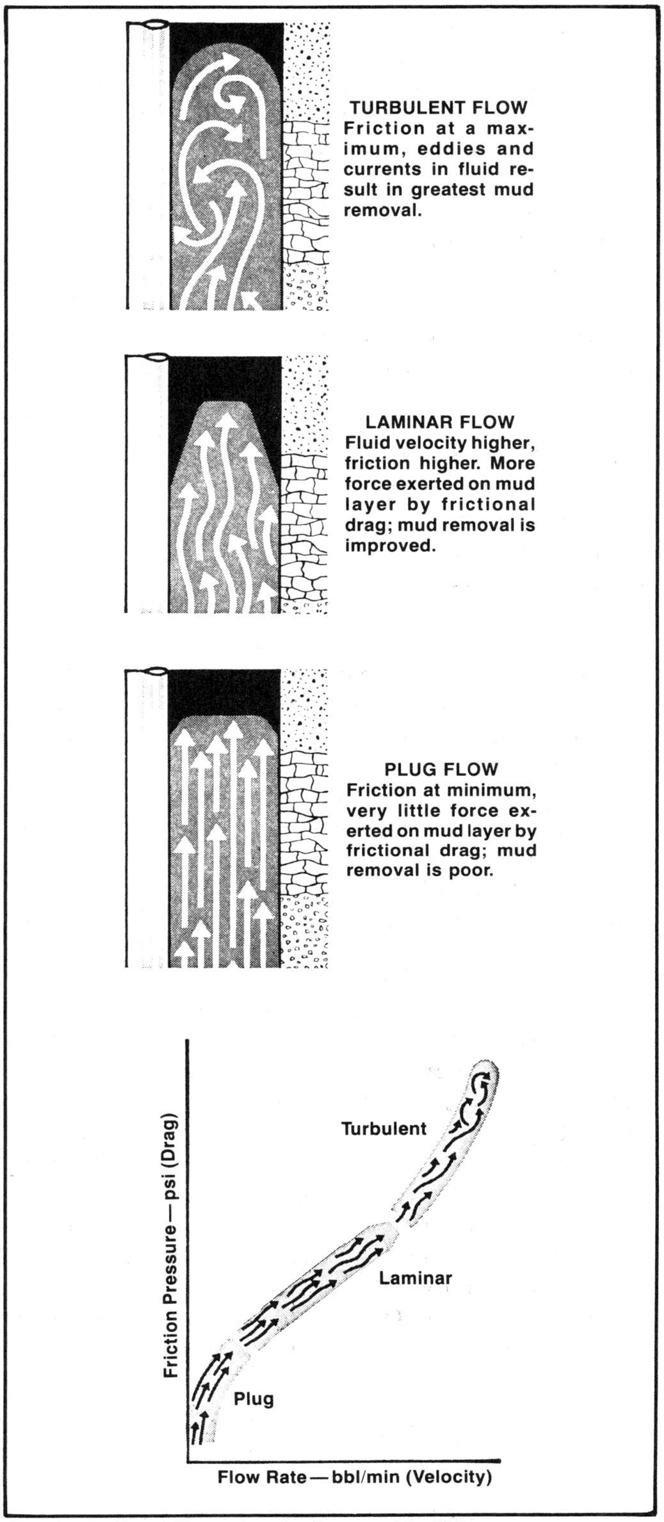

Fig. 21-4. Slurry flow, plug, laminar and turbulent.

An operating practice just as bad or worse than the elimination of the bottom plug is the omission of the float collar (or baffle). This allows the cementing plugs to be pumped all the way to the shoe.

In years gone by it was often erroneously presumed that the cement itself was at fault if the cement failed to set up around the lower joints of the casing string. Contamination most likely was the culprit.

Another common cementing trouble results from the cement channeling through the mud without completely surrounding the casing. This failure becomes more probable as distance above the shoe increases.

There are several aids available that can help alleviate this problem. One is the use of casing centralizers as mentioned above. Another, which has long been used, is to pump a few barrels of water ahead of the cement slurry to wash mud from the outer surface of the casing and possibly dilute the mud filter cake.

Perhaps the most important aid in effective slurry placement is higher pumping rates which move cement up the annulus at a turbulent rate of flow. It has been determined that turbulent flow will most effectively displace the mud in the annulus and that pumping rates above that required to attain turbulent flow will not significantly improve the mud displacement process (Fig. 21-4).

Achievement of turbulent flow rates in the annulus, at normal pumping rates, has been made possible by the use of turbulent flow additives in the cement. Turbulent flow inducers, of whatever chemical composition, tend to increase fluidity of the cement slurry without impairing its other qualities.

A large number of operators now follow the practice, after the top plug reaches the baffle, of immediately releasing the pressure at the surface. If back pressure valves in the floating equipment hold, pressure is left off permanently.

This practice excludes any possibility that pressure left on the casing might be increased to an excessive level by downhole temperatures, causing a ballooning of the casing and the subsequent shrinking away from the cement body when pressure is later released.

In-Casing Logging

After casing has been run and cemented, a temperature survey run soon thereafter will reveal the depth to top of cement column in the annulus. It will also pinpoint intervals where large deposits of cement may have lodged in washed-out cavities or intervals where no cement at all has been placed.

Among the most important guides in perforating and other completion work are radiation logs such as the gamma-ray neutron collar log, run for correlating purposes. Locating and recording the position of casing collars with respect to formation changes and the correlation of the radiation log with resistivity log run in open hole will help locate intervals to be perforated.

Perforating for Production

Prior to development of gun-perforating equipment most wells producing from competent rocks were completed in the open hole. In cases where the pay section was incapable of supporting the well bore, as in loose or poorly consolidated sands, casing was run to give support to the wall of the hole. Generally this casing was slotted or perforated before being placed in the well. This technique, with improvements, is still being used along with gravel packing in areas where unconsolidated sands are prevalent.

Mechanical perforators, which could perforate the casing after it was landed, were available at this time. These tools utilized either a single blade or wheel type knife, which could be opened at the desired level and manipulated to cut vertical slots in the casing.

The disadvantages of mechanical perforators were numerous and their use was greatly limited. This general unreliability undoubtedly hastened the development of gun-perforating equipment.

Advances in cementing and stimulation techniques along with improvements in perforating equipment now have made the perforated completion commonplace even in hard rock areas.

The open-hole completion may have certain inherent advantages but the perforated completion is often more desirable because of such factors as selectivity in thin zones, greater gas/oil and water/oil ratio control, ease of selective treatment and the economics of maintenance and remedial work.

Current methods of perforating casing include two distinct techniques:

1. Use of projectiles or bullets.
2. Employment of the shaped-charge or jet principle.

Bullet Perforating

This form of perforating service came into use in the early 1930s. The principles and overall features of current equipment are essentially the same now as when introduced although considerable improvement has been made in the materials used and in the performance obtained by this equipment.

The bullet-perforator is a multi-barreled firearm designed so it can be lowered into an oil well, positioned at the desired interval and electrically

BULLET TYPE		SIZES AVAILABLE	TYPICAL PERFORATION CHARACTERISTICS (1)		GIVES	USED FOR
NUMBER	NAME		BURR	HOLE DIA.		
1	STEELFLO	1/4" 3/8" 15/32" 9/16"		1/4" 3/8" 15/32" 9/16"	Maximum penetration. Improved rolled bead burr. Strengthened point structure. Flight stability.	Deep Formation penetration.
2	BURRFREE	1/4", 3/8" 15/32" 9/16"		1/4" 3/8" 15/32" 9/16"	Elimination of burr. Less penetration than Steelflo. Otherwise possesses all advantages of Steelflo above.	All jobs where smooth inner surface of pipe is necessary and penetration is of secondary importance.
3	SEMI-MUSHROOM	15/32" 9/16"		9/16" 5/8" (2)	Hole in casing larger than bullet. Considerable neat cement fracturing. Reduced penetration.	Casing removal. Cement fracturing.
4	FULL MUSHROOM	15/32" 9/16"		5/8" 7/8" (2)	Hole in casing considerably larger than bullet. Extensive neat cement fracturing Considerably reduced penetration.	Casing removal. Cement fracturing.
5	NEEDLE	1/16" 1/8"		1/16" 1/8"	Single string tubing penetration.	Establishing circulation in stuck tubing or any other purpose where small orifice is desired.
6	PUNCH	15/32" 9/16"		15/32" 9/16"	Negligible burr. Controlled penetration. Hole same size as bullet.	Perforation of the inner of two strings without damage to outer.

(1) Based on observation of tests in J55 casing. (2) Dependent upon casing thickness and cement support.
Note: Although 15/32" steelflo type bullet is most widely used, other types and sizes are generally available on special order if specified by operator.

Fig. 21-5. Typical bullet types used in perforating guns.

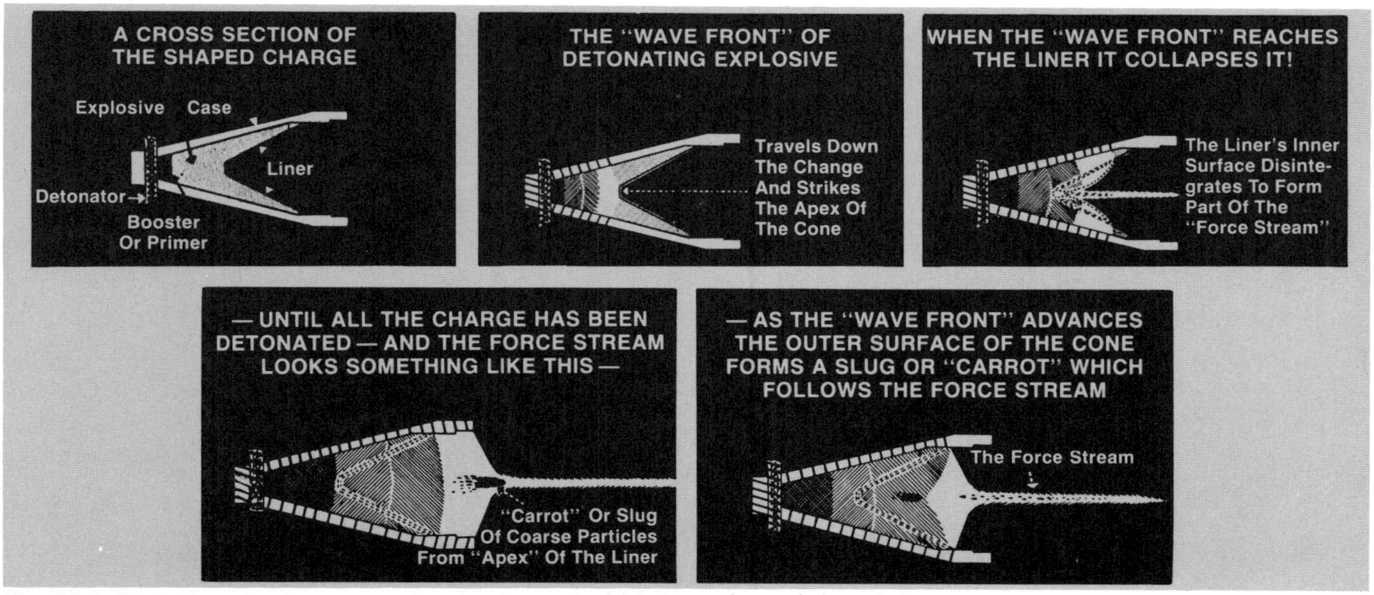

Fig. 21-6. Illustration showing progressive development of jet stream force of shaped charge.

fired from surface controls.

Penetration of the casing, cement and formation is accomplished by a high velocity projectile or bullet. Current equipment allows the selective firing of one bullet at a time, selective firing of independent groups of bullets, or simultaneous firing of all bullets depending on the operator's need.

A number of bullet types are available, each being tailor-made for a particular purpose. A number of these (which are currently available) and their individual characteristics are shown in Fig. 21-5. Bullet guns designed for use in virtually all sizes of casing

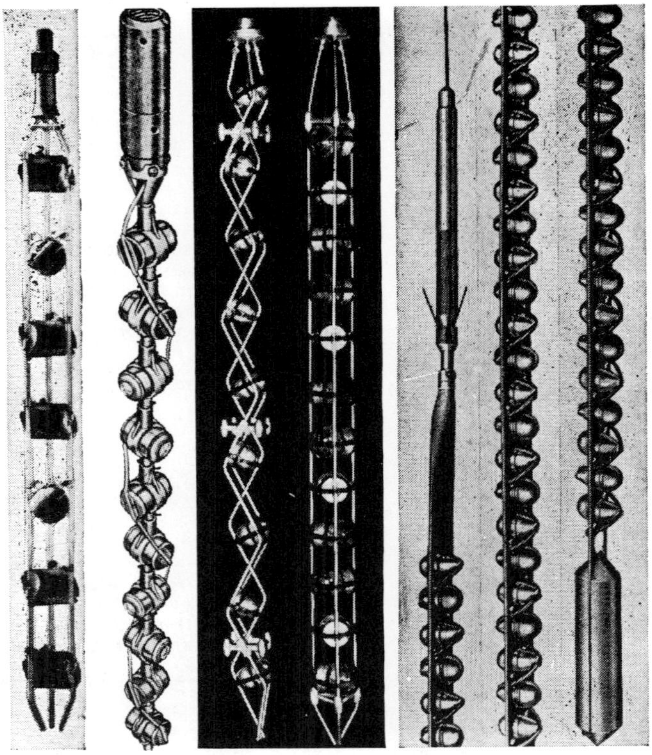

A. B. C. D. E.

A. Horizontal firing gun with cased charges. B. Standard horizontal firing aluminum-cased gun. C. Angle gun where vertical permeability is desired. D. Standard horizontal firing glass gun. E. Steel Strip gun with glass cases showing rope socket and collar indicator at top, and sinker bar at bottom. All these guns disintegrate when fired except the steel strip gun (E) in which the steel carrier is retrieved.

Fig. 21-7. Various configurations of expendable jet perforating guns.

and tubing are readily available from a number of service companies.

Jet Perforating

Adaptation of the Munroe or shaped-charge principle to oil well perforating was made shortly after World War II with the first well being perforated by this technique in 1947. The descriptive term jet perforating is now commonly used to denote this process. Numerous articles describing the theory and mechanism of this phenomenon have appeared in the literature.

Fig. 21-6 is a diagram showing the jet development at succeeding stages of the explosion. Penetration of the target is obtained from the high velocity force stream (jet) formed by the conical liner's inward collapse and partial disintegration. The velocity of this jet is on the order of 30,000 ft/sec, which causes it to exert an impact pressure of some 4.4 million psi on the target.

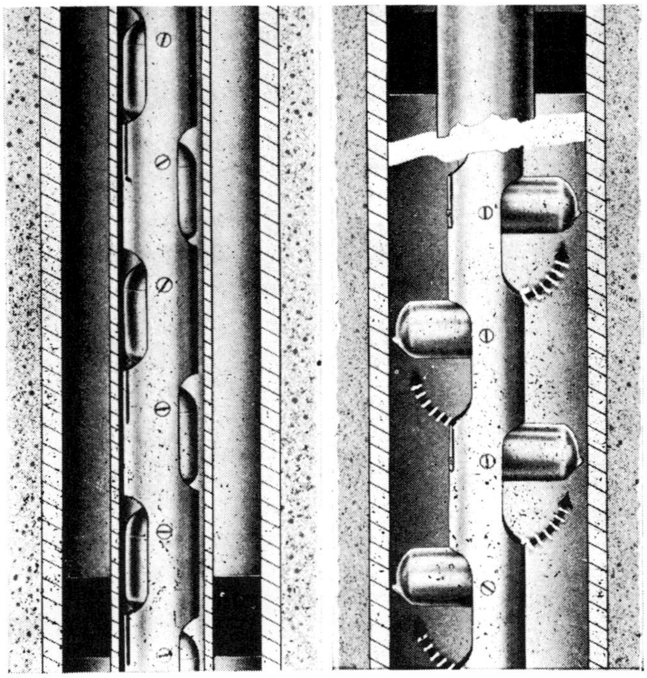

Tool runs smoothly down *2-inch* tubing in closed position. **Full powered jets are snapped into position at firing zone.**

Fig. 21-8. Two views of jet gun, designed to run through tubing and perforate casing below.

Two basic types of jet-perforating equipment are available, namely retrievable and expendable guns. As implied, the retrievable-type guns are composed of a cylindrical steel carrier with the charges opposite ports facing radially from the vertical axis of the carrier.

Expendable guns are composed of materials which disintegrate into small particles when the gun is fired. The materials commonly used for the carrier are aluminum or cast iron with the cases housing the charge being constructed from glass, aluminum, plastic, cast iron or a ceramic material. Typical expendable guns are shown in Fig. 21-7.

A jet gun designed to run through tubing and be fired in the casing below with increased penetration approaching that of standard casing perforators for use in permanent-type completions is shown in Fig. 21-8.

Fig. 21-9 shows a cross section of the new four-way gun which fires four charges in the same horizontal plane as an aid in effective squeeze cementing.

Theoretical Considerations

Prime objective of a perforated completion is to obtain productivity as near that of the uncased hole as possible while still enjoying the advantages of the casing. Principal factors to consider in this respect are:

1. Perforation diameter.
2. Perforation density, i.e., number of holes per foot.
3. Depth of penetration.

Effects of these variables on well productivity have been the subject of considerable study both mathematical and experimental. Partial results of the McDowell and Muskat experiments conducted with electrolytic models are reproduced in Figs. 21-10 and 21-11.

This data might be summarized briefly and qualitatively as follows:

1. The effect of perforation diameter is not critical if the size is kept above ¼-in. at normal perforation densities.
2. For penetration depths of ½ to 1 well diameter very little increase in productivity may be obtained by increasing perforation density above 4 to 6 holes/ft.
3. Greater penetration increases well productivity which may theoretically exceed that of the uncased hole at moderate perforation densities.

The above investigation also reported a decrease in productivity of 5% to 9% if the perforation were tapered rather than uniform in diameter. The depth-of-penetration factor may be more critical than shown if the presence of a damaged or invaded zone is considered.

Operational Aspects

It is often tempting to choose a perforating gun considerably smaller than the inside diameter of the casing in which it is to be run rather than to use the size designed for the job. This will require that the bullet or jet must expend considerable energy in penetrating the drilling fluid prior to reaching the target.

This effect is more critical in jet equipment than bullet and obviously will have an adverse effect on penetration. Standoff is distance the jet or bullet must travel before encountering the target. The effect of standoff on jet penetration under some conditions has been evaluated and is shown in Fig. 21-12.

Accurate depth measurements are essential to any perforating job. The most accurate placement of shots is obtained through the combined use of the collar and radioactivity logs. In this technique the zone to be perforated is picked on the radioactivity log and all measurements are made relative to casing collars,

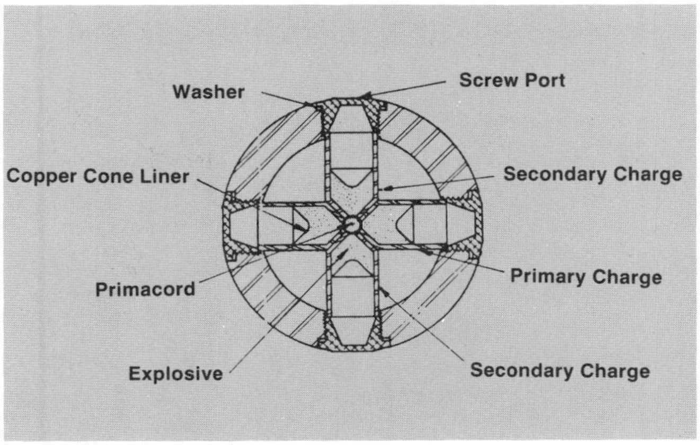

Fig. 21-9. Cross-section of four-way gun.

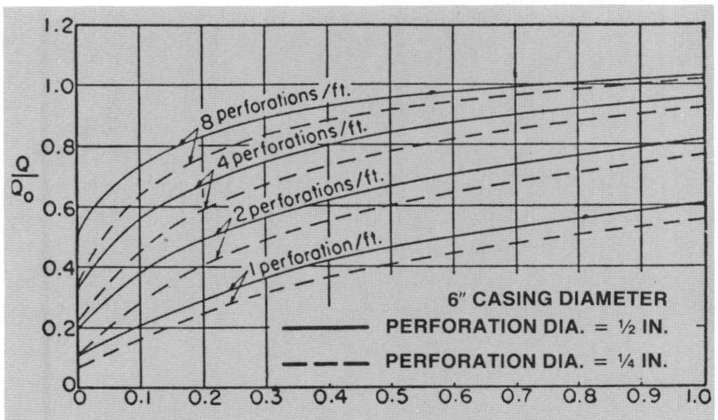

Effect of depth in perforation penetration into formation surrounding casing on well productivity for various perforation densities. Q/Q_O = (Productive capacity of cased and perforated well)/(Productive capacity of uncased well).

Fig. 21-10. Relative productivity vs. depth of formation penetration as measured in multiples of well.

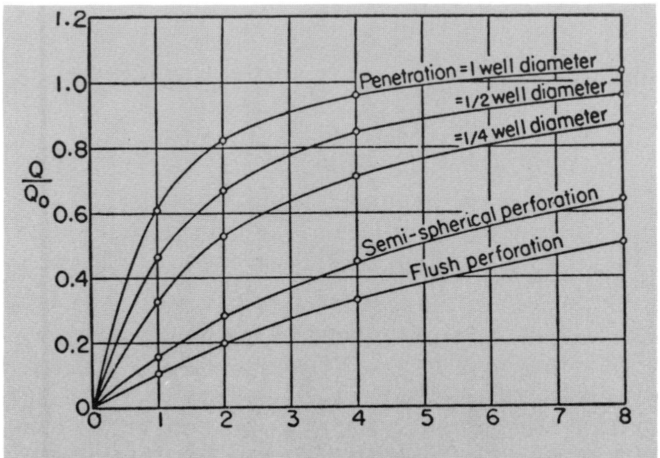

The effect of the perforations density on the productivity of perforated casing completions, with various perforation penetrations into the producing formations. Q/Q_O = (productive capacity of cased and perforated well)/(Productive capacity of uncased well). Casing diameter = 6 inches, perforation diameter = one-half inch.
McDowell and Muskat, A.I.M.E. Petroleum Transactions '50.

Fig. 21-11. Relative productivity vs. density of perforations in holes per foot. Q/QO = Relative productivity of cased and perforated well/productivity of open hole completion.

TABLE 21-1. PERFORATING CONSIDERATIONS

JET	BULLET
1. Deeper penetration in hard rocks and multiple casing strings.	1. Equal penetration in soft rocks†.
2. Minimum burring of casing wall.	2. Maximum fracturing of soft rocks with minimum formation compaction around perforation.
3. Minimum fracturing of cement and soft rocks.	3. Uniform hole diameter.
4. Equal or greater fracturing of hard formations*.	4. Selective firing available.
5. Available for permanent-type well completions.	5. Controlled penetration by bullet selection if applicable.
	6. Burless hole may be obtained by proper bullet selection but only at expense of penetration.

Steel Guns	Expendable Guns
1. Minimum "junk" left in hole.	1. Maximum "junk" left in hole.
2. Semi-slective firing.	2. No "swabbing" action when retrieving line.
3. Less subject to damage while running in hole.	3. Scouring action in open-hole shooting due to gun disintegration.
	4. Large charges for open-hole shooting.
	5. Versatility of make-up, i.e., angle shooting, spacing.

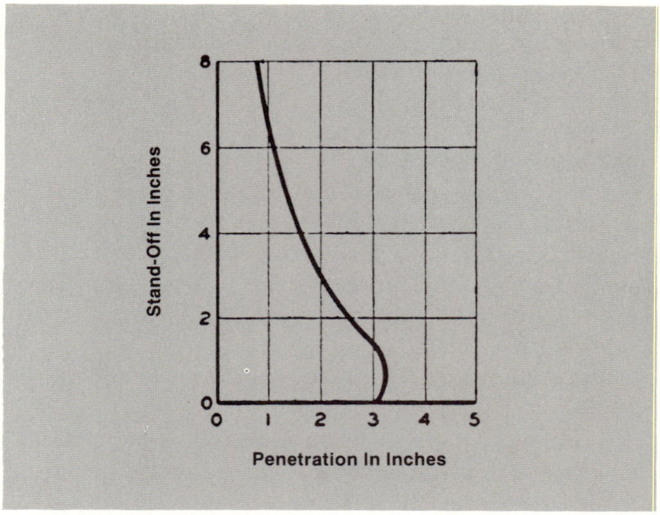

Fig. 21-12. Penetration of formation in inches vs. stand-off in inches.

located by wire detectors attached to the perforating gun.

A critical factor in successful perforating is the type of fluid present in hole when the operation takes place. Considerable restriction to flow may be formed due to the dehydration of some drilling fluids which may form a complete or partial plug over the perforation.

Other fluids may damage the formation by simply contaminating the pay section exposed by the perforations. Oil, oil base mud or salt water spotted opposite the desired interval will in most cases be desirable with general preference in the order listed.

Many other considerations must be evaluated in making the best selection of equipment for a particular job. There are occasions when dual perforating (both bullet and jet) has been used. As a rule the decision to be made by the operator is the type of equipment to use. Table 21-1 attempts to list the most generally accepted characteristics of each type.

This discussion intends to show that no one formula or piece of equipment is best for all perforating jobs but that each situation must be analyzed individually if maximum performance is to be obtained.

Sand Control Measures

There are many important oil and gas fields throughout various parts of the world in which the producing reservoirs consist of semi-consolidated sandstones. Wells completed for production from such reservoirs are likely to present operating problems if produced at other than a minimum rate.

The influx of sand grains may give very little trouble until the well, for some reason, must be shut in. During the shut-in period, sand settles back to the bottom of the tubing string, thus plugging it so that production cannot be reestablished without first washing out the sand grains.

The influx of sand tends to increase significantly if the well starts producing appreciable quantities of water. This is due in part to the increase in the total fluid produced.

Where conditions of this sort are known to exist, it usually is advisable to institute sand control measures at the earliest opportunity.

Two methods have gained prominence in preventing the influx of formation sand into the well through casing perforations. They are: (1) sand or gravel packing and (2) consolidation of the formation with plastics.

In the packing method, the formation is fractured, in effect, with very coarse sand grains or small gravel, sized so that it will bridge over perforations to prevent sand entry.

This technique demands careful analysis of the formation and size of the perforations so that the packing material will hold back formation sand, yet bridge on perforations back of the casing when the well is put on production.

This process is well adapted to multiple zone completions since it leaves clear passage for tools moving past upper zones after having been packed.

Many different plastic compounds are used for consolidation of semi-consolidated sands. Properly applied, they coat each sand grain and bond one to another with only a very slight impairment of permeability. It has been found to be most effective if this type of sand control is instituted immediately upon completion of the well and before appreciable movement of sand has taken place.

Reasons are not fully known but it is presumed that movement of formation sand is accompanied by a graduation of fine sand, which when consolidated by plastics forms a barrier of very low permeability. Early consolidation work, on the other hand, bonds the sand in place as originally deposited.

Testing for Productivity

Unless an open hole test, made prior to running production casing, indicated the probability of prolific production, the operator will almost surely make in-casing production tests after perforating prospective pay zones but before any stimulation treatment has been applied to any such zone.

The operator has at his disposal a variety of drillstem test tools with which he can adequately test the formation without emptying the hole of any fluids that may be in it. In other instances he may elect to run tubing with packer and conduct production tests over a period of several days.

Speed and economy of drillstem testing (DST) give it marked advantages over more protracted formation evaluation methods. These advantages have been enhanced recently by improvement in DST tools, particularly their recording components. Overshadowing even the improvements in pressure and temperature recording instruments has been the industry's ability to interpret test results.

Conventional DST tools with their improved instrumentation and improved interpretation of their findings are making more and more data available to the operator. Acceptance of reservoir information obtained from drillstem test analysis has expanded to such an extent that the drillstem test is now considered as much of an evaluation tool as it is an exploration tool.

Regardless of the type of fluid recovery, most DST results may now be interpreted with reasonable accuracy to throw needed light on reservoir pressure, average effective permeability, productivity index, damage ratio, radius of investigation and absolute open flow potential.

In addition to its aid in formation evaluation and hydrodynamic studies, DST interpretation may be of considerable help to the operator in the design of stimulation treatments, if any.

Well Stimulation

Well stimulation treatments, in one form or another, have been a general practice for many years. During the cable tool era it was often helpful to "shoot" a tight sand with nitroglycerin, the purpose of which was to fracture the rock extensively and thereby increase the surface areas exposed to reduced well bore pressures. Such fractures also provided drainage channels to the well bore.

In the early thirties, acidizing carbonate reservoir rock was done successfully, first in brine wells and next in oil and gas wells. It was equally effective in either cable tool or rotary drilled wells. Hydraulic fracturing of sandstone reservoir rock was first successfully used in the late forties and since that time may have accounted for the greatest single increase in our national petroleum reserves.

One of the principal short-comings of the hydraulic rotary system of drilling oil wells is the probability of damage to the reservoir rock caused by the mud column pressure which causes water to filter into the porous rock and carry with it some very fine solids which tend to plug the pores, thus greatly reducing the effective permeability. For this reason, well stimulation has become one of the important steps in the original well completion.

By acidizing or fracturing, or by some combination of the two, the operator usually can undo any damage to the formation caused by rotary drilling operations and also extend permeability in all directions.

In some cases, notably to remove water blocks, the treatment may include surfactants in oil base or water carriers to change the surface properties of formation liquids.

For stimulation of limestone and dolomite pay zones, hydrochloric acid is most widely used, although acetic acid is being used in instances where slower action and a less corrosive agent is desired.

Acid reaction further back in the formation can be achieved by the inclusion of a retarding agent in the treating fluid. Fluid loss control additives, pumped ahead of the acid, will coat the fractures and delay action of the acid until it can be pumped well back into the formation. Basically, the fluid loss additive approach is a combination of fracturing and acidizing techniques.

It has long been recognized that formations can be broken down or fractured by hydraulic pressure. This phenomenon has too frequently been responsible for loss of circulation in rotary drilling operations.

It was not until 1948 that the industry learned the value of that capacity in well stimulation. It also was learned quickly that sand grains entrained in the fracture fluid served to prop open the fracture and that without a propping agent the treatment was of minimal benefit.

Since those early dates many improvements have been made. Probably the chief one being an understanding of the effects of treatment size and the rate of injection.

As a result of improved techniques, it now is common practice to use many thousands of gallons of fracturing fluid and huge quantities of sand or other propping agents. Although sand (the propping agent) and oil (the fracturing fluid) are most commonly used, water fracs also are frequent.

Originally water with sand was used only in gas well treatments but later water/sand fracs were applied to oil wells. Increased rates of injection, having proved beneficial, required much improvement in pumping capacity. The various service companies have met that challenge.

In addition to sand as a propping agent, many alternate materials are being used, such as ground walnut hulls, aluminum pellets and glass beads of great strength and hardness.

One of the big problems in well stimulation treatments is the problem of creating multiple fractures throughout a thick pay interval. To insure effective treatment throughout the perforated interval, it is necessary to guard against accidental concentration of the treatment in an interval where the formation breaks down first.

One solution to this problem, which has been quite successful, is the use of ball sealers which plug the perforations opposite the zone taking fluid most freely. As pressures increase, some other zone will break down and start taking fluid freely. Subsequently these perforations are sealed off. This sequence is repeated until the entire interval has been treated.

The sealers usually are rubber covered balls made of plastic or other hard material. They are lubricated into the treating fluid stream at pre-selected intervals.

Another popular approach to fracturing thick pay zones is the limited entry technique, which employs a limited number of perforations of pre-selected diameter and spaced a foot or more apart, rather than the usual four holes/foot.

In this plan the pumping rates and pressures are very high so as to achieve near critical flow at each perforation. Each perforation becomes an orifice which will pass a fluid volume proportional to its area. With proper planning of the number and size of the perforations and of the pumping rates to be used, it is possible to distribute the treatment over the entire interval.

Still another innovation is the single entry technique. This plan calls for cutting a single notch through casing and into the formation, rather than the usual perforation. The notch can be cut with a mechanical tool but more often is created by the abrasive action of a sand-laden fluid pumped by a service company.

The basic theory of single entry is that a horizontal fracture will be initiated which may extend over a large area and likely will be confined to the desired section of the pay zone with respect to gas/oil or oil/water contact levels.

To facilitate clean-up after a frac job, it has become common practice to use an inert gas such as carbon dioxide or nitrogen, which is injected into the fluid stream as it is pumped into the well. After pumping all frac fluid, with its entrained gas, into the formation and releasing the pressure, the back flow will be assisted by the expanding gas very much like live oil would flow from the same well. Carbon dioxide has proved effective in removing water or emulsion blocks. Either gas will help in raising fluids to the surface.

Squeeze Cementing

Squeeze cementing, through perforations, usually is undertaken either to seal off or abandon a zone that has been previously exposed or to block squeeze above and/or below an interval scheduled for testing. The purpose of squeeze cementing, in this case, is to protect the prospective zone from communication with adjacent fluid-bearing formations. If the job is successful, it will compensate for an imperfect primary cement job.

Perforating for a block squeeze usually employs three or four holes at as near the same depth as possible. If cement has channeled at this level, the squeeze job will displace the contaminated mud with good cement. Such a job might better be called recementing rather than squeeze cementing.

In the absence of a bad primary cement job, the squeeze cement must create fractures in the formation and be deposited therein. It would be helpful if fractures thus created would follow a horizontal pattern but there has been much evidence to prove this seldom happens. Squeeze cementing can be expected to seal off natural fractures and to plug vugular porosity but it cannot be expected to plug the pores of a normal sandstone.

Achieving a complete shutoff of a long perforated interval is much more difficult than block squeezing. Pumping cement through each of numerous perforations involves many different techniques. It may be desirable to use a chemical wash ahead of the cement to remove filter cake from the formations perforated. Mud acid often is used for this purpose.

It also is desirable that the cement slurry contain fluid loss additives, or at least has low fluid loss characteristics, so that the cement will not quickly dehydrate and form large nodes inside the casing over the perforated holes. Such nodes might hamper subsequent completion operations at lower levels, or possibly require that the casing inner surface be scraped or reamed out.

The hesitation technique has proved a valuable aid in squeezing off perforated intervals and also in recementing mud channels. After a quantity of cement has been pumped through perforations, pumping is stopped. After the cement has had time to filter out some of its water, pumping is resumed.

Perforations previously taking the most cement

will be harder to pump through, therefore cement will start through perforations which have passed lesser cement or possibly none at all. The procedure can be repeated until all perforations have taken cement, which will be indicated by very high pumping pressure.

Fluid loss additives have contributed significantly to the art of squeeze cementing. These chemicals tend to keep the cement slurry pumpable for greater lengths of time so as to permit it to enter a greater number of perforations than it otherwise would.

Ball sealers, which proved helpful in stimulation treatments, have been used for a similar purpose in squeeze cementing by closing off those perforations which have been taking cement so that pressure can be built up opposite perforations which have not.

Tubing Installation

The final completion operation (following the running cementing of production casing, the perforating, stimulating and testing of prospective productive zones) is the installation of tubing.

In single zone completions, of moderate depth, packers are seldom required. In deep wells, packers are often used and in multiple completions, packers are required regardless of well depth.

The impact of multiple completions in cased holes has brought about the slenderization of tubing strings. Instead of the usual 2⅜-in. OD or 2⅞-in. OD tubing commonly used in single zone completions, 2-in. OD and 2 1/16-in. OD tubing are now frequently used in multiple completions.

As a result of this miniaturization, it is possible in 7-in. casing to effect a triple completion with two strings of 2 ⅜-in. integral joint tubing and one string of 2 1/16-in. integral joint tubing and still have adequate clearance. Or the same job can be performed with adequate clearance using three strings of 2-in. tubing with buttress thread couplings on non-upset joints.

Inside 17 lb/ft or smaller 5 ½-in. casing it is possible to run two strings of 2-in. tubing with special (2.30 in. OD) buttress thread couplings or three strings of 1½-in. OD tubing with integral joints.

With these slender tubing strings it was necessary to develop landing nipples, sliding valves, gas lift mandrels and other devices to run on such tubing strings.

Multiple completions in cased holes also stimulated development and miniaturization of a wide range of multi-string packers, both of the permanent and retrievable types.

To insure proper operation, long service and recoverability of tubing, packers and other accessories the operator must devote special attention to packer fluids. Choice of packer fluid in shallow, single zone completions has never been critical. It usually is possible to pull loose, or if failing to do so, the tubing can easily be washed over and thereafter recovered.

Inside casing with two or more strings of tubing with packers there seldom is room for a washover string. It is necessary, in such instances, that all packer fluids be free of solids that settle out and pack around the lower sections of the tubing strings.

Among popular packer fluids are oil compounds and chemical compounds such as calcium chloride and water, weighted with such other chemicals as zinc chloride and treated with some good corrosion inhibitor.

Chapter 22
Marine Drilling Operations

Marine drilling operations might logically be divided into four categories: (1) inland water, (2) close-in offshore, (3) outer continental shelf and (4) deep sea drilling operations.

Inland Water Drilling

Inland water drilling can be defined as all drilling operations in the bays, lakes, rivers, canals and marshes which are reasonably protected from wind and waves associated with the ocean.

In these operations the drilling rig and drilling practices are not unlike operations on land but the foundation for the drilling rig differs materially.

Commencing in the mid-thirties the vast majority of inland water drilling, domestically, has been confined to the marsh lands adjacent to the Gulf of Mexico where submersible barges are used almost exclusively. These barges, complete with rig, equipment and supplies are towed to the drill site through canals dug for that purpose. Subsequently, the canals are used for transportation of personnel and supplies and for laying gathering lines to a centrally located production platform which includes separating, treating and storage facilities. Such platforms have wood decks supported by driven wood piling.

Submersible Barges. Submersible drilling barges are used in inland waters such as bays, lagoons and dredged marsh locations. Equipment needed on such locations and techniques used are different from those on offshore wells. These differences should be kept clearly in mind.

A submersible drilling barge is merely a portable working platform to carry the derrick, drilling rig, drill pipe, mud and casing used to drill a well in relatively shallow water. Procedures and techniques of drilling are the same whether on a land or a water location, except that limitations of space necessarily circumscribe movement of crew members and material handling.

Drilling barges were used first about 1933, shortly after drilling expansion in marshes and bays was begun. Drilling on locations of this type, prior to the advent of the drilling barge, was done by piling platforms. This was costly and construction was slow.It also presented quite a problem in moving the rig as this required the use of a derrick barge to take the rig off the completed well; load it onto the cargo barges, and unload it at the new location.

The Texas Company (Texaco) pioneered in building and using submersible drilling barges. Its first barge of this type was completed in November, 1933 and was named after Captain Louis Gilasso. In 1928 Gilasso filed patent applications both in the United States and Venezuela covering the design of a submersible drilling barge. These patents were issued and Texaco acquired rights to build and use this type barge.

The first Texas Company barge was designed by G. I. McBride, company engineer. It was built in Pittsburgh and towed down the Ohio and Mississippi Rivers. The rig and derrick were then installed and the first well was drilled at Lake Barre, Terrebonne Parish, La. This first barge was a split type, with each side 24-ft wide and 114-ft long, with an 8-ft slot between.

Economy, ease, speed of moving and absence of tearing down and setting up operations on the rig and derrick were apparent after the first well and these advantages soon led to construction of many barges of this type.

The first barges were built for steam rigs. They were about 50-ft to 54-ft wide, 125-ft long, with about 10-ft sides. Many of the early ones were the horseshoe or U type, with boilers mounted on a separate barge and a catwalk connecting the two. Later the preference seemed to be for two barges about 24-ft wide connected by steel trusses with an 8-ft to 10-ft slot between the barges.

Where the location was on a narrow canal or bayou, these barges supposedly could be separated and towed to the location singly and then put back together and the machiney reinstalled. In practice this did not work out very well. The cost of tearing down a rig, loading it on cargo barges, and setting it up again was prohibitive.

Now canals and bayous have, in general, been widened and barges 50-ft to 54-ft wide will generally pass. There are presently some barges only 38 ft wide, which are narrow enough to pass all but a few bridges and locks.

Barges for Power Rigs. The last few years most new barges have been built to accommodate power rigs. These barges have been one piece hulls 50 ft to 54 ft wide, 160-ft to 200-ft long and 11-ft to 14-ft draft. Many

Fig. 22-1. Inboard profile.

Figs. 22-1 to 22-4 are plans of Rowan Drilling Company barge M. G. Rowe, a heavy duty diesel electric rig.

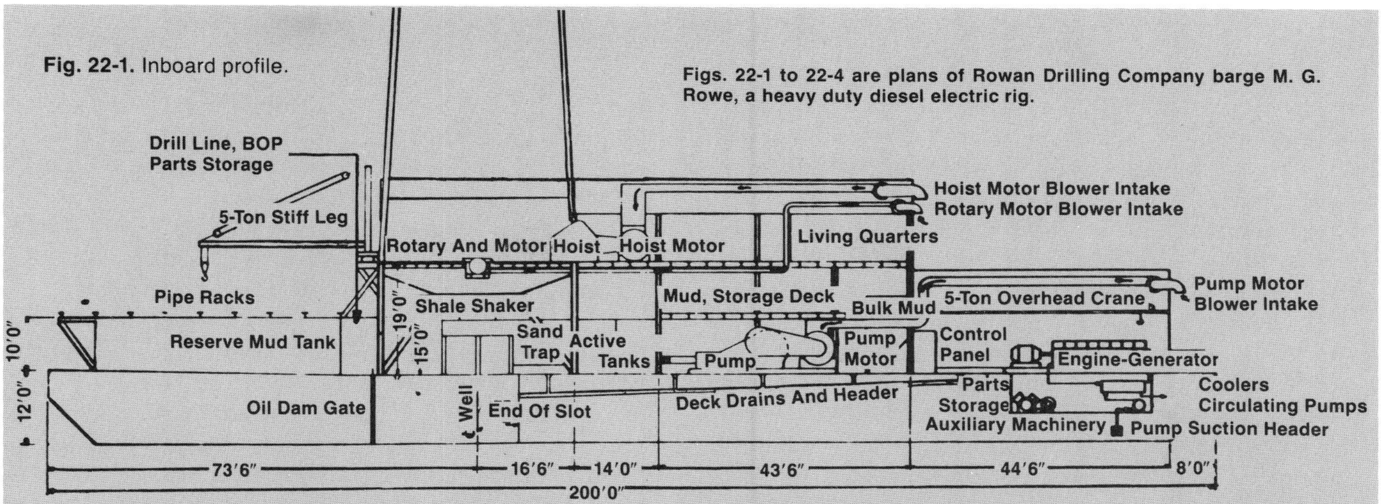

Fig. 22-2. Plan of intermediate deck level.

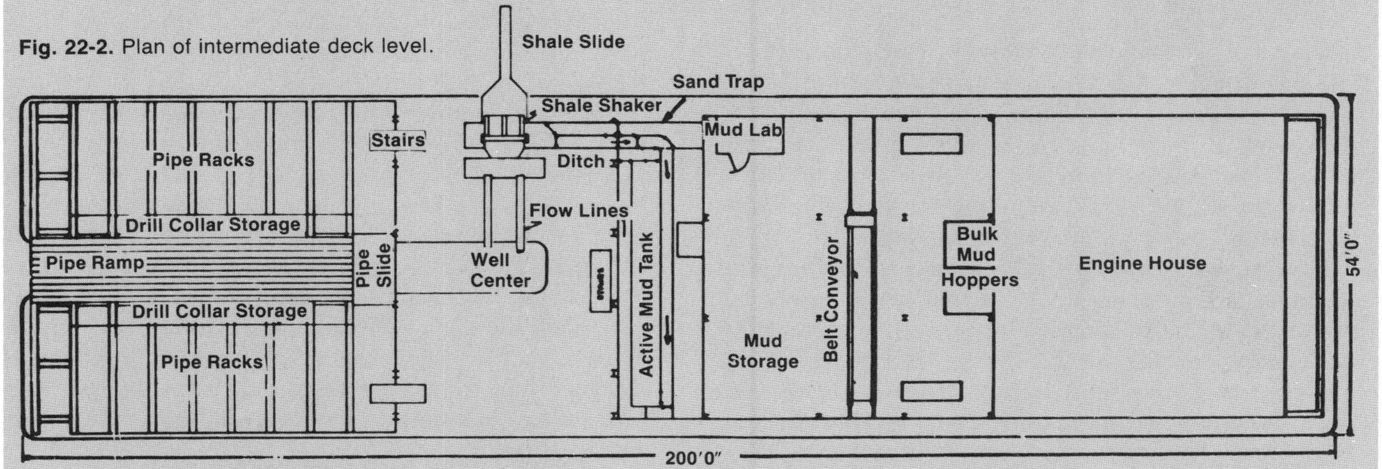

Fig. 22-3. Plan of main deck level.

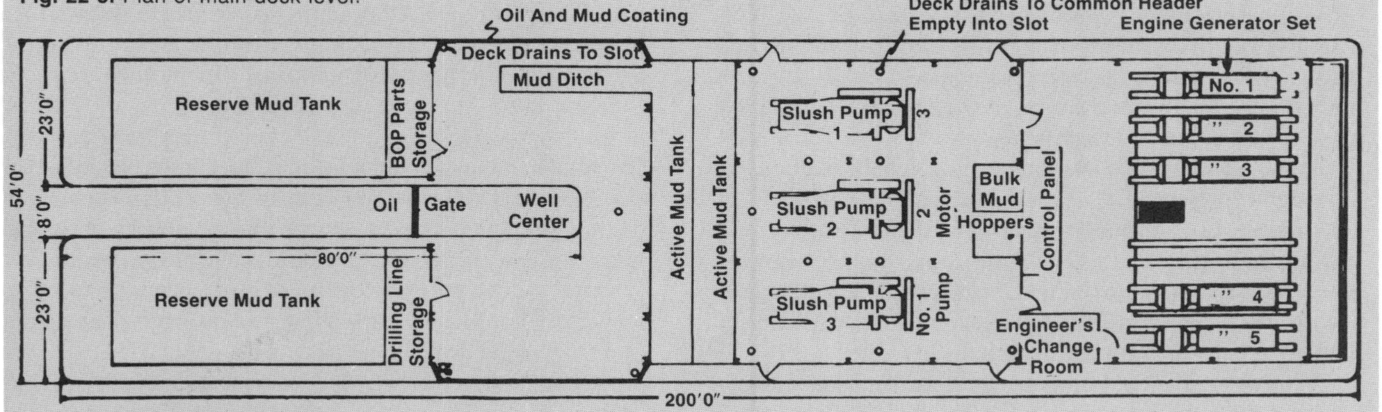

Fig. 22-4. Plan of drill floor level.

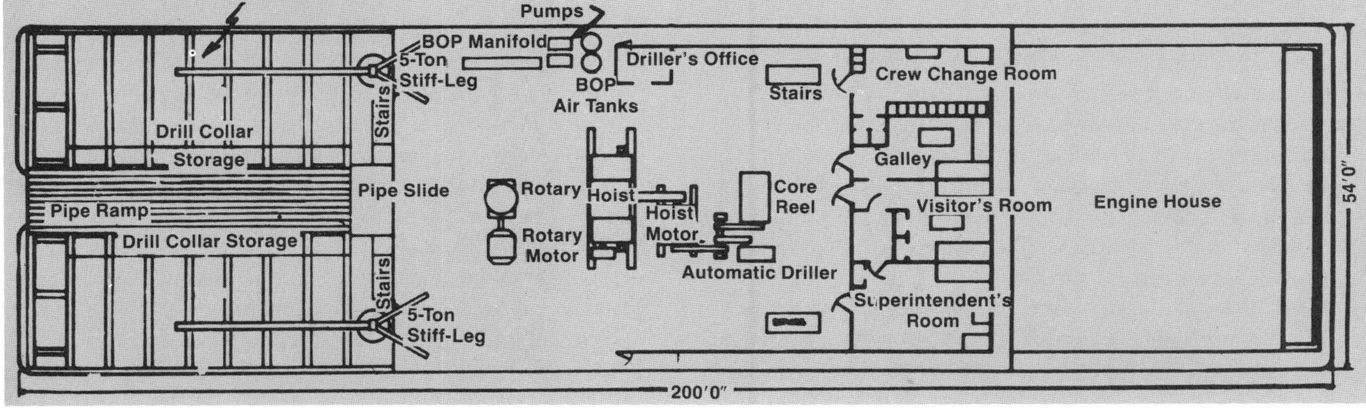

are three stories high. The pumps, active mud pits, reserve pits and deck room around the well connections are on the first deck. The second deck usually carries the engines, dry mud storage and pipe rack. The third story is derrick level and carries drawworks, motor (if electric rig), sand reel, tool house, change rooms and quarters.

The quarters usually consist of a toolpusher's room and office, with ship-to-shore radio telephone, small galley, bath and visitors' room. These quarters, while modest, must be comfortable since the superintendent may have to spend many nights on the job and geologists, engineers and service men may also be compelled to stay overnight on the rig during completions, bad weather or fishing jobs.

Derricks are usually the conventional type and are left standing on short moves except when they interfere with high lines or bridges. They are then dismantled down to the fourth or fifth girth. Some new rigs, particularly the workover rigs, are using jacknife derricks because of the ease and speed with which they can be lowered for frequent moves.

Almost all of the rigs have two mud pumps and many of the deep rigs have three: one large pump about 700 hp for drilling; and two smaller around 350 hp for compounding or paralleling. If the large pump is down or if one small pump is used to mix and condition mud, the other can be used to circulate hole while the larger pump is being repaired.

Many rigs now have electric, paddle-type mud mixers or stirrers. These mixers do a good job, particularly if the pits are designed to take full advantage of their power. Settling pit capacity is limited on barge rigs, so shale shakers are a necessity. All rigs use at least one and many use two.

The spoil goes into the bay or canal, although often during the early part of the well, it is piped around the surface or conductor pipe to fill in and reinforce the outside of this pipe. Shale shakers, of course, eliminate only a small part of the sand. Therefore, a lot of attention should be given to the design of sand traps and mud ditch and settling pit.

Two sand traps with proper bypasses are necessary so that one can be emptied while the other is being used. Mud ditch should go to both settling pit and suction pit so the settling pit can be cleaned as often as necessary.

Keeping the mud level low in the settling pit is important so that a minimum amount of mud will be wasted when this pit is cleaned. A limited amount of surplus mud can be stored in the reserve pits, which usually occupy the space under the pipe racks. These pits usually hold about 1200 bbl each. All mud, before it is stored in reserve pits, should be treated to remove excess sand or entrained gas.

Mud Barge Useful and Economical. In proven fields, a mud barge used in connection with a drilling barge can save large sums of money. The common practice after completing a well is to fill the reserve pits with good mud, then pump the excess mud overboard or, if it is injurious to fish and oysters, into a pit built in the marsh for this purpose.

If this mud is stored in a barge, it can be treated if necessary, then the barge can be towed to the next location and used in place of mixing sacked mud.

Mud in the circulating system at completion of a deep well is usually the very best mud because the deeper the well, the more attention the mud gets. When a barrel of 12-lb mud costs from $9 to $11, it is apparent that there is great waste if this mud is not saved. A considerable amount of time is saved as the mud can be transferred from the mud barge to a drilling well much easier than it could be mixed if it were in dry storage. This is important if a well is losing circulation or trying to blow out.

On barge rigs derrick substructures should have at least 14-ft clearance between rotary support beams and deck level. This is necessary so that the casinghead will be above water level and at the same time afford room for three blowout preventers. Plenty of head room under the derrick also permits use of hoist and trolley to handle blowout preventers. A good overhead system will save many hours and much laborious effort when installing blowout preventers. Choke lines should be manifolded and bolted to a substantial beam and relief lines bolted to the top of the mud pit.

Ample dry mud storage aboard the barge or on a mud barge moored alongside is a must. On deep wells this storage space should provide for a minimum of 2000 sacks and probably a maximum of 5000 sacks. If electric mud mixers are provided, they should be supplemented with a hopper type mixer so that mixing facilities will be available at all times, should the mechanical mixer be down for repair. A conveyor belt for sacked mud is necessary if storage room is on the second deck. It is very convenient in any event.

Some newer barges now have bulk hopper dry mud storage. These hoppers work well. They are fast when moving, clean and no mud is wasted.

Now it is possible to get dry mud in bulk delivered to barge rigs. This mud is moved from mud barge to drilling barge by screw or air conveyors.

A fresh water compartment of several hundred barrels in the hull of the barge should be provided. This is necessary because often the bay, lake or canal water is salty and not fit for cooling engines, mixing mud or personal use. This fresh water supply is replenished by rain water salvaged from the roof of engine and quarter houses or by water transported by tug and water barge.

Moving to Location. The method of getting a drilling barge on location is relatively simple. The barge is towed to the location the same way as a cargo barge. After reaching the location, the barge is spotted so the location stake is exactly under the rotary

Fig. 22-5. Inboard profile of inland drilling barge, the Blind Pass.

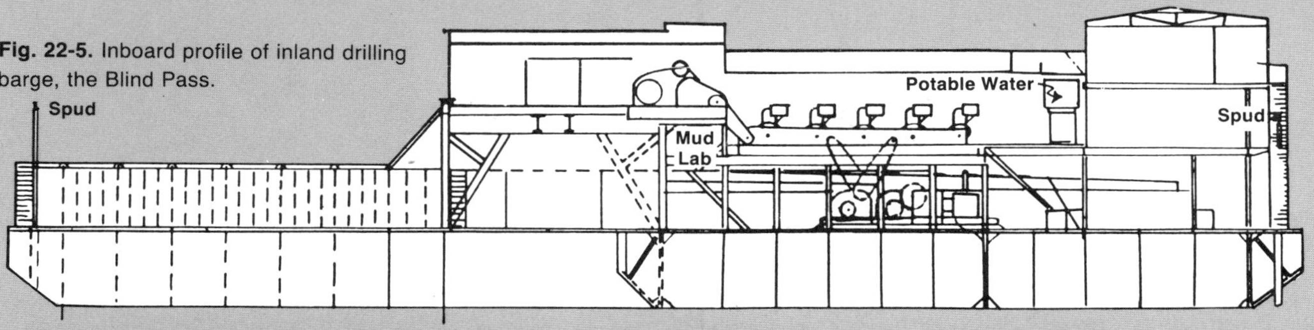

Fig. 22-6. Plan view of drill floor of the Blind Pass.

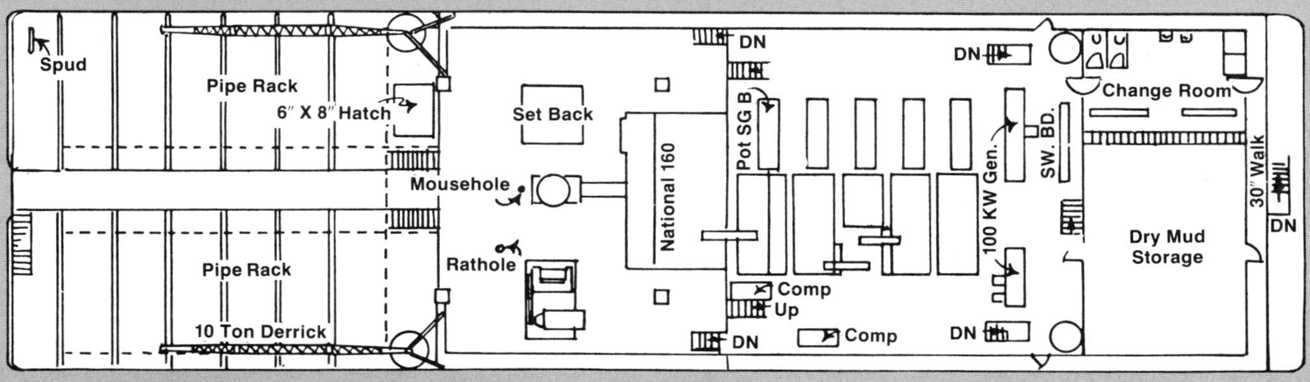

Fig. 22-7. Plan view of main deck of the Blind Pass.

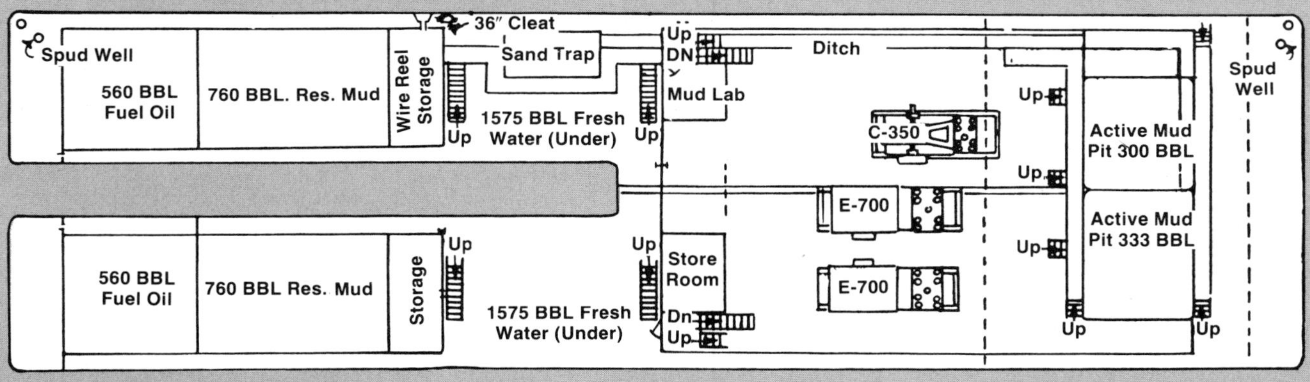

opening. Four piles are driven around the well location and the slot of the barge is spaced by these pilings.

After completion, if the well is a producer, these pilings act as a guard to prevent the barge from bumping against the well connections when it is being towed to the next job.

After the barge is spotted, spuds are then lowered and sea cocks opened in each compartment. The barge level is maintained by the amount of water admitted to each compartment while being sunk.

After the barge is resting on bottom, it can be leveled by decreasing the amount of water in the low side, end or corner. If the barge is not resting on a shell bank, it generally sinks gradually and evenly, due to vibration.

Depth of most barges is from 10 ft to 14 ft. If it is necessary to drill in water deeper than this, a shell fill may be laid to bring the barge higher out of the water. This is not too expensive and in quiet water works very well.

There have been quite a number of drilling barges made from surplus Naval LST vessels. These vessels can be converted to this use very nicely. They have the advantage of about 20-ft sides and therefore can be used in deeper water. They are also longer and have more space for quarters or storage.

Buoyancy in a drilling barge is important. Draft

Fig. 22-8. Photo of self-contained offshore drilling platform.

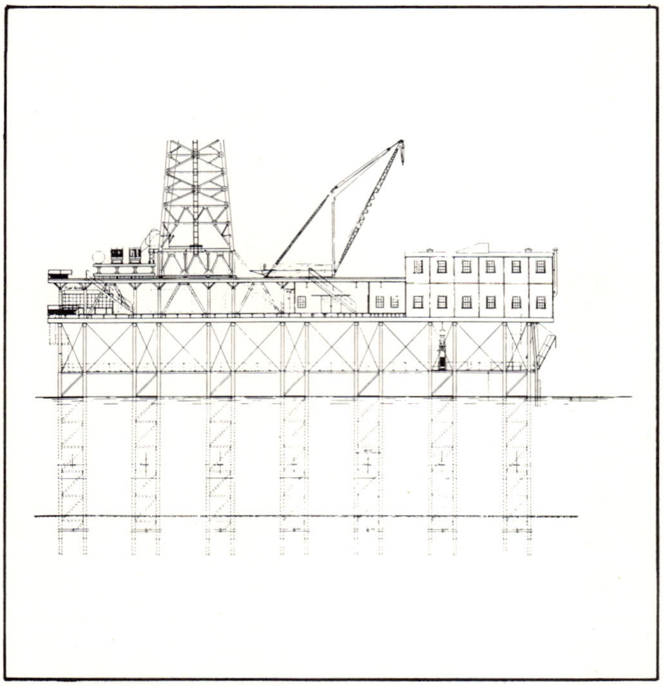

Fig. 22-9. Profile of self contained offshore drilling platform.

when loaded for moving should not be over 5½ ft to 6 ft and a draft of 4 ft to 5 ft can save a lot of expense. This is because many bays are shallow and canals and channels often fill up repeatedly with silt. A strong north wind can cause the water to be so low that moving is delayed for several days.

Buoyancy is, of course, directly related to size: the larger the barge, the more buoyancy. The 20 ft of additional barge would give about 6 in. more buoyancy.

It costs considerable in time, convenience and money to unload drill pipe, mud, fresh water and diesel fuel from a drilling barge onto a cargo barge and back onto the drilling barge to move in low water, so a little more expense when the barge is being built will be easily recouped over the years the barge is in use.

Maintenance Problem Serious. The problem of maintenance on barge rigs is, like on any vessel or structure exposed to salt water, always serious. On barges operating in fresh water lakes and bayous, continual chipping and painting seems to be enough. On barges located in salt water bays and open water and exposed to wave action, the job is much more complicated. These barges should be drydocked every 3 to 5 years, sandblasted and repainted. Some kind of protection against electrolysis also should be provided.

There are now about 130 barge rigs operating along the coasts of Louisiana and Texas. It is estimated that approximately 6000 wells have been drilled by barge rigs. These barges will be used for many years, probably as long as wells are drilled in swamps, marshes and shallow water along the coasts of Texas, Louisiana, Mississippi and Florida. The saturation point in this type barge has been reached.

For offshore drilling an entirely new type drilling barge is needed. There are many new and strange looking barges on the drawing boards of naval engineers and shipyards.

Close-In Offshore Drilling

When it was learned that some producing oil structures along the Gulf coast of Texas and Louisiana and the Pacific coast of California very likely extended offshore, means were conceived to explore these possible reserves.

The most common early practice was to construct a large wood platform, supported by wood piling, on which a conventional derrick and drilling rig was placed. If the platform was not too far from shore a pier or causeway was built connecting the platform to the land area much as board roads are used in some swamp areas. These platforms were large enough to accommodate fresh water storage tanks, mud storage, pipe racks and all other items normally found around a comparable land drill site.

As prospective structures were found further from shore, a variety of supporting structures evolved. The pile-supported platform was enlarged to accommodate living quarters.

These structures were usually about 100-ft wide

Fig. 22-10. Photo of tender-platform offshore rig.

and at least 200-ft long and of double deck arrangement with lower deck 35-ft or 40-ft above mean Gulf level. The gross deck area often exceeded 40,000 sq ft. (Figs. 22-8 and 22-9).

The large self-contained platform was quite expensive and in an effort to reduce the initial cost, the platform-tender combination was devised in the late forties. With this arrangement the platform could be reduced in size to about 50-ft by 150-ft, with a single deck configuration on which the derrick, pipe racks, drawworks with power units and other hoisting equipment were placed. The tenders or floating barges to be used with these smaller platforms were made from surplus Naval LST or YF vessels.

The tender provided space for prime movers, generators, mud pumps, mud tanks, dry mud, cement, fresh water and quarters for 50 to 60 persons. The tender was securely anchored so as to maintain its relative position with respect to the platform so that both high pressure and low pressure mud lines, electric cables and a catwalk extended from tender to platform (Figs. 22-10 and 22-11).

It was estimated, prior to the mid fifties, that perhaps 90% of exploratory and development wells drilled offshore in the Gulf of Mexico employed one or another of the 20 to 30 platform-tenders type of rigs. Some small platforms provided means for drilling four or more development wells (deflected) from the same platform-tender location.

In exploratory drilling where dry holes are probable, the loss in salvaging the small platform is not as great as in the case where a large self contained platform is used. However, neither type platform could be made economical for wildcatting.

This fact led to the development of a submersible barge that could be used in open seas. The submersible barge used on inland waters could not withstand the waves and currents of the open sea, even in shallowest waters.

The advantages of a hull resting on the ocean floor could not long be overlooked. In the late forties John T. Hayward, of Barnsdall Oil Company, designed and directed construction of a special type submersible barge that proved most successful for exploratory

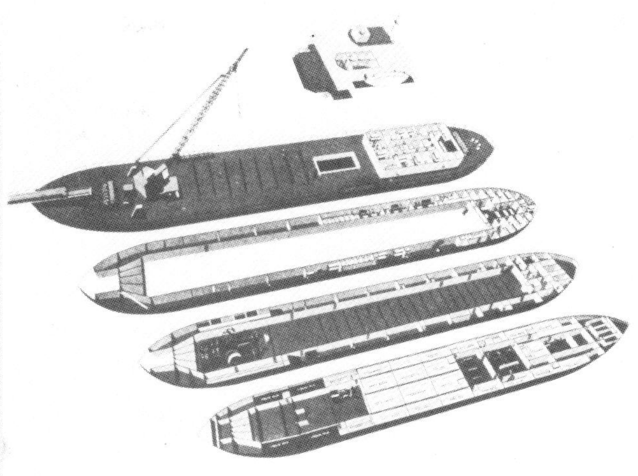

Fig. 22-11. Exploded view of tender showing the layout of various decks.

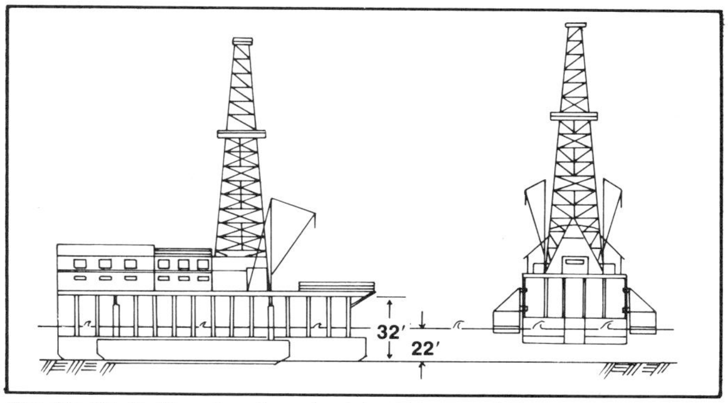

Fig. 22-12. Outline drawing of original Hayward-Barnsdall offshore submersible drilling barge.

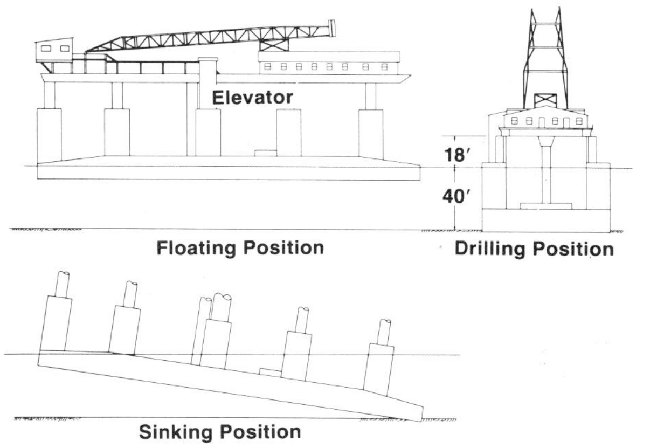

Deck Dimensions: 216 ft by 55 ft
Hull Dimensions: 220 ft by 74 ft by 14 ft
Column Size: 14-ft lower column topped by 6-ft upper column
Operation:
1. By selectively flooding the lower hull, the bow end is sunk to the ocean bottom, the stern end remaining afloat. Stability is maintained by the buoyancy of the columns and small outrigger pontoons which are not shown.
2. The stern is then sunk to the bottom.
3. Bearing pressure on the soil is adjusted by ballasting the hull compartments as necessary.
4. The unit is raised in the reverse manner.

Fig. 2–Ocean Drilling & Exploration Co.
Diesel Electric Submersible Drilling Barge
(For 12 Ft to 40 Ft of Water)

Fig. 22-13. Procedure for submerging a drilling barge.

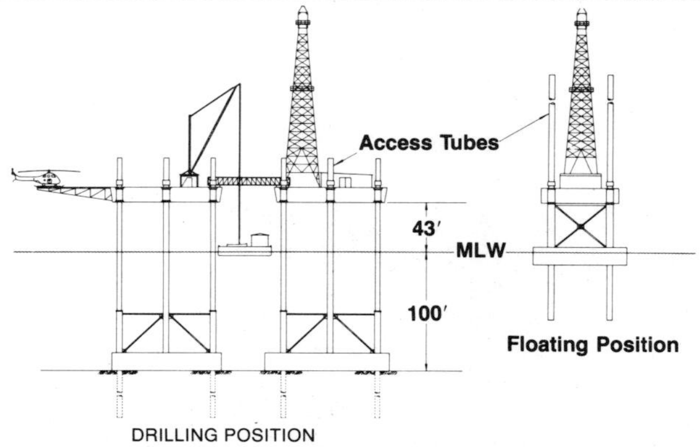

Upper Hull Dimensions: Drilling platform—80 ft by 50 ft by 15 ft
Service platform—80 ft by 50 ft by 15 ft
Lower Hull Dimensions: Drilling platform—106 ft by 80 ft by 15 ft
Service platform—94 ft by 80 ft by 15 ft
Piling Size: 4 ft
Access Tube Size: 5 ft
Operation:
1. The platforms are towed to the drilling location. The piles are dropped to the bottom and forced into the subsoil by means of hydraulic jacks. The weight of the upper hull is applied to two diagonally opposite piles at a time until desired penetration is obtained.
2. The upper hull is clamped to the piles and the lower hull is submerged to bottom by flooding it with water and placing downward pressure on it with the access tubes.
3. Men go down the access tubes and clamp the lower hull to the piles.
4. The two units are raised in a reverse manner with the hydraulic jacks being used to pull the piles, etc.

Fig. 22-14. Outline drawing of an early Jack up platform.

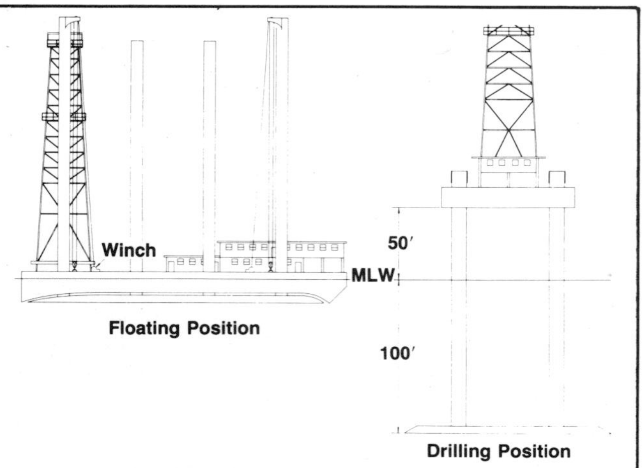

Upper Hull Dimensions: 220 ft by 90 ft by 14 ft
Lower Hull Dimensions: 200 ft by 140 ft by 5 ft
Column Size: 8-ft by 10-ft elliptical columns
Operation:
1. The lower hull is flooded until it is barely buoyant.
2. The lower hull is forced to the bottom by taking up on the cables with power winches.
3. After the stability of the support hull has been established, the upper hull is raised to the required elevation by continuing to take up cable with the winches.
4. The upper hull is then fastened securely to the elliptical columns.
5. During hurricane season, piles may be driven through the columns for additional lateral support.
6. The unit is removed by the reverse procedure.

Fig. 22-15. Second generation of mobile Jack up offshore drilling platform.

drilling in water depths up to approximately 25 ft. This depth was not exceeded in any of the close-in offshore areas then being explored. It is a self-contained drilling barge with quarters and was first used in Breton Sound at a water depth of 22 ft. (Fig. 22-12).

The Hayward barge discussed previously has a working deck with fixed legs resting on a floating hull which can be sunk to rest on bottom. Adjustable outrigger pontoons on each side of the main hull provide stability while sinking the main hull and thereafter they too are sunk to rest on bottom.

The upper or working deck is supported by numerous fixed legs extending from the top of the main hull to the bottom of the working deck, a distance of about 20 ft. The space provided by the legs between hull and deck has proved adequate for wave action in normal seas with water depths less than 25 ft.

The Hayward barge was used extensively for both exploratory and development drilling at close-in offshore locations. Other units of this class were built having more clearance between hull and working deck so as to make them suitable for use in deeper water and rougher seas (Fig. 22-13).

The outstanding advantages of the Hayward type barge are: (1) it is a completely self-contained drilling platform with quarters; (2) it is practically 100% salvageable; and (3) although not self-propelled, it is mobile, except during high seas, without rigging down.

Other mobile drilling platforms were soon to follow. Fig. 22-14 illustrates a unit having an elevated deck maintained at a given height above the water by means of six extendable legs or columns. A service platform is situated nearby and it too has six extendable legs to elevate its deck to same level as the drilling platform.

Figs. 22-15, 22-16 and 22-17 illustrate three different types of barges, each of which has a floating hull which, with all rig elements in place, is lifted out of the water on its extendable legs or columns.

All these jackup types are suitable for use in water deeper than can be explored by submersibles with hull resting on bottom. Some are suitable for use also in what is here termed Outer Continental Shelf (OCS).

Outer Continental Shelf Drilling

The outer continental shelf in the Gulf of Mexico is perceived to extend outward from shore a distance from 25 miles to 200 miles, where the water depth starts rapidly to increase. This anomaly seems to occur at a water depth of about 600 ft.

Drilling exploratory wells or holes in these deeper waters is done mostly with either a Le Tourneau-type jackup barge or a semi-submersible barge. The Le Tourneau barge with slanted legs (Fig. 22-18) is suitable for use in water 300 ft or more in depth. After a reasonable delineation of a productive area the development wells are drilled from fixed, permanent, multi-well, self-contained platforms.

The actual number of wells to be drilled from a single platform may depend on many factors: (1) water depth; (2) nature and depth of the hydrocarbons to be recovered; (3) configuration of the area to

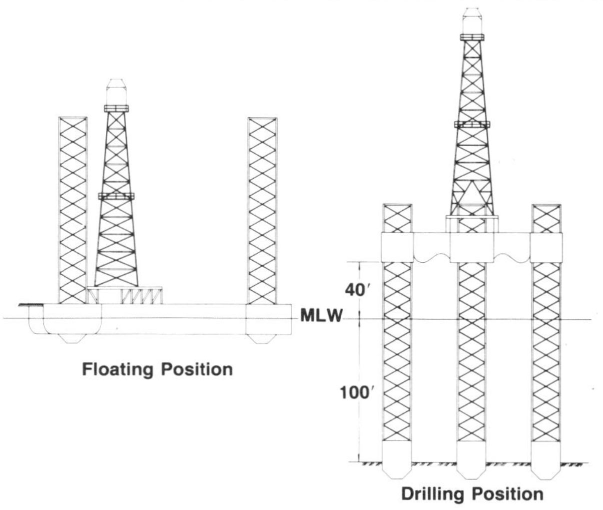

Hull Dimensions: 140 ft by 80 ft by 21 ft
Leg Dimensions: 20 ft by 20 ft
Operation: 1. Unit is towed to location and legs lowered by operating the rock and pinion drives in the corner of each leg. Pinions are powered by alternating-current electric motors (LeTourneau design).
2. When legs penetrate soil deep enough to start carrying load, barge is lifted out of water and elevated to desired height. Springs on motors lock pinion gears when power is shut off.

Fig. 22-16. Le Tourneau tripod mobile Jack up offshore drilling platform.

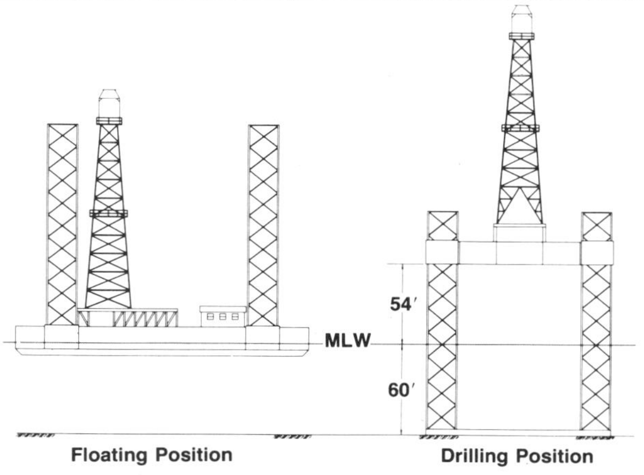

Floating Position
Upper Hull Dimensions: 190 ft by 80 ft by 15 ft
Lower Hull Dimensions: 190 ft by 120 ft by 4 ft
Leg Dimensions: 20 ft by 20 ft
Operation: 1. The lower hull is flooded until it is barely buoyant.
2. Electric motors operating rack and pinion drives on each of the four legs force the lower hull to bottom.
3. The upper hull is then raised to the drilling position with the same electric-motor system.
4. The unit is removed in the reverse manner.

Fig. 22-17. Offshore Drilling Co.'s Mobile rectangular drilling platform with four legs fixed to a submersible section of the hull.

be developed, and many other considerations.

One well may be drilled vertically but all others will be deflected, employing the controlled direction drilling techniques discussed in Chapter 18. Most likely the exploratory wells will have been properly plugged after logging and making conclusive test thereof.

Semi-submersible barges are used for drilling exploratory wells in water depths of about 600 ft or less. They normally are not self-propelled but there have been a few self-propelled designs developed for rectangular types.

Semi-submersibles are built in three basic configurations: (1) rectangular (Fig. 22-19); (2) triangular (Fig. 22-20) and (3) Y-shaped. (Fig. 22-21) Regardless of configuration, each unit employs large buoyancy chambers which, though submerged, provide most of the flotation force. Extending upward from the buoyancy chambers are columns, often in the shape of bottles, which support the working decks and provide supplemental buoyancy.

Semi-submersibles, when drilling operations are in progress, are usually held on location by mooring lines and conventional anchors. If they are intended for use in extremely deep water, provision can be made for dynamic positioning using thrustors.

Operation of the thrustors is controlled by a sensing device receiving signals from some fixed point on the ocean floor. When on location the buoyancy chambers are flooded sufficiently to make them sink below wave action. The columns have sufficient volume to provide stabilizing buoyancy during drilling operations.

Fig. 22-18. Le Tourneau slant leg tripod type Jack up mobile platform for use in deep water.

Where semi-submersibles are used, the BOP stack and other wellhead items are situated at or near the ocean floor. The riser pipe must have some flexibility to adjust to the minor movements of the barge. This requires that a ball joint or some similar flexible connection between wellhead and riser be used.

In many instances the riser pipe incorporates a concentric form of buoyancy chamber which tends to hold the riser in a vertical position. Telescoping risers are used to compensate for tidal action.

Deep Sea Drilling

Deep water drilling operations beyond continental shelves have been limited but are likely to assume

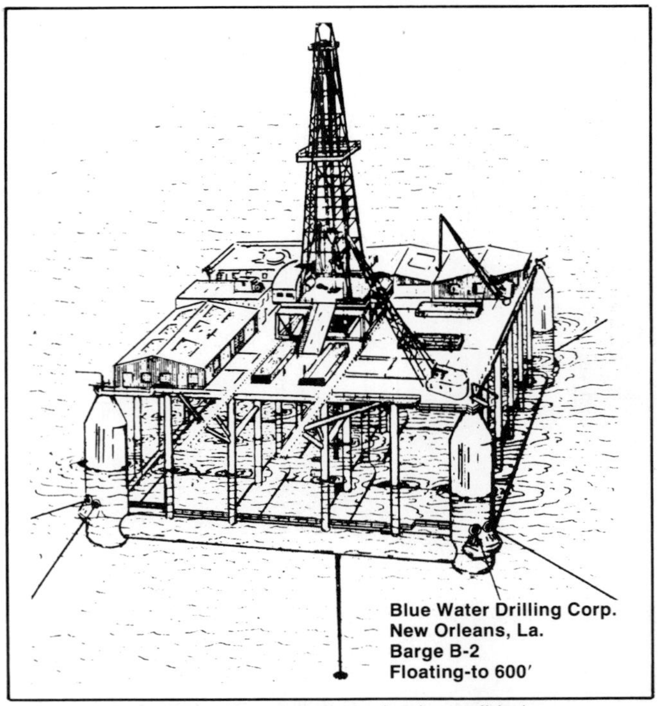

Fig. 22-19. The Blue Water no. 2 semi-submersible barge.

Fig. 22-20. Odeco's triangular configuration type of semi-submersible drilling barge.

Fig. 22-21. Odeco's second generation of triangular or Y shaped semi-submersible offshore drilling barge.

Fig. 22-22. Cuss I drill ship, for deep water drilling.

greater importance in future years. The technology was fairly well demonstrated by project Mohole in successfully drilling in almost 12,000 ft of water near Guadalupe Island in the Pacific Ocean to a depth sufficient to penetrate all sedimentary rocks and 44 ft of basalt, a total interval of 601 ft.

The drilling vessel used was Global Marine's CUSS I which had been modified and equipped especially for this job (Fig. 22-22).

When deep water drilling operations become necessary and are economical, it likely will be done from floating vessels, dynamically positioned and with all wellhead connections at ocean floor level. Such wells will likely be completed without aid of divers as we now know them.

Underwater vessels or miniature submarines have already been developed which can be used as needed in ocean floor completion activities.

In addition to CUSS I there have been several floating barges, converted from surplus naval vessels, used to drill stratigraphic tests in the Gulf of Mexico and elsewhere. The initial drilling from a floating drilling barge occurred offshore California in 1953. The barge used was Zapata Offshore Company's Nola No. 1, which drilled over the side. Fig. 22-23 illustrates the several types of floating drilling barges.

Logistics-Offshore Drilling

Travel and movement of material between an offshore drilling rig and land based terminals is a fair

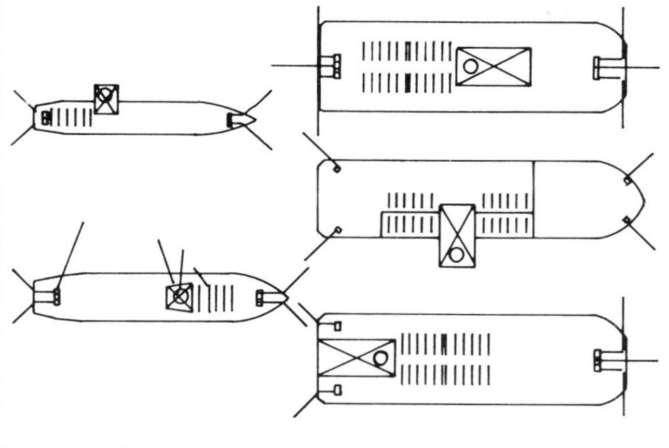

Fig. 22-23a. Various types of floating drilling barges.

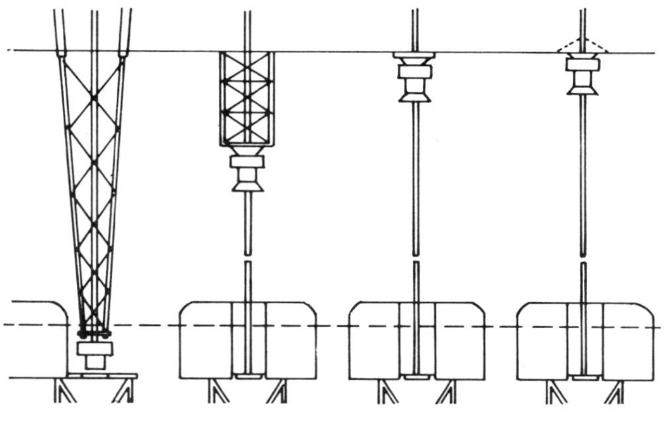

Fig. 22-23b. Several methods of drilling from floating barges.

size job in itself. Conventional workboats, crew boats and helicopters serve a large portion of these needs but some suppliers and service companies have seagoing vessels especially fitted for their service.

Drilling mud supply companies, for example, have their own vessels to deliver dry mud to all rigs in practically all active drilling operations offshore.

During their tour of duty on an offshore rig, drilling crews live and eat well in the quarters provided for them. Each of two crews works 12hr/day for a period of perhaps 10 days, then goes ashore for a like period. If the offshore location is close to shore, the travel is often by boat but sometimes by helicopter if the terminal is inland a considerable distance.

Helicopters are used almost exclusively for movement of crewmen to and from rigs located a distance from shore. Other personnel, such as supervisors, company men and special service personnel, also must be transported.

About the Editor

Wesley W. Moore grew up with the oil business, beginning his career in 1923 with Texaco in Denver, Colorado. He had advanced to district superintendent for southeast New Mexico when he resigned in 1935 to accept employment with Oilwell Supply Co. as machinery sales engineer, a position he held for five years.

In 1940, Mr. Moore joined British-American Oil Producing Co., where he served as Oklahoma City district superintendent and general superintendent of drilling and producing operations at Dallas and Tulsa. He retired in 1962 as drilling analyst.

Mr. Moore worked for Associated Publishers, Inc. of Dallas from 1962 to 1967, as editorial consultant for *Petroleum Equipment and Service* and *Drilling International* magazines. Since 1967, he has done consulting, writing, and research work in the petroleum industry.

Mr. Moore was born March 6, 1897, in Missouri. He served in the U.S. Army during World War I and attended the University of Missouri for three years, studying electrical and mechanical engineering.

A resident of University Village, Tulsa, Oklahoma, Mr. Moore is a member of API and the Society of Petroleum Engineers of AIME.

Tear out this card and mail to order more copies for your business associates or for your company.

ORDER CARD

Or pass this card to a friend who may want to order a personal copy.

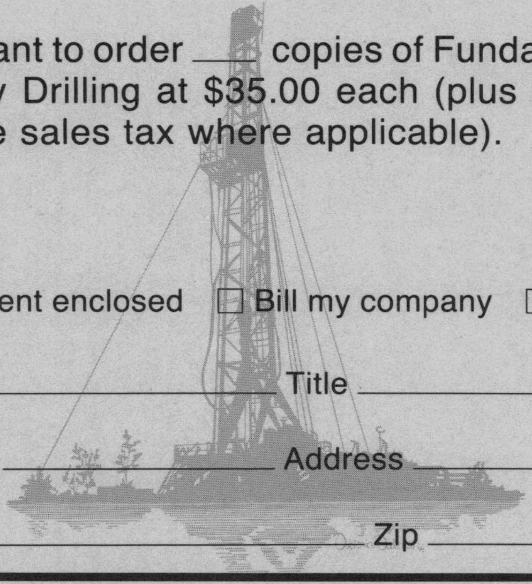

YES! I want to order ___ copies of Fundamentals of Rotary Drilling at $35.00 each (plus shipping and state sales tax where applicable).

☐ Payment enclosed ☐ Bill my company ☐ Bill me

Name ______________ Title ______________

Company ______________ Address ______________

City ______________ Zip ______________

Tear out this card and mail to order more copies for your business associates or for your company.

ORDER CARD

Or pass this card to a friend who may want to order a personal copy.

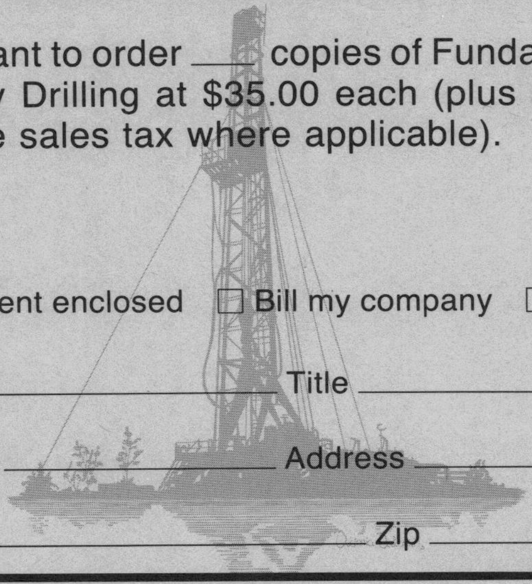

YES! I want to order ___ copies of Fundamentals of Rotary Drilling at $35.00 each (plus shipping and state sales tax where applicable).

☐ Payment enclosed ☐ Bill my company ☐ Bill me

Name ______________ Title ______________

Company ______________ Address ______________

City ______________ Zip ______________

Tear out this card and mail to order more copies for your business associates or for your company.

ORDER CARD

Or pass this card to a friend who may want to order a personal copy.

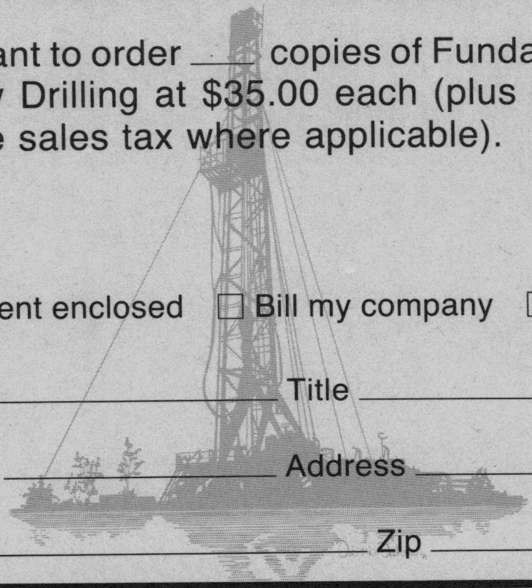

YES! I want to order ___ copies of Fundamentals of Rotary Drilling at $35.00 each (plus shipping and state sales tax where applicable).

☐ Payment enclosed ☐ Bill my company ☐ Bill me

Name ______________ Title ______________

Company ______________ Address ______________

City ______________ Zip ______________

NO POSTAGE
NECESSARY
IF MAILED
IN THE
UNITED STATES

BUSINESS REPLY CARD
FIRST CLASS PERMIT NO. 1571 NEW YORK, N.Y.

POSTAGE WILL BE PAID BY ADDRESSEE

SERVICE R&D DEPARTMENT
ENERGY PUBLICATIONS
a Division of Harcourt Brace Jovanovich
P. O. BOX 1589
DALLAS, TEXAS 75221

Tear out this card and mail to order more copies for your business associates or for your company.

ORDER CARD

Or pass this card to a friend who may want to order a personal copy.

NO POSTAGE
NECESSARY
IF MAILED
IN THE
UNITED STATES

BUSINESS REPLY CARD
FIRST CLASS PERMIT NO. 1571 NEW YORK, N.Y.

POSTAGE WILL BE PAID BY ADDRESSEE

SERVICE R&D DEPARTMENT
ENERGY PUBLICATIONS
a Division of Harcourt Brace Jovanovich
P. O. BOX 1589
DALLAS, TEXAS 75221

Tear out this card and mail to order more copies for your business associates or for your company.

ORDER CARD

Or pass this card to a friend who may want to order a personal copy.

NO POSTAGE
NECESSARY
IF MAILED
IN THE
UNITED STATES

BUSINESS REPLY CARD
FIRST CLASS PERMIT NO. 1571 NEW YORK, N.Y.

POSTAGE WILL BE PAID BY ADDRESSEE

SERVICE R&D DEPARTMENT
ENERGY PUBLICATIONS
a Division of Harcourt Brace Jovanovich
P. O. BOX 1589
DALLAS, TEXAS 75221

Tear out this card and mail to order more copies for your business associates or for your company.

ORDER CARD

Or pass this card to a friend who may want to order a personal copy.